Lecture Notes in Mathematics

1923

Editors:
J.-M. Morel, Cachan
F. Takens, Groningen
B. Teissier, Paris

T0155915

Jaya P.N. Bishwal

Parameter Estimation in Stochastic Differential Equations

Author

Jaya P.N. Bishwal

Department of Mathematics and Statistics
University of North Carolina at Charlotte
376 Fretwell Bldg.
9201 University City Blvd.
Charlotte NC 28223-0001
USA
e-mail: J.Bishwal@uncc.edu
URL: http://www.math.uncc.edu/~jpbishwa

Library of Congress Control Number: 2007933500

Mathematics Subject Classification (2000): 60H10, 60H15, 60J60, 62M05, 62M09

ISSN print edition: 0075-8434
ISSN electronic edition: 1617-9692
ISBN 978-3-540-74447-4 Springer Berlin Heidelberg New York
DOI 10.1007/978-3-540-74448-1

Springer is a part of Springer Science+Business Media
springer.com
© Springer-Verlag Berlin Heidelberg 2008

Typesetting by the authors and SPi using a Springer LaTeX macro package

Cover design: *design & production* GmbH, Heidelberg

Printed on acid-free paper   SPIN: 12111664   41/SPi   5 4 3 2 1 0

To the memory of my late grand father

and to my parents and brothers for their love and affection

Preface

I am indebted to my advisor Arup Bose from who I learned inference for diffusion processes during my graduate studies. I have benefited a lot from my discussions with Yacine Ait-Sahalia, Michael Sørensen, Yuri Kutoyants and Dan Crisan. I am grateful to all of them for their advice.

Charlotte, NC
January 30, 2007   Jaya P.N. Bishwal

Contents

Basic Notations

$(\Omega, \mathcal{F}, P)$	probability space
$\mathbb{R}$	real line
$\mathbb{C}$	complex plane
I_A	indicator function of a set A
$\overset{\mathcal{D}[P]}{\to}$	convergence in distribution under the measure P
$\overset{P}{\to}$	convergence in probability P
a.s. [P]	almost surely under the measure P
P-a.s.	almost surely under the measure P
$a_n = o(b_n)$	$\frac{a_n}{b_n} \to 0$
$a_n = O(b_n)$	$\frac{a_n}{b_n}$ is bounded
$X_n = o_P(b_n)$	$\frac{X_n}{b_n} \overset{P}{\to} 0$
$X_n = O_P(b_n)$	$\frac{X_n}{b_n}$ is stochastically bounded,
	i.e., $\underset{A \to \infty}{\lim} \underset{n}{\sup} P\{|\frac{X_n}{b_n}| > A\} = 0$
$\square$	end of a proof
$A := B$	A is defined by B
$A =: B$	B is defined by A
$\equiv$	identically equal
$\ll$	absolute continuity of two measures
i.i.d.	independent and identically distributed
$\mathcal{N}(a, b)$	normal distribution with mean a and variance b
$\Phi(.)$	standard normal distribution function
$X \sim F$	X has the distribution F
w.r.t.	with respect to
r.h.s.	right hand side
l.h.s.	left hand side
$a \bigvee b$	maximum of a and b
$a \bigwedge b$	minimum of a and b

1

Parametric Stochastic Differential Equations

Stochastic differential equations (SDEs) are a natural choice to model the time evolution of dynamic systems which are subject to random influences (cf. Arnold (1974), Van Kampen (1981)). For example, in physics the dynamics of ions in superionic conductors are modelled via Langevin equations (cf. Dieterich *et al.* (1980)), and in engineering the dynamics of mechanical devices are described by differential equations under the influence of process noise as errors of measurement (cf. Gelb (1974)). Other applications are in biology (cf. Jennrich and Bright (1976)), medicine (cf. Jones (1984)), econometrics (cf. Bergstrom (1976, 1988)), finance (cf. Black and Scholes (1973)), geophysics (cf. Arato (1982)) and oceanography (cf. Adler *et al.* (1996)).

It is natural that a model contains unknown parameters. We consider the model as the parametric Itô stochastic differential equation

$$dX_t \; = \; \mu \left(\theta, \, t, \, X_t \right) dt \; + \; \sigma \left(\vartheta, \, t, \, X_t \right) dW_t, \;\; t \geq 0, \;\; X_0 = \zeta$$

where $\{W_t, t \geq 0\}$ is a standard Wiener process, $\mu : \Theta \times [0,T] \times \mathbb{R} \to \mathbb{R}$, called the *drift coefficient*, and $\sigma : \Xi \times [0,T] \times \mathbb{R} \to \mathbb{R}^+$, called the *diffusion coefficient*, are known functions except the unknown parameters θ and ϑ, $\Theta \subset \mathbb{R}$, $\Xi \subset \mathbb{R}$ and $E(\zeta^2) < \infty$. The drift coefficient is also called the *trend coefficient* or *damping coefficient* or *translation coefficient*. The diffusion coefficient is also called *volatility coefficient*. Under local Lipschitz and the linear growth conditions on the coefficients μ and σ, there exists a unique strong solution of the above Itô SDE, called the *diffusion process* or simply a *diffusion*, which is a continuous strong Markov semimartingale. The drift and the diffusion coefficients are respectively the instantaneous mean and instantaneous standard deviation of the process. Note that the diffusion coefficient is almost surely determined by the process, i.e., it can be estimated without any error if observed continuously throughout a time interval (see Doob (1953), Genon-Catalot and Jacod (1994)). We assume that the unknown parameter in the diffusion coefficient ϑ is known and for simplicity only we shall assume that $\sigma = 1$ and our aim is to estimate the unknown parameter θ.

First we sketch some very popular SDE models.

Bachelier Model
$$dX_t = \beta \, dt + \sigma dW_t$$

Black-Scholes Model
$$dX_t = \beta X_t \, dt + \sigma X_t dW_t$$

Ornstein-Uhlenbeck Model
$$dY_t = \beta X_t \, dt + \sigma dW_t$$

Feller Square root or Cox-Ingersoll-Ross Model
$$dX_t = (\alpha - \beta X_t)dt + \sigma \sqrt{X_t} dW_t$$

Radial Ornstein-Uhlenbeck Process
$$dX_t = (\alpha X_t^{-1} - X_t) \, dt + \sigma dW_t$$

Squared Radial Ornstein-Uhlenbeck Process
$$dX_t = (1 + 2\beta X_t) \, dt + 2\sigma \sqrt{X_t} dW_t$$

Note that X_t the square of the Ornstein-Uhlenbeck process Y_t

$$dY_t = \beta Y_t \, dt + \sigma dW_t$$

Chan-Karloyi-Logstaff-Sanders Model
$$dX_t = \kappa(\theta - X_t) \, dt + \sigma \, X_t^{\gamma} \, dW_t$$

Hyperbolic Diffusion
$$dX_t = \alpha \frac{X_t}{\sqrt{1 + X_t^2}} \, dt + \sigma dW_t$$

Gompertz Diffusion
$$dX_t = (\alpha X_t - \beta X_t \log X_t) \, dt + \sigma X_t dW_t$$

Here X_t is the tumor volume which is measured at discrete time, α is the intrinsic growth rate of the tumor, β is the tumor growth acceleration factor, and σ is the diffusion coefficient.

The knowledge of the distribution of the estimator may be applied to evaluate the distribution of other important growing parameters used to access tumor treatment modalities. Some of these parameters are the plateau of the model $X_\infty = \exp(\frac{\alpha}{\beta})$, tumor growth decay, and the first time the growth curve of the model reaches X_∞.

Logistic Diffusion

Consider the stochastic analogue of the logistic growth model

$$dX_t = (\alpha X_t - \beta X_t^2)\, dt + \sigma X_t dW_t$$

This diffusion is useful for modeling the growth of populations.

Kessler-Sørensen Model

$$dX_t = -\theta \tan(X_t)\, dt + \sigma dW_t$$

By applying Itô formula, a diffusion process with some diffusion coefficient can be reduced to one with unit diffusion coefficient.

Following are most popular short term interest rate models.

Vasicek Model

$$dX_t = (\alpha + \beta X_t)dt + \sigma dW_t$$

Cox-Ingersoll-Ross Model

$$dX_t = (\alpha + \beta X_t)dt + \sigma \sqrt{X_t}dW_t$$

Dothan Model

$$dX_t = (\alpha + \beta X_t)dt + \sigma X_t dW_t$$

Black-Derman-Toy Model

$$dX_t = \beta(t)X_t dt + \sigma(t)X_t dW_t$$

Black-Karasinksi Model

$$d(\log X_t) = (\alpha(t) + \beta(t)\log X_t)dt + \sigma_t dW_t$$

Ho-Lee Model

$$dX_t = \alpha(t)dt + \sigma dW_t^H$$

Hull-White (Extended Vasicek) Model

$$dX_t = (\alpha(t) + \beta(t)X_t)dt + \sigma_t dW_t$$

Hull-White (Extended CIR) Model

$$dX_t = (\alpha(t) + \beta(t)X_t)dt + \sigma_t \sqrt{X_t}dW_t$$

Cox-Ingersoll-Ross 1.5 model

$$dX_t = \sigma X_t^{3/2}dW_t$$

Inverse Square Root Model or Ahn-Gao Model

$$dX_t = \beta(\mu - X_t)X_t dt + \sigma X_t^{3/2}dW_t$$

Ait-Sahalia Model

$$dX_t = (\alpha + \beta X_t + \gamma X_t^{-1} + \delta X_t^2)dt + \sigma X_t^\gamma dW_t$$

This a nonlinear interest rate model.

For existence and uniqueness of solutions of finite dimensional stochastic differential equations, properties of stochastic integrals, and diffusion and diffusion type processes see e.g., the books by McKean (1969), Gikhman and Skorohod (1972), Itô and McKean (1974), McShane (1974), Arnold (1974), Friedman (1975), Stroock and Varadhan (1979), Elliot (1982), Ikeda and Watanabe (1989), Rogers and Williams (1987), Karatzas and Shreve (1987), Liptser and Shiryayev (1977, 1989), Kunita (1990), Protter (1990), Revuz and Yor (1991), Øksendal (1995), Krylov (1995), Mao (1997). For numerical analysis and approximations of SDEs see the books by Gard (1988), Kloeden, Platen and Schurz (1994), Kloeden and Platen (1995) and Milshtein (1995). For existence, uniqueness of solutions and other properties of infinite dimensional SDEs see the books by Curtain and Pritchard (1978), Metivier and Pellaumail (1980), Itô (1984), Walsh (1986) and Kallianpur and Xiong (1995).

The asymptotic approach to statistical estimation is frequently adopted because of its general applicability and relative simplicity. In this monograph we study the asymptotic behaviour of several estimators of the unknown parameter θ appearing in the drift coefficient based on observations of the diffusion process $\{X_t, t \geq 0\}$ on a time interval $[0, T]$. Note that the observation of diffusion can be continuous or discrete. Continuous observation of diffusion is a mathematical idealization and has a very rich theory, for example Itô stochastic calculus, stochastic filtering, inference for continuously observed diffusions and much more behind it. But the path of the diffusion process is very kinky and no measuring device can follow a diffusion trajectory continuously. Hence the observation is always discrete in practice. Research on discretely observed diffusions is growing recently with a powerful theory of simulation schemes and numerical analysis of SDEs behind it.

The asymptotic estimation of θ, based on continuous observation of $\{X_t\}$ on $[0, T]$ can be studied by different limits, for example, $T \to \infty$, $\sigma(\vartheta, t, X_t) \to 0$, $\mu(\theta, t, X_t) \to \infty$, or any combination of these conditions that provide the increase of the integrals $\int_0^T [\mu(\theta, t, X_t)\sigma^{-1}(\vartheta, t, X_t)]^2 dt$ and $\int_0^T [\mu'(\theta, t, X_t)\sigma^{-1}(\vartheta, t, X_t)]^2 dt$, where prime denotes derivative with respect to θ. Parameter estimation in SDE was first studied by Arato, Kolmogorov and Sinai (1962) who applied it to a geophysical problem. For long time asymptotics $(T \to \infty)$ of parameter estimation in stochastic differential equations see the books by Liptser and Shiryayev (1978), Basawa and Prakasa Rao (1980), Arato (1982), Linkov (1993), Küchler and Sørensen (1997), Prakasa Rao (1999) and Kutoyants (1999). For small noise asymptotics $(\sigma \to 0)$ of parameter estimation see the books by Ibragimov and Khasminskii (1981) and Kutoyants (1984a, 1994a).

If $\{X_t\}$ is observed at $0 = t_0 < t_1 < t_2 < ... < t_n = T$ with $\Delta_n = \max\limits_{1 \leq i \leq n} |t_i - t_{i-1}|$ the asymptotic estimation of θ can be studied by different limits, for example, $\Delta_n \to 0, n \to \infty$ and $T \to \infty$ (or $\sigma \to 0$) or $\Delta_n = \Delta$ remaining fixed and $n \to \infty$. See Genon-Catalot (1987).

In the infinite dimensional diffusion models there are even different asymptotic frameworks. For example, in a stochastic partial differential equation, based on continuous observation, asymptotics can also be obtained when the intensity of noise and the observation time length remain fixed, but the number of Fourier coefficients in the expansion of the solution random field increases to infinity. Based on discrete observations, asymptotics can be obtained by this condition along with some sampling design conditions of discrete observations as in the finite dimensional case.

Our asymptotic framework in this monograph is long time for continuous observation and decreasing lag time along with increasing observation time for discrete observations.

The monograph is broadly divided into two parts. The first part (Chapters 2-6) deals with the estimation of the drift parameter when the diffusion process is observed continuously throughout a time interval. The second part (Chapters 7-10) is concerned with the estimation of the drift parameter when the diffusion process is observed at a set of discrete time points.

Asymptotic properties such as weak or strong consistency, asymptotic normality, asymptotic efficiency etc. of various estimators of drift parameter of Itô SDEs when observed continuously throughout a time interval, has been studied extensively during the last three decades. In linear homogeneous SDEs, maximum likelihood estimation was studied by Taraskin (1974), Brown and Hewitt (1975a), Kulinich (1975), Lee and Kozin (1977), Feigin (1976, 1979), Le Breton (1977), Tsitovich (1977), Arato (1978), Bellach (1980, 1983), Le Breton and Musiela (1984), Musiela (1976, 1984), Sørensen (1992), Küchler and Sørensen (1994a,b), Jankunas and Khasminskii (1997) and Khasminskii et al. (1999). In nonlinear homogeneous SDEs maximum likelihood estimation was studied by Kutoyants (1977), Bauer (1980), Prakasa Rao and Rubin (1981), Bose (1983a, 1986b), Bayes estimation was studied by Kutoyants (1977), Bauer (1980), Bose (1983b, 1986b), maximum probability estimation was studied by Prakasa Rao (1982), minimum contrast estimation was studied by Lanska (1979), M-estimation was studied by Yoshida (1988, 1990), minimum distance estimation was studied by Dietz and Kutoyants (1997). In nonlinear nonhomogeneous SDEs maximum likelihood estimation was studied by Kutoyants (1978, 1984a), Borkar and Bagchi (1982), Mishra and Prakasa Rao (1985), Dietz (1989) and Levanony, Shwartz and Zeitouni (1993, 1994), Bayes estimation was studied by Kutoyants (1978, 1984a). For survey of work in continuously observed diffusions, see Bergstrom (1976), Prakasa Rao (1985), Barndorff-Neilson and Sørensen (1994) and Prakasa Rao (1999). The following is a summary of Chapters 1-5.

In Chapter 2 we start with the historically oldest example of stochastic differential equation called the Langevin equation and whose solution is called the Ornstein-Uhlenbeck (O-U) process. In this case $\mu(\theta, t, X) = \theta X_t$. The first order theory like consistency, asymptotic normality etc. is well known for this case, see Le Breton (1977), Liptser and Shiryayev (1978). We study the rate of convergence in consistency and asymptotic normality via the large deviations probability bound and the Berry-Esseen bound for the minimum contrast estimator (MCE) of the drift parameter when the process is observed continuously over $[0, T]$. Then we study more general nonlinear ergodic diffusion model and study the Berry-Esseen bound for Bayes estimators. We also posterior large deviations and posterior Berry-Esseen bound. Mishra and Prakasa Rao (1985a) obtained $O(T^{-1/5})$ Berry-Esseen bound and $O(T^{-1/5})$ large deviation probability bound for the MLE for the Ornstein-Uhlenbeck model. For the MLE, Bose (1986a) improved the Berry-Esseen bound to $O(T^{-1/2}(\log T)^2)$. (The main result in Bose (1986a) has a misprint and gives the rate as $O(T^{-1/2})$, but by following the proof given there it is clear that the rate is $O(T^{-1/2}(\log T)^2)$.) Bose (1985) obtained the rate $O(T^{-1/2} \log T)$. Bishwal and Bose (1995) improved this rate to $O(T^{-1/2}(\log T)^{1/2})$. For the MLE, Bishwal (2000a) obtained $O(T^{-1})$ bound on the large deviation probability and the Berry-Esseen bound of the order $O(T^{-1/2})$ using nonrandom norming. This bound is consistent with the classical i.i.d. situation. Next we consider nonlinear diffusion model and obtain exponential rate of concentration of the posterior distribution, suitably normalized and centered at the MLE, around the true value of the parameter and also $O(T^{-1/2})$ rate of convergence of posterior distribution to normal distribution. We then establish $o(T^{-1/2})$ bound on the equivalence of the MLE and the Bayes estimator, thereby improving the $O(T^{-3/20})$ bound in Mishra and Prakasa Rao (1991). We obtain $O(T^{-1/2})$ Berry-Esseen bound and $O(T^{-1})$ bound on the large deviation probability of the BEs. This chapter is adapted from Bishwal (2004a) and Bishwal (2005a).

In Chapter 3 we deal with estimation in nonlinear SDE with the parameter appearing nonlinearly in the drift coefficient, based on continuous observation of the corresponding diffusion process over an interval $[0, T]$. In this case $\mu(\theta, t, X_t) = f(\theta, X_t)$. We obtain exponential bounds on large deviation probability for the MLE and regular BEs. The method of proof is due to Ibragimov and Khasminskii (1981). Some examples are presented. This chapter is adapted from Bishwal (1999a).

In Chapter 4 we study the asymptotic properties of various estimators of the parameter appearing nonlinearly in the nonhomogeneous drift coefficient of a functional stochastic differential equation when the corresponding solution process, called the diffusion type process, is observed over a continuous time interval $[0, T]$. We show that the maximum likelihood estimator, maximum probability estimator and regular Bayes estimators are strongly consistent and when suitably normalised, converge to a mixture of normal distribution and are locally asymptotically minimax in the Hajek-Le Cam sense as $T \to \infty$

under some regularity conditions. Also we show that posterior distributions, suitably normalised and centered at the maximum likelihood estimator, converge to a mixture of normal distribution. Further, the maximum likelihood estimator and the regular Bayes estimators are asymptotically equivalent as $T \to \infty$. We illustrate the results through the exponential memory Ornstein-Uhlenbeck process, the nonhomogeneous Ornstein-Uhlenbeck process and the Kalman-Bucy filter model where the limit distribution of the above estimators and the posteriors is shown to be Cauchy. This chapter is adapted from Bishwal (2004b).

In Chapter 5 we study estimation of a real valued parameter in infinite dimensional SDEs based on continuous observation of the diffusion. This area is relatively young and in our opinion is exciting and difficult. A few contributions in the existing literature are devoted to parameter (finite or infinite dimensional) estimation in infinite dimensional SDEs see, e.g., Aihara (1992, 1994, 1995), Aihara and Bagchi (1988, 1989, 1991), Bagchi and Borkar (1984), Loges (1984), Koski and Loges (1985, 1986), Huebner, Khasminskii, Rozovskii (1992), Huebner (1993), Huebner and Rozovskii (1995), Kim (1996). We consider the drift coefficient as $\theta A X_t$ with A being the infinitesimal generator of a strongly continuous semigroup acting on a real separable Hilbert space H and θ real valued. We obtain the Bernstein-von Mises theorem concerning the normal convergence of the posterior distributions and and strong consistency and asymptotic normality of the BEs of a parameter appearing linearly in the drift coefficient of Hilbert space valued SDE when the solution is observed continuously throughout a time interval $[0, T]$ and $T \to \infty$. It is also shown that BEs, for smooth priors and loss functions, are asymptotically equivalent to the MLE as $T \to \infty$. Finally, the properties of sequential maximum likelihood estimate of θ are studied when the corresponding diffusion process is observed until the observed Fisher information of the process exceeds a predetermined level of precision. In particular, it is shown that the estimate is unbiased, uniformly normally distributed and efficient. This chapter is adapted from Bishwal (1999b) and Bishwal (2002a).

In Chapter 6 we consider non-Markovian non-semimartingale models. Recently long memory processes or stochastic models having long range dependence phenomena have been paid a lot of attention in view of their applications in finance, hydrology and computer networks (see Beran (1994), Mandelbrot (1997), Shiryaev (1999), Rogers (1997), Dhehiche and Eddahbi (1999)). While parameter estimation in discrete time models having long-range dependence like the autoregressive fractionally integrated moving average (ARFIMA) models have already been paid a lot of attention, this problem for continuous time models is not well settled. Here we study estimation problem for continuous time long memory processes. This chapter is adapted from Bishwal (2003a).

Parameter estimation in diffusion processes based on observations at discrete time points is of much more practical importance due to the impossibility of observing diffusions continuously throughout a time interval. Note that diffusion process can be observed either at deterministic or at random sampling instants. For random sampling, e.g., from a point process, the sampling process may be independent of the observation process or may be dependent on it. Also in random sampling scheme, e.g., Poisson sampling (see Duffie and Glynn (2004)) the inverse estimation problem arises i.e., the estimation of the parameters of the sampling process when the parameters of the observation process is known. For a survey of earlier works on inference in continuous time processes based on observations at random sampling schemes see Stoyanov (1984). Jacod (1993) studied random sampling from a process with independent increments. Later on this scheme was used by Genon-Catalot and Jacod (1994) for the estimation of the parameter of the diffusion coefficient of a diffusion process. Duffie and Glynn (2004) studied the asymptotics of generalized method of moments estimators for a continuous time Markov process from observations at random sampling instants. In this monograph we will only be dealing with deterministic sampling scheme. We assume that the diffusion process $\{X(t)\}$ is observed at $\{0 = t_0 < t_1 < \ldots < t_n = T\}$ with $t_i - t_{i-1} = \frac{T}{n} = h \to 0, i = 1, 2, \ldots, n$.

Note that when one observes the process continuously throughout a time interval the diffusion coefficient is almost surely determined by the process. But when one has discrete observations, the problem of estimation of the diffusion coefficient also arises. Estimation of the parameter in the diffusion coefficient from discrete observations has been studied by Penev (1985), Dohnal (1987), Genon-Catalot and Jacod (1993), Florens-Zmirou (1993), and others. However, we will not deal with estimation of the diffusion coefficient in this monograph.

Drift parameter estimation in diffusion processes based on discrete observations has been studied by many authors. Le Breton (1976) and Dorogovcev (1976) appear to be the first persons to study estimation in discretely observed diffusions. While Le Breton (1976) used approximate maximum likelihood estimation, Dorogovcev (1976) used conditional least squares estimation. Robinson (1977) studied exact maximum likelihood estimation in discretely observed Ornstein-Uhlenbeck process. Other works on approximate maximum likelihood estimation (where the continuous likelihood is approximated), also called the maximum contrast estimation, are Bellach (1983), Genon-Catalot (1987, 1990), Yoshida (1992), Bishwal and Mishra (1995), Harison (1996), Clement (1993, 1995, 1997a,b) and Kessler (1997). Dacunha-Castelle and Florens-Zmirou (1986) studied consistency and asymptotic normality of MLE by using an expansion of the transition density of an ergodic diffusion. While ideally one should of course use maximum likelihood, in practice it is difficult because only in a few cases the transition densities are available. Pedersen (1995a,b) used numerical approximations based on iterations of the Gaussian transition densities emanating from the Euler-Maruyama scheme and studied approximate maximum likelihood estimation. Ait-Sahalia (2002) used Hermite function expansion of the transition density giving an accurate theoretical approximation and studied approximate maximum likelihood estimation.

In the conditional least squares estimation method (see Hall and Heyde (1981)) one minimizes the quadratic

$$Q_n(\theta) = \sum_{i=1}^{n} \frac{\left[X_{t_i} - X_{t_{i-1}} - \mu(\theta, t_{i-1}, X_{t_{i-1}})\Delta t_i\right]^2}{\sigma^2(t_{i-1}, X_{t_{i-1}})\Delta t_i}.$$

For equally spaced partition $0 = t_0 < t_1 < \ldots < t_n = T$ with $\Delta t_i = t_i - t_{i-1} = \frac{T}{n}, i = 1, 2, \ldots, n$, for the homogeneous stationary ergodic diffusion, Dorogovcev (1976) proved the weak consistency of the CLSE $\hat{\theta}_{n,T}$ as $T \to \infty$ and $\frac{T}{n} \to 0$ which we call the *slowly increasing experimental design* (SIED) condition. Kasonga (1988) proved the strong consistency of $\hat{\theta}_{n,T}$ as $T \to \infty$ and $\frac{T}{n} \to 0$. Prakasa Rao (1983) proved the asymptotic normality and asymptotic efficiency of $\hat{\theta}_{n,T}$ as $T \to \infty$ and $\frac{T}{\sqrt{n}} \to 0$, called the *rapidly increasing experimental design* (RIED) condition (see, Prakasa Rao (1988b)). Penev (1985) studied the consistency and asymptotic normality of a multidimensional parameter extending Dorogovcev (1976) and Prakasa Rao (1983). Florens-Zmirou (1989) proved the weak consistency of the minimum contrast estimator as $T \to \infty$ and $\frac{T}{n} \to 0$. He proved the asymptotic normality of this estimator as $T \to \infty$ and $\frac{T}{n^{2/3}} \to 0$ which we call the *moderately increasing experimental design* (MIED) condition. The properties of AMLE based on Itô type approximation of the Girsanov density are studied by Yoshida (1992) by studying the weak convergence of the approximate likelihood ratio random field when $T \to \infty$ and $\frac{T}{n^{2/3}} \to 0$. Genon-Catalot (1990) studied the asymptotic properties of maximum contrast estimators (using contrast function related to an approximate likelihood) in the nonlinear SDE as the intensity of noise becomes small and $T \to \infty$. Genon-Catalot et al. (1998a,b) studied the asymptotic properties of minimum contrast estimator in a stochastic volatility model that is a partially observed diffusion process. Note that the estimation methods discussed in this paragraph through different approaches are equivalent.

There are many other approaches to drift estimation for discretely observed diffusions. Prakasa Rao (1988b, 1999) gave a survey of estimation in discretely observed stochastic processes. Because of the difficulty in performing accurate maximum likelihood, much research has focussed on finding alternatives in the form of various estimating functions. Bibby and Sørensen (1995a, b) allowed $T \to \infty$ and $n \to \infty$ letting $t_i - t_{i-1} = \Delta(i = 1, 2, \ldots, n)$ fixed and found approximate martingale estimating functions based on approximate log-likelihood function and showed that the estimators based on these estimating functions are consistent and asymptotically normal as $n \to \infty$. Other estimating functions have also been proposed in the literature, e.g., estimating functions based on eigenvalues (see Kessler and Sørensen (1999)) and simple, explicit estimating functions (see Kessler (2000)). Sørensen, H. (2001, 2004), Sørensen, M. (1997, 1999) and Jacobsen (2002) also studied discretely observed diffusions through estimating functions. Bibby and Sørensen

(1995b) gave a review of martingale estimating functions based on discretely observed diffusions. Pedersen (1994) used quasi-likelihood approach for martingale estimating functions. McLeish and Kolkiewicz (1997) proposed method of estimation based on higher order Itô-Taylor expansion.

Lo (1988) studied the maximum likelihood estimation (both based on exact likelihood function based on transition densities and also on approximate discretized model) of a jump-diffusion process. Laredo (1990) studied the asymptotic sufficiency property of incomplete observations of a diffusion process which include discrete observations and studied consistency and asymptotic normality of the minimum contrast estimator. Sørensen (1992) studied the properties of estimates based on discrete observations from a linear SDE by embedding the discrete process into the continuous one. Kloeden *et al.* (1992) studied the effect of discretization on drift estimation. Gouriéroux *et al.* (1993), Gouriéroux and Monfort (1994) and Broze *et al.* (1998) studied the properties of estimates by indirect inference method. Overbeck and Ryden (1997) studied the asymptotics of the MLE and the CLSE in the Cox-Ingersoll-Ross model whose solution is a Bessel process. Gallant and Long (1997) showed the asymptotic properties of minimum chi-squared estimator as the moment function entering the chi-squared criterion and the number of past observations entering each moment function increase. Elerian *et al.* (2001) studied Bayesian estimation in nonlinear SDEs through a MCMC method using the Euler-Maruyama discretization scheme. The following is a summary of Chapters 7-10.

In Chapter 7 we assume $\mu(\theta, t, X_t) = \theta f(X_t)$. If complete (continuous) observation were available, a desirable estimate would be the maximum likelihood estimate. Based on discrete observations, one approach of finding a good estimate would to try to find an estimate which is as close as possible to the continuous MLE. For further refinement, one should measure the loss of information due to discretization. For the Ornstein-Uhlenbeck process, Le Breton (1976) proposed an approximate maximum likelihood estimator (AMLE) and showed that the difference between his AMLE and the continuous MLE is of the order $O_P((\frac{T^2}{n})^{1/2})$. We obtain several AMLEs with faster rates of convergence than that of Le Breton (1976). For this purpose first we obtain several higher order discrete approximations of the Fisk-Stratonovich integral. We use these approximations and the rectangular rule ordinary integral approximation to obtain different AMLEs. *Interalia* we introduce a new stochastic integral which will be of independent interest.

In Chapter 8 we return to the Ornstein-Uhlenbeck process, i.e., $\mu(\theta, t, X_t) = \theta X_t$ and we investigate the asymptotic properties of the conditional least squares estimator (CLSE) (see Hall and Heyde (1981)) as $T \to \infty$ and $\frac{T}{n} \to 0$. For the homogeneous nonlinear stationary ergodic case, i.e., $\mu(\theta, t, X) = f(\theta, X_t)$ under some regularity conditions Dorogovcev (1976) showed that the weak consistency and Kasonga (1988) showed the strong consistency of the CLSE as $T \to \infty$ and Prakasa Rao (1983) showed that one needs $T \to \infty$ and $\frac{T}{\sqrt{n}} \to 0$ as $n \to \infty$. This means that one needs larger number of observations to obtain asymptotic normality of the CLSE than for consistency. Till date no approximation results are known for this estimator. We obtain Berry-Esseen bound of the order $O(\max(T^{-1/2}(\log T)^{1/2}, \frac{T^2}{n}(\log T)^{-1},$ $\frac{T^2}{n}(\log T)^{-1}))$ for this estimator using nonrandom and parameter free random normings. Using parameter dependent random norming, we obtain the rate $O(T^{-1/2} \bigvee (\frac{T^2}{n})^{1/3})$. We also obtain large deviation probability bound for the CLSE. Its rate of convergence to the continuous MLE for fixed T, is of the order $O_P(\frac{T^2}{n})^{1/2}$. We study another approximate MLE here whose Berry-Esseen bound is of the order $O(T^{-1/2}(\log T)^{1/2} \bigvee \frac{T^4}{n^2}(\log T)^{-1})$ using nonrandom and sample dependent random normings. With a random norming which is parameter dependent the Berry-Esseen bound is shown to be of the order $O(T^{-1/2} \bigvee (\frac{T}{n})^{2/3}))$ and its rate of convergence to the continuous MLE is of the order $O_P(\frac{T^2}{n})$. From the above result it is clear that one needs $T \to \infty$ and $\frac{T}{n^{2/3}} \to 0$ as $n \to \infty$ for the asymptotic normality of the AMLE and the AMLE has a faster rate of convergence than CLSE. This chapter is adapted from Bishwal and Bose (2001) and Bishwal (2006a).

In Chapter 9 we consider discretely observed SDE with homogeneous stationary ergodic solution where the parameter and the process appear nonlinearly in the drift coefficient, i.e., $\mu(\theta, t, X_t) = f(\theta, X_t)$. Asymptotic normality of approximate MLEs, approximate Bayes estimators and approximate maximum probability estimators of the drift parameter based on two different approximate likelihoods are obtained via the study of weak convergence of the approximate likelihood ratio random fields. Also the *Bernstein-von Mises type theorems* with the two approximate likelihoods are studied. Asymptotic properties of conditional least squares estimator are studied via the weak convergence of the least squares random field. We relax to some extent the regularity conditions and the RIED condition used by Prakasa Rao (1983) who obtained asymptotic normality through the usual normal equations and Cramer's approach. Instead we use the *moderately increasing experimental design* (MIED) condition, i.e., $T \to \infty$ and $\frac{T}{n^{2/3}} \to 0$ as $n \to \infty$. This chapter is adapted from Bishwal (1999c) and Bishwal (2005b).

In Chapter 10 we show that discretization after the application of Itô formula in the Girsanov likelihood produces estimators of the drift which have faster rates of convergence than the Euler estimator for stationary ergodic diffusions and is free of approximating the stochastic integral. The discretization schemes are related to the Hausdorff moment problem. We show strong consistency, asymptotic normality and a Berry-Esseen bound for the corresponding approximate maximum likelihood estimators of the drift parameter from high frequency data observed over a long time period. This chapter is adapted from Bishwal (2007a).

Part I

Continuous Sampling

2

Rates of Weak Convergence of Estimators in Homogeneous Diffusions

2.1 Introduction

Let $(\Omega, \mathcal{F}, \{\mathcal{F}_t\}_{t\geq 0}, P)$ be a complete filtered probability space where the filtration $\{\mathcal{F}_t\}_{t\geq 0}$ satisfies the *usual hypotheses*:

(i) nondecreasing, i.e., $\mathcal{F}_s \subseteq \mathcal{F}_t \subset \mathcal{F}$ for $s \leq t$;

(ii) right continuity i.e., $\mathcal{F}_t = \mathcal{F}_{t_+} = \underset{u>t}{\cap} \mathcal{F}_u$;

(iii) completeness, i.e., $\mathcal{F}_0$ contains all the P-null sets of $\mathcal{F}$.

Such a complete filtered probability space satisfying the usual hypotheses is called a *stochastic basis*.

On the stochastic basis $(\Omega, \mathcal{F}, \{\mathcal{F}_t\}_{t\geq 0}, P)$ define the Ornstein-Uhlenbeck process $\{X_t\}$ satisfying the stochastic differential equation

$$dX_t = -\theta X_t dt + dW_t, \quad X_0 = 0 \tag{1.1}$$

where $\{W_t\}$ is a standard Wiener process with respect to the filtration $\{\mathcal{F}_t\}_{t\geq 0}$. Here $\theta \in \Theta \subseteq \mathbb{R}^+$ is the unknown parameter to be estimated on the basis of continuous observation of the process $\{X_t\}$ on the time interval $[0, T]$.

Model (1.1) is historically the oldest example of SDE known as the *Langevin equation*. One may think of X as being the speed of a particle of mass m which, at time t, is subjected to a force composed of two parts, a frictional force $-m\theta X_t$ and a fluctuating force, formally written as $m\frac{dW_t}{dt}$, so that equation (1.1) is, formally, nothing else but Newton's law. Alternatively, one may think of X as the prevailing short term interest rate in a term-structure model. It is also known as *Vasicek model* in financial literature. This is a homogeneous linear model and is the continuous version of the first order Gaussian autoregressive process. If X_0 is normally distributed or a constant, then the solution $\{X_t\}$ is a Gaussian process.

It is well known that the maximum likelihood estimator (MLE) and Bayes estimators (BEs) (for smooth prior and loss functions) are strongly consistent and asymptotically normally distributed as $T \to \infty$ (see Basawa and Prakasa Rao (1980), Prakasa Rao (1999)). We obtain the rates of convergence in asymptotic normality, i.e., Berry-Esseen bound for the pseudo MLE in Theorem 2.5. The asymptotic normality of the posterior distribution, i.e., the Bernstein-von Mises theorem is also well known for this model (see Basawa and Prakasa Rao (1980)). We obtain rate of convergence in posterior consistency in Theorem 4.2 and Theorem 4.3. Rate of convergence in posterior asymptotic normality is obtained in Theorem 4.9. We obtain a bound on the equivalence of the MLE and the Bayes estimators in Theorem 5.10. Finally for Bayes estimators we obtain large deviation bound in Theorem 5.7 and Berry-Esseen bound in Theorem 5.11. Our results show that the MLE and the Bayes estimators have the same rates of convergence.

The Chapter is organized as follows: In Section 2.2 we obtain different rates of convergence of the MCE to the normal distribution with nonrandom norming extending the methods of Pfanzagl (1971) developed in the context of i.i.d. situation for minimum contrast estimates. Also we obtain large deviation bound for the MLE in this Section. In Section 2.3 we obtain the rate of convergence in posterior consistency and posterior asymptotic normality extending the methods of Strasser (1977) developed for the i.i.d. case. In Section 2.4 we obtain bounds on the large deviation probability for the Bayes estimator, bound on the asymptotic equivalence of the MLE and the Bayes estimators and the Berry-Esseen bound for the Bayes estimator extending the methods of Strasser (1977) who obtained the results for the i.i.d. case.

We begin with some preliminary lemmas. Lemma 1.1 (a) is well known and a proof of part (b) may be found in Michel and Pfanzagl (1971). Throughout the Chapter C denotes a generic constant (perhaps depending on θ, but not on anything else).

Lemma 1.1 *Let X, Y and Z be arbitrary random variables on a probability space and $P(Z > 0) = 1$. Then for all $\epsilon > 0$,*

(a) $\displaystyle\sup_{x \in \mathbb{R}} |P\{X + Y \leq x\} - \varphi(x)| \leq \sup_{x \in \mathbb{R}} |P\{X \leq x\} - \Phi(x)| + P\{|Y| > \epsilon\} + \epsilon.$

(b) $\displaystyle\sup_{x \in \mathbb{R}} \left| P\left(\frac{Y}{Z} \leq x\right) - \Phi(x) \right| \leq \sup_{x \in \mathbb{R}} |P(Y \leq x) - \Phi(x)| + P(|Z - 1| > \epsilon) + \epsilon.$

This chapter is adapted from Bishwal (2004a) and Bishwal (2005a).

2.2 Berry-Esseen Bounds for Estimators in the Ornstein-Uhlenbeck Process

Let us denote the realization $\{X_t, 0 \leq t \leq T\}$ by X_0^T. Let P_θ^T be the measure generated on the space (C_T, B_T) of continuous functions on $[0, T]$ with the associated Borel σ-algebra B_T generated under the supremum norm by the process X_0^T and P_0^T be the standard Wiener measure. P_θ^T is absolutely continuous with respect to P_0^T and the Radon-Nikodym derivative (likelihood) of P_θ^T with respect to P_0^T based on X_0^T is given by

$$L_T(\theta) := \frac{dP_\theta^T}{dP_0^T}(X_0^T) = \exp\left\{-\theta \int_0^T X_t dX_t - \frac{\theta^2}{2} \int_0^T X_t^2 dt\right\}. \qquad (2.1)$$

Maximizing the log-likelihood with respect to θ provides the maximum likelihood estimate (MLE)

$$\theta_T = -\frac{\int_0^T X_t dX_t}{\int_0^T X_t^2 dt}. \qquad (2.2)$$

θ_T is strongly consistent and $T^{1/2}(\theta_T - \theta)$ asymptotically $\mathcal{N}(0, \frac{1}{2\theta})$ distributed as $T \to \infty$ (see Le Breton (1977) and Basawa and Prakasa Rao (1980)).

Note that

$$\left(\frac{T}{2\theta}\right)^{1/2} (\theta_T - \theta) = -\frac{\left(\frac{2\theta}{T}\right)^{1/2} Z_T}{\left(\frac{2\theta}{T}\right) I_T}. \qquad (2.3)$$

where

$$Z_T = \int_0^T X_t dW_t \quad \text{and} \quad I_T = \int_0^T X_t^2 dt.$$

I_T is called the energy of the O-U process. In (2.3), the numerator of the normalized MLE is a normalized martingale which converges in distribution to the standard normal variable and the denominator is its corresponding normalized increasing process which converges to one almost surely as $T \to \infty$.

The Berry-Esseen rate for the ratio of two processes can be split up into two components: the Berry-Esseen rate for the numerator and, the rate of convergence of the denominator to one (see Lemma 1.1 (b)). For the Berry-Esseen bound of the MLE θ_T using norming given in (2.3), Mishra and Prakasa Rao (1985a) used this approach. For the normal approximation of the numerator, they embedded it in a Brownian motion by Kunita-Watanabe theorem and used Lemma 3.2 of Hall and Heyde (1980) on the Berry-Esseen bound for the Brownian motion with random time. For the convergence of the denominator to one, they used Chebyshev's inequality. These led to the rate $O(T^{-1/5})$. One can use Burkholder inequality for the convergence of the denominator to one to improve this rate to $O(T^{-1/4} + \epsilon)$, $\epsilon > 0$. Note that using Skorohod embedding method, one can not obtain a rate better than $O(T^{-1/4}$. Bose (1986a) used characteristic function followed by Esseen's lemma for the numerator. The denominator was linked with the numerator via Itô formula. He obtained the rate $O(T^{-1/2}(\log T)^2)$. (Theorem 3.4 in Bose (1986a) has a misprint and gives the rate as $O(T^{-1/2})$, but by following the proof given there it is clear that the rate is $O(T^{-1/2}(\log T)^2)$.)

A slight improvement in this bound is possible in the following way. Bose (1985) linked the denominator with the numerator by Itô formula and using

Lemma 1.1 obtained the rate $O(T^{-1/2}(\log T))$. Bose (1986a) has shown that the rate of convergence to normality of the numerator is $O(T^{-1/2})$. For the convergence of the denominator to one, in Bishwal and Bose (1995) we obtain exponential bound using the moment generating function of the denominator. Using these two together and Lemma 1.1, in Bishwal and Bose (1995) we improved the rate to $O(T^{-1/2}(\log T)^{1/2})$. Note that using the *splitting technique*, i.e., Lemma 1.1, one can not obtain a rate better than $O(T^{-1/2}(\log T)^{1/2})$ as is evident from the i.i.d. case of Michel and Pfanzagl (1971), who obtained the rate $O(n^{-1/2}(\log n)^{1/2})$ where n is the number of observations.

Note that the norming in (2.3) depends on the unknown parameter. One may also use random normings, for two reasons. First, this may lead to an improved rate of convergence. Second, if the purpose is to obtain an approximate confidence interval then note that the pivot (2.3) when inverted will yield a confidence set but not necessarily a confidence interval. Bishwal (2001) the rate of convergence with appropriate random normings. The same rate as above is also obtained using different random normings.

Consider the score function, the derivative of the log-likelihood function, which is given by

$$Y_T(\theta) := \int_0^T X_t dX_t - \theta \int_0^T X_t^2 dt. \tag{2.4}$$

A solution of $Y_T(\theta) = 0$ provides the maximum likelihood estimate θ_T.

Strictly speaking, θ_T is not the maximum likelihood estimate of θ since $\hat{\theta}_T$ may take negative values where as the parameter θ is assumed to be strictly positive. For an exact definition of MLE, see Kutoyants (1994). Nevertheless, we use this terminology. It is well known that θ_T is strongly consistent and $T^{1/2}(\theta_T - \theta)$ asymptotically $\mathcal{N}(0, 2\theta)$ distributed as $T \to \infty$ (see Kutoyants (1994) and Prakasa Rao (1999)). Beyond the first order asymptotics, Bishwal (2000a) obtained the rate of normal approximation of the order $O(T^{-1/2})$ for the MLE thereby improving $O(T^{-1/5})$ rate of Mishra and Prakasa Rao (1985) and $O(T^{-1/2}(\log T)^{1/2})$ rate of Bishwal and Bose (1995). As far as the order of T is concerned, $O(T^{-1/2})$ rate is the sharpest rate one could expect since the norming is $T^{1/2}$. Bishwal (2000b) obtained rate of normal approximation of the order $O(T^{-1/2})$ for posterior distributions and Bayes estimators, with smooth priors and loss functions. Bishwal (2000a, b) used parameter dependent nonrandom norming. Bishwal (2001) obtained normal approximation of the order $O(T^{-1/2})$ for the MLE and Bayes estimators using two different random normings which are useful for computation of a confidence interval.

Using Itô formula (see Friedman (1975)), the score function $Y_T(\theta)$ can be written as

$$Y_T(\theta) = -\frac{1}{2}X_T^2 - \int_0^T (\theta X_t^2 - \frac{1}{2}) dt. \tag{2.5}$$

Consider the estimating function

$$M_T(\theta) = \int_0^T (\theta X_t^2 - \frac{1}{2}) dt \qquad (2.6)$$

and the minimum contrast estimate (MCE)

$$\hat{\theta}_T := \frac{T}{2 \int_0^T X_t^2 dt.} \qquad (2.7)$$

This estimator is simpler than the maximum likelihood estimator. The MCE does not involve stochastic integral unlike the MLE. The M-estimator is reduced to the minimum contrast estimator. It is well known that $\hat{\theta}_T$ is strongly consistent and asymptoticaly $\mathcal{N}(0, 2\theta)$ distributed as $T \to \infty$ (see Lanksa (1979)). The large deviations of $\hat{\theta}_T$ and $\hat{\theta}_T$ were obtained in Florens-Landais and Pham (1999). In particular, it was shown the large deviation probabilities of $\hat{\theta}_T$ are identical to those of $\hat{\theta}_T$ for $\theta \geq \theta_0/3$ but weaker for $\theta < \theta_0/3$ where θ_0 is the true value of the parameter. However, as far as the rate of normal approximation is concerned, we show that $\hat{\theta}_T$ and $\hat{\theta}_T$ have the same normal approximation rate of the order $O(T^{-1/2})$.

Note that using the Skorohod embedding of martingale, which has, since long, been the basic tool for normal approximation of martingales, will not give a rate better than $O(T^{-1/4})$ (see Borokov (1973)). To obtain the rate of normal approximation of the order $O(T^{-1/2})$, we adopt the Fourier method followed by the squeezing technique of Pfanzagl (1971) and Michel (1973) developed for the minimum contrast estimator in the i.i.d. case.

Observe that

$$\left(\frac{T}{2\theta}\right)^{1/2} (\hat{\theta}_T - \theta) = \frac{\left(\frac{2\theta}{T}\right)^{1/2} N_T}{\left(\frac{2\theta}{T}\right) I_T} \qquad (2.8)$$

where

$$N_T := \theta I_T - \frac{T}{2} \quad \text{and} \quad I_T := \int_0^T X_t^2 dt.$$

Our aim in this Section is to obtain the rate of convergence to normality, i.e., the bounds for $\sup_{x \in \mathbb{R}} |P\{r_T(\hat{\theta}_T - \theta) \leq x\} - \Phi(x)|$ where r_T is a suitable norming increasing to ∞ as $T \to \infty$. Here $\Phi(\cdot)$ is the standard normal cummulative distribution function. Such bounds are called Berry-Esseen bounds.

In order to obtain the optimum rate, we use the *squeezing technique* developed by Pfanzagl (1971) for the minimum contrast estimators in the i.i.d. case. Using this technique, we improve the Berry-Esseen bound for θ_T to $O(T^{-1/2})$. This is the sharpest rate one could hope for, since in the case of n i.i.d. observations the rate of convergence in the usual central limit theorem is $O(n^{-1/2})$.

Let $\Phi(\cdot)$ denote the standard normal distribution function. Throughout the chapter, C denotes a generic constant (which does not depend on T and x).

We start with some preliminary lemmas.

Lemma 2.1 (Bishwal (2000a)) For every $\delta > 0$,

$$P\left\{\left|\left(\frac{2\theta}{T}\right)I_T - 1\right| \geq \delta\right\} \leq CT^{-1}\delta^{-2}.$$

Lemma 2.2 (Bishwal (2000a))

$$\sup_{x \in \mathbb{R}}\left|P\left\{\left(\frac{2\theta}{T}\right)^{1/2} N_T \leq x\right\} - \Phi(x)\right| \leq C\, T^{-1/2}.$$

The following lemma gives the error rate on the difference of the characteristic function of the denominator of the MCE and the normal characteristic function.

Lemma 2.3 (a) (Bishwal (2000a)) Let $\phi_T(z_1) := E\exp(z_1 I_T), z_1 \in \mathbb{C}$. Then $\phi_T(z_1)$ exists for $|z_1| \leq \delta$, for some $\delta > 0$ and is given by

$$\phi_T(z_1) = \exp\left(\frac{\theta T}{2}\right)\left[\frac{2\gamma}{(\gamma - \theta)e^{-\gamma T} + (\gamma + \theta)e^{\gamma T}}\right]^{1/2}$$

where $\gamma = (\theta^2 - 2z_1)^{1/2}$ and we choose the principal branch of the square root.
(b) Let $H_{T,x} := \left(\frac{2\theta}{T}\right)^{1/2} N_T - \left(\frac{2\theta}{T}I_T - 1\right)x$.
   Then for $|x| \leq 2(\log T)^{1/2}$ and for $|u| \leq \epsilon T^{1/2}$, where ϵ is sufficiently small

$$\left|E\exp(iuH_{T,x}) - \exp(\frac{-u^2}{2})\right| \leq C\exp(\frac{-u^2}{4})(|u| + |u|^3)T^{-1/2}.$$

Proof: Note that part (a) is a Cameron-martin type formula. To prove (b), observe that

$$E\exp(iuH_{T,x})$$
$$= E\exp\left[-iu\left(\frac{2\theta}{T}\right)^{1/2}N_T - iu\left(\frac{2\theta}{T}I_T - 1\right)x\right]$$
$$= E\exp\left[-iu\left(\frac{2\theta}{T}\right)^{1/2}\left\{\theta I_T - \frac{T}{2}\right\} - iu\left(\frac{2\theta}{T}I_T - 1\right)x\right]$$
$$= E\exp(z_1 I_T + z_3)$$
$$= \exp(z_3)\phi_T(z_1)$$

where $z_1 := -iu\theta\delta_{T,x}$, and $z_3 := \frac{iuT}{2}\delta_{T,x}$ with $\delta_{T,x} := \left(\frac{2\theta}{T}\right)^{1/2} + \frac{2x}{T}$.
Note that $\phi_T(z_1)$ satisfies the conditions of (a) by choosing ϵ sufficiently small. Let $\alpha_{1,T}(u), \alpha_{2,T}(u), \alpha_{3,T}(u)$ and $\alpha_{4,T}(u)$ be functions which are

$O(|u|T^{-1/2}), O(|u|^2 T^{-1/2}), O(|u|^3 T^{-3/2})$ and $O(|u|^3 T^{-1/2})$ respectively. Note that for the given range of values of x and u, the conditions on z_1 for part (a) of Lemma are satisfied. Further, with $\beta_T(u) := 1 + iu\frac{\delta_{T,x}}{\theta} + \frac{u^2 \delta_{T,x}^2}{2\theta^2}$,

$$\gamma = (\theta^2 - 2z_1)^{1/2}$$
$$= \theta\left[1 - \frac{z_1}{\theta^2} - \frac{z_1^2}{2\theta^4} + \frac{z_1^3}{2\theta^8} + \cdots\right]$$
$$= \theta\left[1 + iu\frac{\delta_{T,x}}{\theta} + \frac{u^2\delta_{T,x}^2}{2\theta^2} + \frac{iu^3\delta_{T,x}^3}{2\theta^3} + \cdots\right]$$
$$= \theta[1 + \alpha_{1,T}(u) + \alpha_{2,T}(u) + \alpha_{3,T}(u)]$$
$$= \theta\beta_T(u) + \alpha_{3,T}(u)$$
$$= \theta[1 + \alpha_{1,T}(u)].$$

Thus
$$\gamma - \theta = \alpha_{1,T}, \ \ \gamma + \theta = 2\theta + \alpha_{1,T}.$$

Hence the above expectation equals

$$\exp\left(z_3 + \frac{\theta T}{2}\right)[2\theta\beta_T(u) + \alpha_{3,T}(u)]^{1/2}$$
$$\times [\alpha_{1,T}\exp\{-\theta T\beta_T(u) + \alpha_{4,T}(u)\}$$
$$+ (2\theta + \alpha_{1,T}(u))\exp\{\theta T\beta_T(u) + \alpha_{4,T}(u)\}]^{-1/2}$$
$$= [1 + \alpha_{1,T}(u)]^{1/2}[\alpha_{1,T}\exp(\chi_T(u)) + (1 + \alpha_{1,T}(u))\exp(\psi_T(t))]^{-1/2}$$

where

$$\chi_T(u) =: -\theta T\beta_T(u) + \alpha_{4,T}(u) - 2z_3 - \theta T$$
$$= -2\theta T + \alpha_{1,T}(u) + t^2\alpha_{1,T}(u).$$

and

$$\psi_T(u) =: \theta T\beta_T(u) + \alpha_{4,T}(u) - 2z_3 - \theta T$$
$$= \theta T\left[1 + iu\frac{\delta_{T,x}}{\theta} + \frac{u^2\delta_{T,x}^2}{2\theta^2}\right] + \alpha_{4,T}(u) - iuT\delta_{T,x} - \theta T$$
$$= \frac{u^2 T}{2\theta}\left[\left(\frac{2\theta}{T}\right)^{1/2} + \frac{2x}{T}\right]^2$$
$$= u^2 + u^2\alpha_{1,T}(u).$$

Hence, for the given range of values of u, $\chi_T(t) - \psi_T(u) \leq -\theta T$.

Hence the above expectation equals

$$\exp(-\frac{u^2}{2})(1 + \alpha_{1,T})^{1/2} \left[\alpha_{1,T} \exp\{-2\theta T + \alpha_{1,T} + t^2 \alpha_{1,T}\}\right.$$

$$+ (1 + \alpha_{1,T}(t)) \exp\{t^2 \alpha_{1,T}(t)\}\big]^{-1/2}$$

$$= \exp(-\frac{u^2}{2}) \left[1 + \alpha_{1,T})(1 + \alpha_{1,T}(1 + \alpha_{1,T}) \exp\{-\theta T + \alpha_{1,T} + t^2 \alpha_{1,T}\}\right]$$

$$\exp(u^2 \alpha_{1,T}(u)).$$

This completes the proof of the lemma. □

To obtain the rate of normal approximation for the MCE, we need the following estimate on the tail behavior of the MCE.

Lemma 2.4

$$P\left\{(\frac{T}{2\theta})^{1/2}|\hat{\theta}_T - \theta| \geq 2(\log T)^{1/2}\right\} \leq CT^{-1/2}.$$

Proof: Observe that

$$P\left\{(\frac{T}{2\theta})^{1/2}|\hat{\theta}_T - \theta| \geq 2(\log T)^{1/2}\right\}$$

$$= P\left\{\left|\frac{(\frac{2\theta}{T})^{1/2} N_T}{(\frac{2\theta}{T}) I_T}\right| \geq 2(\log T)^{1/2}\right\}$$

$$\leq P\left\{\left|(\frac{2\theta}{T})^{1/2} N_T\right| \geq (\log T)^{1/2}\right\} + P\left\{\left|\frac{2\theta}{T} I_T\right| \leq \frac{1}{2}\right\}$$

$$\leq \left|P\left\{(\frac{2\theta}{T})^{1/2}|N_T| \geq (\log T)^{1/2}\right\} - 2\Phi(-(\log T)^{1/2})\right|$$

$$+ 2\Phi(-(\log T)^{1/2}) + P\left\{\left|\frac{2\theta}{T} I_T - 1\right| \geq \frac{1}{2}\right\}$$

$$\leq \sup_{x \in \mathbb{R}} \left|P\left\{(\frac{2\theta}{T})^{1/2}|N_T| \geq x\right\} - 2\Phi(-x)\right|$$

$$\leq \sup_{x \in \mathbb{R}} \left|P\left\{(\frac{2\theta}{T})^{1/2}|N_T| \geq x\right\} - 2\Phi(-x)\right|$$

$$+ 2\Phi(-(\log T)^{1/2}) + P\left\{\left|(\frac{2\theta}{T}) I_T - 1\right| \geq \frac{1}{2}\right\}$$

$$\leq CT^{-1/2} + C(T \log T)^{-1/2} + CT^{-1}$$

$$\leq CT^{-1/2}.$$

The bounds for the first and the third terms come from Lemma 2.2 and Lemma 2.1 respectively and that for the middle term comes from Feller (1957, p. 166).

$\square$

Now we are ready to obtain the uniform rate of normal approximation of the distribution of the MCE.

Theorem 2.5

$$\sup_{x \in \mathbb{R}} \left| P\left\{ (\frac{T}{2\theta})^{1/2}(\hat{\theta}_T - \theta) \leq x \right\} - \Phi(x) \right| \leq C_\theta T^{-1/2}.$$

Proof: We shall consider two possibilities (i) and (ii).
(i) $|x| > 2(\log T)^{1/2}$.
We shall give a proof for the case $x > 2(\log T)^{1/2}$. The proof for the case $x < -2(\log T)^{1/2}$ runs similarly. Note that

$$\left| P\left\{ (\frac{T}{2\theta})^{1/2}(\hat{\theta}_T - \theta) \leq x \right\} - \Phi(x) \right| \leq P\left\{ (\frac{T}{2\theta})^{1/2}(\hat{\theta}_T - \theta) \geq x \right\} + \Phi(-x).$$

But $\Phi(-x) \leq \Phi(-2(\log T)^{1/2}) \leq CT^{-2}$. See Feller (1957, p. 166).
Moreover by Lemma 2.4, we have

$$P\left\{ (\frac{T}{2\theta})^{1/2}(\hat{\theta}_T - \theta) \geq 2(\log T)^{1/2} \right\} \leq CT^{-1/2}.$$

Hence

$$\left| P\left\{ (\frac{T}{2\theta})^{1/2}(\hat{\theta}_T - \theta) \leq x \right\} - \Phi(x) \right| \leq CT^{-1/2}.$$

(ii) $|x| \leq 2(\log T)^{1/2}$.

Let $A_T := \left\{ (\frac{T}{2\theta})^{1/2}|\hat{\theta}_T - \theta| \leq 2(\log T)^{1/2} \right\}$ and $B_T := \left\{ \frac{I_T}{T} > c_0 \right\}$

where $0 < c_0 < \frac{1}{2\theta}$. By Lemma 2.4, we have

$$P(A_T^c) \leq CT^{-1/2}. \tag{2.9}$$

By Lemma 2.1, we have

$$P(B_T^c) = P\left\{ \frac{2\theta}{T}I_T - 1 < 2\theta c_0 - 1 \right\} < P\left\{ |\frac{2\theta}{T}I_T - 1| > 1 - 2\theta c_0 \right\} \leq CT^{-1}. \tag{2.10}$$

Let b_0 be some positive number. On the set $A_T \cap B_T$ for all $T > T_0$ with $4b_0(\log T_0)^{1/2}(\frac{2\theta}{T_0})^{1/2} \leq c_0$, we have

$$(\frac{T}{2\theta})^{1/2}(\hat{\theta}_T - \theta) \leq x$$

$$\Rightarrow I_T + b_0 T(\hat{\theta}_T - \theta) < I_T + (\frac{T}{2\theta})^{1/2} 2b_0 \theta x$$

$$\Rightarrow (\frac{T}{2\theta})^{1/2}(\hat{\theta}_T - \theta)[I_T + b_0 T(\theta_T - \theta)] < x[I_T + (\frac{T}{2\theta})^{1/2} 2b_0 \theta x]$$

$$\Rightarrow (\hat{\theta}_T - \theta)I_T + b_0 T(\theta_T - \theta)^2 < (\frac{2\theta}{T})^{1/2} I_T x + 2b_0 \theta x^2$$

$$\Rightarrow -N_T + (\hat{\theta}_T - \theta)I_T + b_0 T(\hat{\theta}_T - \theta)^2 < -N_T + (\frac{2\theta}{T})^{1/2} I_T x + 2b_0 \theta x^2$$

$$\Rightarrow 0 < -N_T + (\frac{2\theta}{T})^{1/2} I_T x + 2b_0 \theta x^2$$

since

$$I_T + b_0 T(\theta_T - \theta)$$
$$> T c_0 + b_0 T(\theta_T - \theta)$$
$$> 4b_0 (\log T)^{1/2} (\frac{2\theta}{T})^{1/2} - 2b_0 (\log T)^{1/2} (\frac{2\theta}{T})^{1/2}$$
$$= 2b_0 (\log T)^{1/2} (\frac{2\theta}{T})^{1/2} > 0.$$

On the other hand, on the set $A_T \cap B_T$ for all $T > T_0$ with $4b_0(\log T_0)^{1/2}$ $(\frac{2\theta}{T_0})^{1/2} \leq c_0$, we have

$$(\frac{T}{2\theta})^{1/2}(\theta_T - \theta) > x$$

$$\Rightarrow I_T - b_0 T(\hat{\theta}_T - \theta) < I_T - (\frac{T}{2\theta})^{1/2} 2b_0 \theta x$$

$$\Rightarrow (\frac{T}{2\theta})^{1/2}(\hat{\theta}_T - \theta)[I_T - b_0 T(\theta_T - \theta)] > x[I_T - (\frac{T}{2\theta})^{1/2} 2b_0 \theta x]$$

$$\Rightarrow (\hat{\theta}_T - \theta)I_T - b_0 T(\hat{\theta}_T - \theta)^2 > (\frac{2\theta}{T})^{1/2} I_T x - 2b_0 \theta x^2$$

$$\Rightarrow -N_T + (\hat{\theta}_T - \theta)I_T - b_0 T(\theta_T - \theta)^2 > -N_T + (\frac{2\theta}{T})^{1/2} I_T x - 2b_0 \theta x^2$$

$$\Rightarrow 0 > -N_T + (\frac{2\theta}{T})^{1/2} I_T x - 2b_0 \theta x^2$$

since

$$I_T - b_0 T(\hat{\theta}_T - \theta)$$
$$> T c_0 - b_0 T(\hat{\theta}_T - \theta)$$
$$> 4b_0 (\log T)^{1/2} (\frac{2\theta}{T})^{1/2} - 2b_0 (\log T)^{1/2} (\frac{2\theta}{T})^{1/2}$$
$$= 2b_0 (\log T)^{1/2} (\frac{2\theta}{T})^{1/2} > 0.$$

Hence

$$0 < -N_T + (\frac{2\theta}{T})^{1/2}I_T x - 2b_0\theta x^2 \Rightarrow (\frac{T}{2\theta})^{1/2}(\theta_T - \theta) \le x.$$

Letting $D_{T,x}^{\pm} := \{-N_T + (\frac{2\theta}{T})^{1/2}I_T x \pm 2b_0\theta x^2 > 0\}$, we obtain

$$D_{T,x}^{-} \cap A_T \cap B_T \subseteq A_T \cap B_T \cap \left\{(\frac{T}{2\theta})^{1/2}(\theta_T - \theta) \le x\right\} \subseteq D_{T,x}^{+} \cap A_T \cap B_T. \quad (2.11)$$

If it is shown that

$$\left|P\left\{D_{T,x}^{\pm}\right\} - \Phi(x)\right| \le CT^{-1/2} \quad (2.12)$$

for all $T > T_0$ and $|x| \le 2(\log T)^{1/2}$, then the theorem would follow from (2.9) - (2.12).

We shall prove (2.12) for $D_{T,x}^{+}$. The proof for $D_{T,x}^{-}$ is analogous. Observe that

$$\begin{aligned}
&\left|P\left\{D_{T,x}^{+}\right\} - \Phi(x)\right| \\
=\ &\left|P\left\{(\frac{2\theta}{T})^{1/2}N_T - \left(\frac{2\theta}{T}I_T - 1\right)x < x + 2(\frac{2\theta}{T})^{1/2}b_0\theta x^2\right\} - \Phi(x)\right| \\
\le\ &\sup_{y\in\mathbb{R}}\left|P\left\{(\frac{2\theta}{T})^{1/2}N_T - \left(\frac{2\theta}{T}I_T - 1\right)x \le y\right\} - \Phi(y)\right| \\
&+ \left|\Phi\left(x + (\frac{2\theta}{T})^{1/2}b_0\theta x^2\right) - \Phi(x)\right| \\
=:\ &\Delta_1 + \Delta_2.
\end{aligned} \quad (2.13)$$

Lemma 2.3 (b) and Esseen's lemma immediately yield

$$\Delta_1 \le CT^{-1/2}. \quad (2.14)$$

On the other hand, for all $T > T_0$,

$$\Delta_2 \le 2(\frac{2\theta}{T})^{1/2}b_0\theta x^2(2\pi)^{-1/2}\exp(-\bar{x}^2/2)$$

where

$$|\bar{x} - x| \le 2(\frac{2\theta}{T})^{1/2}b_0\theta x^2.$$

Since $|x| \le 2(\log T)^{1/2}$, it follows that $|\bar{x}| > |x|/2$ for all $T > T_0$ and consequently

$$\begin{aligned}
\Delta_2 &\le 2(\frac{2\theta}{T})^{1/2}b_0\theta x^2(2\pi)^{-1/2}x^2\exp(-x^2/8) \\
&\le CT^{-1/2}.
\end{aligned} \quad (2.15)$$

From (2.13) - (2.15), we obtain

$$\left|P\left\{D_{T,x}^{+}\right\} - \Phi(x)\right| \le CT^{-1/2}.$$

This completes the proof of the theorem.   □

Remarks

(1) The rate in Theorem 2.5 is uniform over compact subsets of the parameter space.
(2) The rates of normal approximation of the conditional least squares estimator and an approximate maximum likelihood estimator when the O-U process is observed at discrete time points in $[0, T]$ has been studied in Bishwal and Bose (2001). The rate of normal approximation for some approximate minimum contrast estimators based on discrete observations of the O-U process has been studied in Bishwal (2006a).
(3) It remains to investigate the nonuniform rates of convergence to normality which are more useful.

2.3 Rates of Convergence in the Bernstein-von Mises Theorem for Ergodic Diffusions

In this section and the next section rates of convergence in consistency and asymptotic normality of the posterior distributions and regular Bayes estimators, for smooth priors and loss functions, of the drift parameter in an ergodic diffusion process satisfying a nonlinear Itô stochastic differential equation observed continuously over $[0, T]$ are obtained as $T \to \infty$ using the uniform nondegeneracy of the Malliavin covariance which plays the analogous role as the Cramer condition plays in the independent observation case.

First order asymptotic theory of estimation in diffusion processes is well developed now, see Liptser and Shiryayev (1977, 1978), Basawa and Prakasa Rao (1980), Kutoyants (1984, 1994, 1999) and Prakasa Rao (1999). Second order theory for diffusion processes is still in its infancy.

Let $(\Omega, \mathcal{F}, \{\mathcal{F}_t\}_{t \geq 0}, P)$ be a stochastic basis satisfying the usual conditions on which is defined a diffusion process X_t satisfying the Itô stochastic differential equation

$$dX_t = f(\theta, X_t)dt + dW_t, t \geq 0, \quad X_0 = \xi \tag{3.1}$$

where $\{W_t\}$ is a standard Wiener process with respect to the filtration $\{\mathcal{F}_t\}_{t \geq 0}$,
$f : \Theta \times \mathbb{R} \to \mathbb{R}$ is a known function except $\theta \in \Theta$ open in $\mathbb{R}$ the unknown parameter which is to be estimated on the basis of the observation $\{X_t, 0 \leq t \leq T\}$.

Let the realization $\{X_t, 0 \leq t \leq T\}$ be denoted by X_0^T. Let P_θ^T be the measure generated on the space (C_T, B_T) of continuous functions on $[0, T]$ with the associated Borel σ-algebra B_T generated under the supremum norm by the process X_0^T and P_0^T be the standard Wiener measure. Assume that when θ is the true value of the parameter P_θ^T is absolutely continuous with respect to P_0^T. Then the Radon-Nikodym derivative (likelihood) of P_θ^T with respect to P_0^T based on X_0^T is given by

$$L_T(\theta) := \frac{dP_\theta^T}{dP_0^T}(X_0^T) = \exp\left\{\int_0^T f(\theta, X_t)dX_t - \frac{1}{2}\int_0^T f^2(\theta, X_t)dt\right\}. \quad (3.2)$$

(See Liptser and Shiryayev (1977)). Denote $l_T(\theta) := \log L_T(\theta)$. The maximum likelihood estimate (MLE) is defined as

$$\theta_T := \arg\max_{\theta \in \Theta} l_T(\theta). \quad (3.3)$$

Under stationarity and ergodicity of the diffusion and some regularity conditions, it is well known that θ_T is strongly consistent and asymptotically normally distributed as $T \to \infty$ (see Prakasa Rao and Rubin (1981)). Beyond the first order asymptotics, Yoshida (1997) obtained the asymptotic expansion of the distribution of the normalized MLE using the Malliavin calculus techniques. In particular, it gives a Berry-Esseen bound of the order $O(T^{-1/2})$ for the MLE.

In this chapter, we study Bayesian asymptotics beyond the first order. We proceed as follows: Let Λ be a prior probability measure on $(\Theta, \mathcal{B})$ where $\mathcal{B}$ is the Borel σ-algebra of subsets of Θ. We assume the following regularity conditions on the prior:

(A1) For every $\delta > 0$ and every compact $K \subseteq \Theta$,

$$\inf_{\theta \in K} \Lambda\{\phi \in \Theta : |\phi - \theta| < \delta\} > 0.$$

(A2) For every compact $K \subseteq \Theta$, there exists $\alpha \geq 0$ (depending on K) such that

$$\liminf_T \inf_{\theta \in K} T^\alpha \Lambda\{\phi \in \Theta : |\phi - \theta| < T^{-1/2}\} > 0.$$

(A3) Λ has a density $\lambda(\cdot)$ with respect to Lebesgue measure on $\mathbb{R}$ which is continuous and positive on Θ and for every $\theta \in \Theta$ there exists a neighborhood U_θ of θ such that

$$\left|\frac{\lambda(\theta_1)}{\lambda(\theta_2)} - 1\right| \leq C|\theta_1 - \theta_2| \text{ for all } \theta_1, \theta_2 \in U_\theta.$$

(A4)$_m$ Λ_θ possesses m-th order absolute moments in $\theta \in \Theta$.

The posterior probability measure of $B \in \mathcal{B}$ given X_0^T is defined as

$$R_T(B) := \frac{\int_B L_T(\theta)\lambda(\theta)d\theta}{\int_\Theta L_T(\theta)\lambda(\theta)d\theta}. \quad (3.4)$$

In the i.i.d. case, it is well established that, for large sample size the mass of the posterior distribution concentrates in arbitrarily small neighbourhoods of the true parameter value (see Doob (1948), Schwarz (1965), Le Cam (1973)). This phenomenon is known as *posterior consistency*. Strasser (1976) estimated the rate at which the posterior distribution concentrates around the true parameter value. Recently, rates of convergence in posterior consistency in infinite dimensional parameter situation was studied by Ghoshal, Ghosh and Van der Vaart (2000), Shen and Wasserman (1998) and Huang (2000) in the i.i.d. case. We obtain rates of convergence in posterior consistency for the nonlinear ergodic diffusion process in Theorems 2.2 and 2.3 extending Strasser (1976).

The Bernstein-von Mises theorem (rather called the Bernstein-von Mises phenomenon in the terminology of Le Cam and Yang (1990)) or the Bayesian central limit theorem states that for large number of observations, suitably centered and normalized posterior distribution behaves like the normal distribution. In particular cases, the theorem was shown by Laplace, Bernstein and von-Mises in the early period of last century. See Le Cam (1986), Le Cam and Yang (1990) for a history. A complete proof in the i.i.d. case was given by Le Cam (1953). Since then many authors have extended the theorem to many dependent observations cases or refined the theorem in the i.i.d. case. Borwanker, Kallianpur and Prakasa Rao (1973) extended the theorem to discrete time stationary Markov process. Strasser (1976) and Hipp and Michel (1976) estimated the speed of convergence in this theorem in the i.i.d. case. Prakasa Rao (1978) obtained the rate of convergence in the Bernstein - von Mises theorem for Markov processes extending the work of Hipp and Michel (1976) in the i.i.d. case. Note that Bernstein-von Mises theorem fails in infinite dimensional parameter case even in i.i.d. case, see Freedman (1999).

The Bernstein-von Mises theorem for linear homogeneous diffusion processes was proved by Prakasa Rao (1980). Bose (1983) extended the work of Prakasa Rao (1980) to nonlinear homogeneous diffusion processes and Mishra (1989) and Harison (1992) extended the work of Bose (1983) to nonhomogeneous diffusion processes. Prakasa Rao (1981) extended his work to linear homogeneous diffusion field. Bishwal (1999b) proved Bernstein-von Mises theorem for Hilbert space valued diffusions. Bishwal (2000c) studied Bernstein-von Mises theorem and asymptotics of Bayes estimators for parabolic stochastic partial differential equations through the Fourier method.

Not much is known beyond the first order asymptotics of posteriors for diffusions. For linear homogeneous diffusion processes, Mishra and Prakasa Rao (1987) studied the rate of convergence in the Bernstein-von Mises theorem, i.e., they obtained bound in the approximation of suitably normalized and centered posterior density to normal density. The posterior density of the normalized parameter (centered at the maximum likelihood estimate) being a ratio of two measurable functions, its difference with the normal density can be expressed as the ratio of two measurable functions. The rate of convergence to zero of the ratio of two processes can be decomposed as the rate of convergence of the numerator to zero and the rate of convergence of the denominator to one. Mishra and Prakasa Rao (1987) used this method to obtain the rate of convergence of the normalized posterior density. This technique does not give the best possible rate which is evident when their result is applied to the Ornstein-Uhlenbeck process. Their results give $O(T^{-1/5})$ order of approximation of posterior density by normal density and $O(T^{-1/4}(\log T)^{1/4})$ order for the P_θ^T probabilities of exceptional set. Bishwal (2000b) obtained $O(T^{-1/2})$ rate of convergence of normalized posterior density to normal density and $o(T^{-1/2})$ rate for the P_θ^T probabilities of the exceptional set by using the method of Strasser (1977) for the i.i.d. case. In Theorem 2.9, we obtain same rate of convergence for posterior distributions of the drift parameter in stationary ergodic diffusion process X_t satisfying (3.1).

A $\mathcal{B} \times \mathcal{B}$ measurable function $D : \overline{\Theta} \times \overline{\Theta} \to \mathbb{R}$ is called a loss function if

$$D(\theta, \theta) < D(\phi, \theta) \quad \text{for all } \theta \in \Theta, \phi \in \overline{\Theta}, \theta \neq \phi.$$

A Bayes estimator (relative to a prior distribution Λ and a loss function D) is defined as

$$\widetilde{\theta}_T := \arg\min_{\theta \in \overline{\Theta}} \int_\Theta D(\theta, \phi) R_T(d\phi). \tag{3.5}$$

For the quadratic loss function $\widetilde{\theta}_T$ is the the posterior mean

$$E(\theta | X_0^T) = \frac{\int_\Theta \phi L_T(\phi) \lambda(\phi) d\phi}{\int_\Theta L_T(\phi) \lambda(\phi) d\phi}. \tag{3.6}$$

For nonlinear ergodic diffusion model (3.1), Bose (1983) showed that, for smooth priors and loss functions, Bayes estimator $\widetilde{\theta}_T$ of θ is consistent and asymptotically normally distributed as $T \to \infty$. Bose (1983) also showed that $\sqrt{T}(\widetilde{\theta}_T - \theta_T) \to 0$ a.s. as $T \to \infty$. In particular, it was shown that the asymptotic behaviour of the Bayes estimator $\widetilde{\theta}_T$ and the MLE θ_T is the same as $T \to \infty$. For linearly parametrized (but in general nonlinear) diffusion, Mishra and Prakasa Rao (1991) obtained a bound on the convergence to zero of the probability $P\{\sqrt{T}(\widetilde{\theta}_T - \theta_T) > r(T)\}$ where $r(T) \downarrow 0$ as $T \to \infty$, that is, bound on the equivalence of the MLE and the BE. Their bound is not sharp which is evident when applied to the Ornstein-Uhlenbeck process, it gives only a bound $O(T^{-1/5})$. For the Ornstein-Uhlenbeck process, Bishwal (2000b) improved the bound on asymptotic equivalence of the above two estimators to $o(T^{-1/2})$ and Berry-Esseen bound of the Bayes estimator to $O(T^{-1/2})$. We generalize the result to stationary ergodic nonlinear diffusion process and obtain the same bound on the asymptotic equivalence of the MLE and the BE and same Berry-Esseen bound for the Bayes estimators as in the Ornstein-Uhlenbeck process. Note that asymptotic expansion of distribution of the bias corrected Bayes estimator for quadratic loss function, i.e., the posterior mean in diffusion model when the intensity of noise decreases to zero, was studied by Yoshida (1993) using the Malliavin calculus technique.

Let $\Phi(\cdot)$ denote the standard normal distribution function. Throughout the chapter C denotes a generic constant (perhaps depending on θ, but not on anything else). *Throughout the paper prime denotes derivative with respect to θ and dot denotes derivative with respect to x.*

We assume the following conditions:

(C1) The diffusion process $\{X_t, t \geq 0\}$ is stationary and ergodic with invariant measure ν at $\theta = \theta_0$ given by

$$\nu_\theta(dx) = \frac{n(\theta, x)}{\int_{-\infty}^{\infty} n(\theta, u)du} dx$$

where $n(\theta, x) = \exp(2 \int_0^x f(\theta, v)dv)$.

Further, ξ obeys the stationary distribution $\nu = \nu_{\theta_0}$.

(C2) (i) $|f(\theta, x) - f(\theta, y)| \leq C|x - y|$,
(ii) $|f(\theta, x) - f(\phi, x)| \leq C|\theta - \phi|$,
(iii) There exists $\gamma > 0$ such that

$$|f'(\theta, x) - f'(\phi, x)| \leq C|\theta - \phi|^\gamma.$$

(iv) There exists positive constant $C > 0$ such that for all u,

$$E_\theta \left[\exp \left\{ -u^2 (3T)^{-1} \int_0^T \inf_{\phi \in \Theta} (f'(\phi, X_t))^2 dt \right\} \right] \leq C \exp(-u^2 C).$$

(v) $f'(\theta, x)$ has polynomial majorant, i.e., $|f'(\theta, x)| \leq C(1 + |x|^p), p \geq 1$, C does not depend on θ.
(C3) $f(., x) \in C^4(\mathbb{R})$ with derivatives of at most polynomial growth order. Further $\sup_{x \in \mathbb{R}} \dot{f}(\theta_0, x) < 0$.

(C4) There exist constants $A > 0$ and $\gamma > 0$ such that for all $|x| \geq A$ and all $\theta \in \Theta$,

$$\mathrm{sgn}(x)f(\theta, x) \leq -\gamma.$$

(C5) (i) $f(\theta, x)$ is continuously differentiable with respect to θ.
(ii) $f'(\theta, x)$ is uniformly continuous in the following sense: For any compact $K \subset \Theta$,

$$\lim_{\delta \to 0} \sup_{\theta_0 \in K} \sup_{|\theta - \theta_0| < \delta} E_{\theta_0}|f'(\theta, X_0) - f'(\theta_0, X_0)|^2 = 0.$$

(C6) $|\dot{f}'(\theta, x)| \leq C(1 + |x|^c)$ for any $x \in \mathbb{R}$ and $\theta \in \Theta$.
The Fisher information satisfies

$$0 < I(\theta) := \int_{-\infty}^{\infty} (f'(\theta, x))^2 d\nu(x) < \infty$$

and for any $\delta > 0$, for any compact $K \subset \Theta$,

$$\inf_{\theta_0 \in K} \sup_{|\theta - \theta_0| > \delta} E_{\theta_0}|f'(\theta, X_0) - f'(\theta_0, X_0)|^2 > 0.$$

(C7) The Malliavin covariance of the process is nondegenerate.
We need the following results in the sequel.

Theorem 3.1 Under the conditions (C1) – (C7), for every $\epsilon > 0$,

$$P_{\theta_0}\left\{\left|\frac{1}{T}\int_0^T (f'(\theta_0, X_t))^2 dt - I(\theta_0)\right| > \epsilon\right\} \le CT^{-1}\epsilon^{-2}.$$

Proof: Let

$$g(\theta, x) := (f'(\theta, x))^2 - I(\theta).$$

Thus $E_{\theta_0}g(\theta, X_0) = 0$. Let

$$h(\theta, x) := G(\theta)^{-1}\exp\left\{2\int_0^x g(\theta, v)dv\right\}, x \in \mathbb{R}$$

where

$$G(\theta) := \int_{-\infty}^{\infty}\exp\left\{\int_0^y g(\theta, v)dv\right\}dy < \infty.$$

Let

$$F(\theta, x) := \int_0^x \frac{1}{g(\theta, y)}\int_{-\infty}^y h(\theta, z)g(\theta, z)dzdy.$$

By Itô formula,

$$dF(\theta, X_t) = g(\theta, X_t)dt + F'(\theta, X_t)dW_t.$$

Hence

$$\int_0^T g(\theta, X_t)dt = F(\theta, X_t) - F(\theta, X_0) - \int_0^T \frac{1}{g(\theta, X_t)}\int_{-\infty}^{X_t} h(\theta, v)g(\theta, v)dvdW_t.$$

By the martingale central limit theorem

$$\frac{1}{\sqrt{T}}\int_0^T \frac{1}{g(\theta, X_t)}\int_{-\infty}^{X_t} h(\theta, v)g(\theta, v)dvdW_t \xrightarrow{D} \mathcal{N}(0, \nu(F'^2)) \text{ as } T \to \infty$$

and

$$\frac{1}{\sqrt{T}}[F(\theta, X_t) - F(\theta, X_0)] \xrightarrow{P} 0 \text{ as } T \to \infty.$$

Observe that

$$|F(\theta, \xi)|$$
$$\le \left|\int_0^x \frac{1}{g(\theta, y)}\int_{-\infty}^y h(\theta, z)g(\theta, z)dzdy\right|$$
$$= \left|\int_0^x \int_y^{\infty} g(\theta, z)\exp\left\{2\int_y^z f(\theta, v)dv\right\}dy\right|$$
$$\le C\int_0^x \int_y^{\infty} |g(\theta, z)|\exp\{-2\gamma(z - y)\}dzdy$$

$$\leq C \int_0^x \int_y^\infty |1 + z^p| \exp\{-2\gamma(z-y)\} dz dy$$

$$\leq C \int_0^x \int_0^\infty |1 + (y+v)^p| \exp\{-2\gamma v\} dv dy$$

$$\leq C \int_0^x (1 + y^p) dy.$$

This estimate holds for $y > A$. We have a similar estimate for $y < -A$. Thus

$$E_\theta |F(\theta, \xi)|^2 \leq C_1 + C_2 \int_{|y|>A} (1 + |y|^p) \exp\{-2\gamma|y-A|\} dy \leq C.$$

Similarly it can be shown that

$$E_\theta \left| \frac{g(\theta, v)h(\theta, v)}{h(\theta, \xi)} dv \right|^2 < C.$$

Thus

$$E_\theta \left| \frac{1}{T} \int_0^T g(\theta, X_t) dt \right|^2$$

$$\leq CT^{-2} E_\theta |F(\theta, \xi)|^2 + CT^{-1} E_\theta \left| \int_{-\infty}^\xi \frac{h(\theta, v)g(\theta, v)}{g(\theta, \xi)} dv \right|^2$$

$$\leq CT^{-1}.$$

Now the result follows by Chebyshev inequality.   □

Theorem 3.2 (Bishwal (1999a)) Under the conditions (C1) and (C2), for every $\epsilon > 0$ and every compact $K \subseteq \Theta$,

$$\sup_{\theta \in K} P_\theta \{|\theta_T - \theta| > \epsilon\} \leq C_K \exp(-CT\epsilon^2).$$

The following theorem is a consequence of Theorem 8 in Yoshida (1997).

Theorem 3.3 Under the conditions (C1) – (C7), for every compact $K \subseteq \Theta$,

$$\sup_{\theta \in K} \sup_{x \in \mathbb{R}} \left| P_\theta \left\{ (TI(\theta))^{1/2}(\theta_T - \theta) \leq x \right\} - \Phi(x) \right| = O(T^{-1/2}).$$

The rest of the chapter is organized as follows: In Section 2.4 we obtain rates of convergence in consistency and asymptotic normality of the posterior distributions and in Section 2.5 we obtain bound on the asymptotic equivalence of the MLE and the Bayes estimators and Berry-Esseen bounds for the Bayes estimators.

2.4 Rates of Convergence of the Posterior Distributions in Ergodic Diffusions

Introduce some notations:
Let $K \subseteq \Theta$ be compact and $M_{T,\theta} \in \mathcal{F}_T, \theta \in K$ be sets satisfying

$$\sup_{\theta \in K} P_\theta^T (M_{T,\theta}^c) = o(T^{-1/2}). \tag{4.1}$$

For every Borel set $B \in \mathcal{B}$, define

$$B_T(\theta) := \left\{ \phi \in \mathbb{R}^+ : (TI(\theta))^{1/2}(\phi - \theta_T) \in B \right\}, \tag{4.2}$$

$$J_T(\theta) := \frac{I_T(\theta)}{TI(\theta)} - 1 \tag{4.3}$$

where $I_T(\theta) := \int_0^T f'^2(\theta, X_t)dt$.
Let further $\Phi_m, m \in \mathbb{N}$ be signed measures on $\mathcal{B}$ defined by

$$\Phi_m(B) := \frac{1}{\sqrt{2\pi}} \int_B t^m \exp(-\frac{t^2}{2})dt, \ B \in \mathcal{B}. \tag{4.4}$$

Define
$$u_T(\phi) := (TI(\theta))^{1/2}(\phi - \theta_T), \ \phi \in \Theta, \tag{4.5}$$

$$V_T(s) := \left\{ \phi \in \mathbb{R}^+ : (TI(\theta))^{1/2}|\phi - \theta_T| \leq (s \log T)^{1/2} \right\} \tag{4.6}$$

where $s > 0, \theta \in K$, and

$$A_T(\phi) := \frac{1}{\sqrt{2\pi}}(TI(\theta))^{1/2} \frac{L_T(\phi)}{L_T(\theta_T)}. \tag{4.7}$$

where $\phi \in \Theta$ and $\theta \in K$.

Lemma 4.1(a) For every $\delta > 0$ and every compact $K \subseteq \Theta$ there exists $\epsilon_K > 0$ such that

$$\sup_{\theta \in K} P_\theta \left\{ - \sup_{|\phi - \theta| \geq \delta} \frac{1}{T} l_T(\phi) + E_\theta \left[l_T(\theta) \right] \leq \epsilon_K \right\} \leq C_K T^{-1}.$$

(b) For every $\epsilon > 0$ and every compact $K \subseteq \Theta$ there exists $\delta_K > 0$ such that

$$\sup_{\theta \in K} P_\theta \left\{ - \inf_{|\phi - \theta| < \delta_K} \frac{1}{T} l_T(\theta) + E_\theta \left[l_T(\theta) \right] > \epsilon \right\} \leq C_K T^{-1}.$$

Proof: Similar to Lemmas 1 and 2 in Strasser (1976). Details are omitted. $\square$

The next two theorems give the rates of convergence in posterior consistency.

Theorem 4.2 Let $B_\delta(\theta) = \{\phi \in \Theta : |\phi - \theta| < \delta\}$. Then for every $\delta > 0$ and every compact $K \subseteq \Theta$ there exists $\eta_K > 0$ such that

$$\sup_{\theta \in K} P_\theta\{R_T(B_\delta^c(\theta)) > \exp(-\eta_K T)\} \le C_K T^{-1}.$$

Proof: From (3.3), we have

$$R_T(B_\phi^c(\theta)) = \frac{\int_{|\phi-\theta|\ge\delta} L_T(\phi)\lambda(\phi)d\phi}{\int_\Theta L_T(\phi)\lambda(\phi)d\phi}.$$

Thus

$$R_T(B_\delta^c(\theta)) > \exp(-\eta_K T)$$

is equivalent to

$$\frac{1}{T} \log R_T(B_\delta^c(\theta)) > -\eta_K$$

which is further equivalent to

$$\frac{1}{T} \log \int_{|\phi-\theta|\ge\delta} L_T(\phi)\lambda(\phi)d\phi - \frac{1}{T} \log \int_\Theta L_T(\phi)\lambda(\phi)d\phi > -\eta_K$$

which implies for arbitrary $\delta_1 > 0$ and $\theta \in K$

$$- \inf_{|\phi-\theta|<\delta_1} \frac{1}{T} \log L_T(\phi) + \sup_{|\phi-\theta|\ge\delta} \frac{1}{T} \log L_T(\phi) > -\eta_K + \frac{a}{T}$$

where

$$a := \inf_{\theta \in K} \log \Lambda\{\phi \in \Theta : |\phi - \theta| < \delta_1\}$$

From Lemma 4.1(a), we obtain

$$\sup_{\theta \in K} P_\theta \left\{ E_\theta[l_T(\theta)] - \epsilon_K \le \sup_{|\phi-\theta|\ge\delta} \frac{1}{T} l_T(\phi) \right\} \le C_K T^{-1}.$$

Therefore we only have to show that

$$\sup_{\theta \in K} P_\theta \left\{ - \inf_{|\phi-\theta|<\delta_1} \frac{1}{T} l_T(\theta) + E_\theta(l_T(\theta)) > \epsilon_K - \eta_K + \frac{a}{T} \right\} \le C_K T^{-1}.$$

But this follows from Lemma 4.1(b) choosing δ_1 sufficiently small. Combination of the last two inequalities completes the proof. □

The next theorem shows that a result similar to Theorem 4.2 is true when the radius of the neighbourhoods of the MLE θ_T decreases of the order $T^{-1/2}(\log T)^{1/2}$.

Theorem 4.3 For every $r > 0$ and every compact $K \subseteq \Theta$, there exist $s_K \geq 0$ and $C_K \geq 0$ such that

$$P_\theta \left\{ R_T \left(V_T^c(s_K) \right) \geq C_K T^{-r} \right\} \leq C_K T^{-1}.$$

Proof: Using Theorem 3.2 and Theorem 4.2, we have for every $\delta > 0$ there exists $\eta_K > 0$ such that

$$\sup_{\theta \in K} P_\theta \left\{ R_T \left\{ \phi \in \Theta : |\phi - \theta_T| \geq \delta \right\} > \exp(-\eta_K T) \right\} \leq C_K T^{-1}.$$

Hence it is sufficient to prove that for every $r > 0$ and every compact $K \subseteq \Theta$ there exists $s_K > 0, C_K$ and $\delta_K > 0$ such that

$$\sup_{\theta \in K} P_\theta \left[R_T \left\{ \phi \in \Theta : \delta_K > |\phi - \theta_T| > T^{-1/2} \left(\frac{s_K \log T}{I(\theta)} \right)^{1/2} \right\} \geq C_K T^{-r} \right]$$
$$\leq C_K T^{-1}.$$

From Theorem 3.1, for every $\epsilon > 0$,

$$P_\theta \left\{ \left| \frac{I_T}{T} - I(\theta) \right| > \epsilon \right\} \leq C T^{-1}.$$

Define the sets

$$N_T(\epsilon, \theta) := \left\{ \left| \frac{I_T}{T} - I(\theta) \right| > \epsilon \right\},$$

$$S_T^\theta(s) := \left\{ \phi \in \Theta : \delta_K > |\phi - \theta_T| > T^{-1/2} \left(\frac{s \log T}{I(\theta)} \right)^{1/2} \right\},$$

$$H_T := \left\{ \phi \in \Theta : |\phi - \theta_T| < T^{-1/2} \right\}.$$

By Taylor expansion, we have

$$l_T(\phi) = l_T(\theta_T) + (\phi - \theta_T) l'_T(\theta_T) + \frac{(\phi - \theta_T)^2}{2} l''_T(\theta^*) \tag{4.8}$$

where $|\theta^* - \theta| \leq |\theta_T - \theta|$. Since $l'_T(\theta_T) = 0$, we obtain

$$\exp\left[l_T(\phi) - l_T(\theta_T) \right] = \exp\left\{ -\frac{T}{2}(\phi - \theta_T)^2 \frac{l''_T(\theta^*)}{T} \right\} \tag{4.9}$$
$$=: G_T(\phi)$$

It follows that in $N_T^c(\epsilon, \theta), T > T_K$ implies

$$R_T(S_T^\theta(s)) = \frac{\int_{S_T^\theta(s)} G_T(\phi) \lambda(\phi) d\phi}{\int_\Theta G_T(\phi) \lambda(\phi) d\phi}$$
$$\leq \frac{\int_{S_T^\theta(s)} \exp\left\{ -\frac{T}{2}(\phi - \theta_T)^2 (I(\theta) - \epsilon) \right\} \lambda(\phi) d\phi}{\int_{H_T} \exp\left\{ -\frac{T}{2}(\phi - \theta_T)^2 (I(\theta) + \epsilon) \right\} \lambda(\phi) d\phi}$$
$$\leq \frac{\exp\left\{ -\frac{1}{2} s (\log T)(1 - \epsilon I^{-1}(\theta)) \right\}}{\exp\left\{ -\frac{1}{2}(I(\theta) + \epsilon) \right\} \Lambda(H_T)}$$

Let $\rho > 0$ be such that $K_1 := \{\phi \in \Theta : dist(\phi, K) \le \rho\}$ is a compact subset of Θ. Let

$$\tilde{N}_T(\rho, \theta) := \{|\theta_T - \theta| > \rho\}.$$

Then from Theorem 3.2,

$$\sup_{\theta \in K} P_\theta \left(\tilde{N}_T(\rho, \theta) \right) \le CT^{-1}.$$

Now in $(\tilde{N}_T(\rho, \theta))^c$, by assumption (A2)

$$\liminf_T T^\alpha \Lambda(H_T)$$

$$= \liminf_T T^\alpha \inf_{\theta \in K_1} \Lambda \left\{ \phi \in \Theta : |\phi - \theta| < T^{-1/2} \right\} > 0.$$

Thus it follows that there exists $C_K \ge 0$ with

$$P_\theta \left\{ R_T(S_T^\theta(s)) > C_K T^{(2\alpha - s(1 - \epsilon I^{-1}(\theta)))/2} \right\} \le C_K T^{-1}.$$

It is obvious that for every $r > 0, s > 0$ can be chosen such that the assertion of the Theorem holds. $\square$

Lemma 4.4 Let $g_\theta : \Theta \to \bar{\mathbb{R}}, \theta \in \Theta$, be a class of Λ-integrable functions satisfying

$$\sup_{\theta \in K} \int_\Theta |g_\theta(\phi)| \Lambda(d\phi) < \infty$$

for every compact $K \subseteq \Theta$. Then for every $\delta > 0$ and every compact $K \subseteq \Theta$ there exist $\eta_K > 0$ such that

$$P_\theta \left\{ \int_{|\phi - \theta| \ge \delta} |g_\theta(\phi)| R_T(d\phi) > \exp(-\eta_K T) \right\} \le C_K T^{-1}$$

Proof: Since g_θ is Λ-integrable for every $\theta \in \Theta$ it is also R_T integrable. The proof of the lemma is completely same as the proof of Theorem 4.2 if the constant a is replaced by

$$\tilde{a} := \inf_{\theta \in K} \left\{ \log \Lambda \left\{ \phi \in \Theta : |\phi - \theta| < \delta_1 \right\} - \log \int_\Theta |g_\theta(\phi)| \Lambda(d\phi) \right\}.$$ $\square$

Lemma 4.5 Let $s > 0$. Let $K \subseteq \Theta$ be compact and choose $\epsilon > 0$ such that $\bar{K}^\epsilon \subseteq \Theta$ where $K^\epsilon = \{\phi \in \mathbb{R}^+ : |\phi - \theta| < \epsilon \text{ if } \theta \in K\}$. Then there exist sets $M_{T,\theta} \in \mathcal{F}_T, \theta \in K$ satisfying (2.2) such that $V_T(s) \subseteq K^\epsilon$ on the set $M_{T,\theta}$.
Proof: The proof follows from Theorem 3.2 which implies that we can assume $|\theta_T - \theta| < \epsilon/2$ on the set $M_{T,\theta}$. $\square$

Lemma 4.6 Under (A1) and (A4)$_m$, for $K \subseteq \Theta$ there exist $M_{T,\theta} \in \mathcal{F}_T, \theta \in K$ satisfying (4.2) such that on the set $M_{T,\theta}$

$$\left| \int_\Theta |u_T(\phi)|^m R_T(d\phi) - \int_{V_T(s_K)} |u_T(\phi)|^m R_T(d\phi) \right| \leq C_K T^{-1}.$$

Proof: Lemma 4.5 shows that we can assume $\theta_T \in K^\epsilon$ for $\epsilon > 0$ on the set $M_{T,\theta}$ and $\bar{K}^\epsilon \subseteq \Theta$. From Lemma 2.4 we obtain $\delta_K > 0$ and $\eta_K > 0$ such that in $M_{T,\theta}$,

$$\int_{|\phi - \theta| \geq \delta} |u_T(\phi)|^m R_T(d\phi) \leq C_K T^{m/2} \exp(-\eta_K T).$$

Hence it follows that

$$\int_{\Theta \setminus V_T(s_K)} |u_T(\phi)|^m R_T(d\phi) \leq C_K T^{m/2} \exp(-\eta_K T) + C_K T^{m/2} R_T(V_T^c(s_K))$$

in $M_{T,\theta}$. Now for sufficiently large $s_K > 0$ the assertion of the theorem follows from Theorem 4.3. $\square$

Lemma 4.7 Under (A1) and (A4)$_m$, for every $K \subseteq \Theta$ there exist sets $M_{T,\theta}, \theta \in K$ satisfying (4.1) such that on the set $M_{T,\theta}$

$$\sup_{B \in \mathcal{B}} \left| \int_{B_T(\theta) \cap \Theta} [u_T(\phi)]^m R_T(d\phi) - \frac{\int_{B_T(\theta) \cap V_T(s_K)} [u_T(\phi)]^m A_T(\phi) \lambda(\phi) d\phi}{\int_{V_T(s_K)} A_T(\phi) \lambda(\phi) d\phi} \right|$$
$$\leq C_K T^{-1/2}.$$

Proof: Lemma 4.5 shows that we may restrict our attention to the case $V_T(s_K) \subseteq \Theta$. Now

$$\sup_{B \in \mathcal{B}} \left| \int_{B_T(\theta) \cap \Theta} [u_T(\phi)]^m R_T(d\phi) - \frac{\int_{B_T(\theta) \cap V_T(s_K)} [u_T(\phi)]^m A_T(\phi) \lambda(\phi) d\phi}{\int_{V_T(s_K)} A_T(\phi) \lambda(\phi) d\phi} \right|$$
$$\leq \int_{\Theta \setminus V_T(s_K}} [u_T(\phi)]^m R_T(d\phi) + \frac{|R_T(V_T(s_K)) - 1|}{R_T(V_T(s_K))} \int_{V_T(s_K)} [u_T(\phi)]^m R_T(d\phi).$$

By Theorem 4.3 we have on the set $M_{T,\theta}$

$$\int_{V_T(s_K)} [u_T(\phi)]^m R_T(d\phi) \leq C_K < \infty$$

and

$$R_T(V_T(s_K)) \geq \epsilon_K > 0.$$

By Theorem 4.3 on the set $M_{T,\theta}$

$$|R_T(V_T(s_K)) - 1| \geq C_K T^{-r}.$$

By Lemma 4.6
$$\int_{\Theta \setminus V_T(s_K)} |u_T(\phi)|^m R_T(d\phi) \le C_K T^{-1}.$$

This completes the proof of the lemma. $\qquad\qquad\square$

Theorem 4.8 Under (A1) and (A4)$_m$, for every compact $K \subseteq \Theta$ there exist sets $M_{T,\theta}, \theta \in K$ satisfying (4.1) such that on the set $M_{T,\theta}$

$$\sup_{B \in \mathcal{B}} \left| \int_{B_T(\theta) \cap V_T(s_K)} [u_T(\phi)]^m A_T(\phi) \lambda(\phi) d\phi - \lambda(\theta_T) \left[\Phi_m(B) + \frac{1}{2} J_{T,\theta} \Phi_{m+2}(B) \right] \right|$$
$$\le C_K T^{-1/2}.$$

Proof: According to Lemma 4.5 we can assume that $V_T(s_K) \subseteq \Theta$ on the set $M_T(\theta)$. Note that
$$\exp[l_T(\phi) - l_T(\theta_T)] = \frac{L_T(\phi)}{L_T(\theta_T)}.$$

Hence from (4.7) and (4.9), we obtain

$$\begin{aligned} A_T(\phi) &= \frac{1}{\sqrt{2\pi}} (TI(\theta))^{1/2} \frac{L_T(\phi)}{L_T(\theta_T)} \\ &= \frac{1}{\sqrt{2\pi}} (TI(\theta))^{1/2} \exp\left\{ -\frac{T}{2} (\phi - \theta_T)^2 \frac{I_T}{T} \right\} \\ &= \frac{1}{\sqrt{2\pi}} (TI(\theta))^{1/2} \exp\left\{ -\frac{TI(\theta)}{2} (\phi - \theta_T)^2 (1 + J_T(\theta)) \right\} \quad \text{(from (4.3))} \end{aligned}$$

Define
$$Q_T := \left\{ y \in \mathbb{R} : y \in B, y^2 \le s_K \log T \right\}, \quad B \in \mathcal{B}.$$

Then for every $B \in \mathcal{B}$, substituting $u_T(\phi) = y$, on the set $M_T(\theta)$, we have

$$\int_{B_T(\theta) \cap V_T(s_K)} [u_T(\phi)]^m A_T(\phi) d\phi = \int_{Q_T} \exp\left(-\frac{y^2}{2} J_{T,\theta} \right) \Phi_n(dy).$$

Let us now use the inequality
$$\left| e^{-\beta} - 1 + \beta \right| \le \beta^2 e^{|\beta|}, \quad \beta \in \mathbb{R}$$

and put $\beta = \frac{y^2}{2} J_{T,\theta}$.

From Theorem 3.1 it follows that on the set $M_{T,\theta}$

$$|J_{T,\theta}| \le C_K T^{-1/2} (\log T)^{1/2}.$$

Thus we obtain

$$\sup_{B \in \mathcal{B}} \left| \int_{B_T(\theta) \cap V_T(s_K)} [u_T(\phi)]^m A_T(\phi) d\phi - \int_{Q_T} \left(1 - \frac{y^2}{2} J_{T,\theta} \right) \Phi_m(dy) \right| \le C_K T^{-1/2},$$

i.e.,

$$\sup_{B \in \mathcal{B}} \left| \int_{B_T(\theta) \cap V_T(s_K)} [u_T(\phi)]^m A_T(\phi) d\phi - \Phi_m(Q_T) + \frac{1}{2} J_{T,\theta} \Phi_{m+2}(Q_T) \right| \leq C_K T^{-1/2}.$$

But it is easily seen that

$$\sup_{B \in \mathcal{B}} |\Phi_m(B) - \Phi_m(Q_T)| \leq C_K T^{-1/2}$$

for sufficiently large $s_K > 0$. Hence

$$\sup_{B \in \mathcal{B}} \left| \int_{B_T(\theta) \cap V_T(s_K)} [u_T(\phi)]^m A_T(\phi) d\phi - \Phi_m(B) + \frac{1}{2} J_{T,\theta} \Phi_{m+2}(B) \right| \leq C_K T^{-1/2}.$$

According to the uniform cover theorem for every $\epsilon > 0$ satisfying $\bar{K}^\epsilon \subseteq \Theta$ there exists $e_K > 0$ such that $|\phi - \phi'| < e_K$ implies $|\lambda(\phi) - \lambda(\phi')| \leq C_K |\phi - \phi'|$ for every $\phi \in K^\epsilon$. Hence on the set $M_{T,\theta}$ we can assume that $V_T(s_K)$ is contained in an ϵ-neighbourhood of θ for every $\theta \in K$.
Then on the set $M_{T,\theta}$

$$\left| \sup_{B \in \mathcal{B}} \int_{B_T(\theta) \cap V_T(s_K)} [u_T(\phi)]^m A_T(\phi) \lambda(\phi) d\phi \right.$$

$$\left. - \lambda(\theta_T) \int_{B_T(\theta) \cap V_T(s_K)} [u_T(\phi)]^m A_T(\phi) d\phi \right|$$

$$\leq C_K \lambda(\theta_T) \int_{V_T(s_K)} |u_T(\phi)|^m A_T(\phi) |\phi - \theta_T| d\phi.$$

Using a similar substitution argument as before we obtain on the set $M_{T,\theta}$

$$\int_{V_T(s_K)} |u_T(\phi)|^m A_T(\phi) |\phi - \theta_T| d\phi$$

$$\leq C_K T^{-1/2} \int_{y^2 \geq s_K \log T} |y|^{m+1} \exp\left(-\frac{y^2}{2}(1 - \delta_K)\right) dy$$

since on the set $M_{T,\theta}$ we may assume that $J_{T,\theta} \geq -\delta_K$ if $y^2 \leq s_K \log T$. Now using (A1) and Theorem 3.2, we obtain $\lambda(\theta_T) \leq C_K < \infty$ on the set $M_{T,\theta}$. Hence the assertion of the theorem follows. $\qquad \square$

The next theorem gives the rate of convergence in the Bernstein-von Mises theorem.

Theorem 4.9 Under (A1) and (A4)$_m$, for every compact $K \subseteq \Theta$ there exist sets $M_{T,\theta}, \theta \in K$ satisfying (4.1) such that on the set $M_{T,\theta}$

$$\sup_{B \in \mathcal{B}} \left| \int_{B_T(\theta) \cap \Theta} [u_T(\phi)]^m R_T(\phi) d\phi - \Phi_m(B) + \frac{1}{2} J_{T,\theta} [\Phi_m(B) + \Phi_{m+2}(B)] \right|$$
$$\leq C_K T^{-1/2}.$$

Proof: Let us consider Lemma 4.7. The following inequality holds for $(\alpha, \beta, \gamma, \delta) \in \mathbb{R}^4, \beta, \delta \neq 0$.

$$\left| \frac{\alpha}{\beta} - \frac{\gamma}{\delta} \right| \leq \frac{1}{|\delta|} |\alpha - \gamma| + \left| \frac{\gamma}{\beta \delta} \right| |\beta - \delta|. \tag{4.10}$$

For fixed $B \in \mathcal{B}$, on the set $M_{T,\theta}$ define

$$\alpha := \int_{B_T(\theta) \cap V_T(s_K)} [u_T(\phi)]^m A_T(\phi) \lambda(\phi) d\phi, \quad \beta := \int_{V_T(s_K)} A_T(\phi) \lambda(\phi) d\phi$$

$$\gamma := \lambda(\theta_T) \left[\Phi_m(B) - \frac{1}{2} J_{T,\theta} \Phi_{m+2}(B) \right], \quad \delta := \lambda(\theta_T) \left[1 - \frac{1}{2} J_{T,\theta} \right].$$

Observe that

$$\sup_{B \in \mathcal{B}} \left| \int_{B_T(\theta) \cap \Theta} [u_T(\phi)]^m R_T(d\phi) - \frac{\Phi_m(B) - \frac{1}{2} J_{T,\theta} \Phi_{m+2}(B)}{1 - \frac{1}{2} J_{T,\theta}} \right|$$

$$\leq \sup_{B \in \mathcal{B}} \left| \int_{B_T(\theta) \cap \Theta} [u_T(\phi)]^m R_T(d\phi) - \frac{\int_{B_T(\theta) \cap V_T(s_K)} [u_T(\phi)]^m A_T(\phi) \lambda(\phi) d\phi}{\int_{V_T(s_K)} A_T(\phi) \lambda(\phi) d\phi} \right|$$

$$+ \sup_{B \in \mathcal{B}} \left| \frac{\int_{B_T(\theta) \cap V_T(s_K)} [u_T(\phi)]^m A_T(\phi) \lambda(\phi) d\phi}{\int_{V_T(s_K)} A_T(\phi) \lambda(\phi) d\phi} - \frac{\lambda(\theta_T) \left[\Phi_m(B) - \frac{1}{2} J_{T,\theta} \Phi_{m+2} \right]}{\lambda(\theta_T) \left[1 - \frac{1}{2} J_{T,\theta} \right]} \right|$$

$$=: H_1 + H_2$$

$$\leq C_K T^{-1/2} + \sup_{B \in \mathcal{B}} \frac{1}{\lambda(\theta_T) \left[1 - \frac{1}{2} J_{T,\theta} \right]} \times$$

$$\left| \int_{B_T(\theta) \cap V_T(s_K)} [u_T(\phi)]^m A_T(\phi) \lambda(\phi) d\phi - \lambda(\theta_T) \left[\Phi_m(B) - \frac{1}{2} J_{T,\theta} \Phi_{m+2}(B) \right] \right|$$

$$+ \frac{\left| \lambda(\theta_T) \left[\Phi_m(B) - \frac{1}{2} J_{T,\theta} \Phi_{m+2}(B) \right] \right|}{\left| \int_{V_T(s_K)} A_T(\phi) \lambda(\phi) d\phi \right| \left| \lambda(\theta_T) \left[1 - \frac{1}{2} J_{T,\theta} \right] \right|} \times$$

$$\left| \int_{V_T(s_K)} A_T(\phi) \lambda(\phi) d\phi - \lambda(\theta_T) \left[1 - \frac{1}{2} J_{T,\theta} \right] \right|$$

$$=: C_K T^{-1/2} + N_1 + N_2$$

$$\leq C_K T^{-1/2} \qquad \text{(since } \lambda(\theta_T) > 0 \text{ for every } \theta_T \in \Theta\text{)}.$$

Here the bound for H_1 comes from Lemma 4.7 and the second term H_2 is split up by using the inequality (4.10). The bounds for N_1 and N_2 come from Theorem 4.8.

Now we shall use the following inequality which holds for $\alpha \in \mathbb{R}, |\alpha| < 1$,

$$\left| \frac{1}{1-\alpha} - (1+\alpha) \right| \le \frac{\alpha^2}{1 - |\alpha|} \tag{4.11}$$

It justifies replacing division by $1 - \frac{1}{2} J_{T,\theta}$ with multiplying by $1 + \frac{1}{2} J_{T,\theta}$. Another application of the inequality $J_{T,\theta}^2 \le C_K T^{-1/2}$ on the set $M_{T,\theta}$ proves the assertion completely since terms containing $J_{T,\theta}^2$ can be omitted.   □

2.5 Berry-Esseen Bound for the Bayes Estimator

We obtain $o(T^{-1/2})$ bound for the asymptotic equivalence of the Bayes estimators and the maximum likelihood estimator. Using this bound and the Berry-Esseen bound for the maximum likelihood estimator finally we obtain $O(T^{-1/2})$ Berry-Esseen bound for the normalized Bayes estimator. Thus the maximum likelihood estimator and the Bayes estimators are *asymptotically efficient of order* $T^{-1/2}$.

We assume the following conditions on the loss function D.

(B1) $D : \bar{\Theta} \times \Theta \to \mathbb{R}$ is continuous.

(B2) (a) The mapping $\phi \to D(\theta, \phi), \theta \in \Theta$ is extended Λ-integrable for all $\theta \in \bar{\Theta}$ and has finite Λ- expectation if $\theta \in \Theta$.

(b) For every $\theta \in \bar{\Theta}$ there exists a neighbourhood U_θ^λ of θ such that for every neighbourhood $U \subseteq U_\theta^\lambda$ of θ, the mapping $\phi \to \inf L(U, \phi)$ is Λ-integrable.

(B3) (a) For every $\phi \in \Theta$ the mapping $\theta \to D(\theta, \phi)$ is twice differentiable in Θ. We denote

$$D_{10}(\theta, \phi) := \frac{\partial}{\partial \theta} D(\theta, \phi), D_{20}(\theta, \phi) := \frac{\partial^2}{\partial \theta^2} D(\theta, \phi).$$

(b) For every $\theta \in \Theta$, the mapping $\phi \to D_{10}(\theta, \phi)$ is differentiable in Θ. We denote

$$D_{11}(\theta, \phi) := \frac{\partial}{\partial \phi} D_{10}(\theta, \phi).$$

(B4) For every compact $K \subseteq \Theta$
  (a) $\sup_{\theta \in K} |D_{10}(\theta, \theta)| < \infty$
  (b) $\sup_{\theta \in K} |D_{20}(\theta, \theta)| < \infty$.

(B5) For every compact $K \subseteq \Theta$, $\inf_{\theta \in K} D_{20}(\theta, \theta) > 0$

(B6) For every compact $K \subseteq \Theta$,

  (a) $\sup_{\theta \in K} \int |D_{10}(\theta, \phi)| \Lambda(d\phi) < \infty$

  (b) $\sup_{\theta \in K} \int |D_{20}(\theta, \phi)| \Lambda(d\phi) < \infty$.

(B7) (a) For every $\theta \in \Theta$ there exists a neighbourhood $\widetilde{V}_\theta$ of θ and a constant $\widetilde{C}_\theta \ge 0$ such that

$$|D_{20}(\phi, \phi_1) - D_{20}(\phi, \phi_2)| \leq \tilde{C}_\theta |\phi_1 - \phi_2| \text{ for all } \phi, \phi_1, \phi_2 \in \tilde{V}_\theta.$$

(b) For every $\theta \in \Theta$, there exist a neighbourhood $\tilde{\tilde{V}}_\theta$ of θ and a constant $\tilde{\tilde{C}}_\theta \geq 0$ such that

$$|D_{11}(\phi, \phi_1) - D_{11}(\phi, \phi_2)| \leq \tilde{\tilde{C}}_\theta |\phi_1 - \phi_2| \text{ for all } \phi, \phi_1, \phi_2 \in \tilde{\tilde{V}}_\theta.$$

(B8) For every $\theta \in \Theta$ there exist a neighbourhood V_θ^λ of θ and a continuous function $k_\theta : \Theta \to \mathbb{R}$ such that

$$|D_{20}(\phi_1, \phi) - D_{20}(\phi_2, \phi)| \leq k_\theta(\phi)|\phi_1 - \phi_2| \text{ for all } \phi_1, \phi_2 \in V_\theta^\lambda,$$

and all $\theta \in \Theta$ and $\int k_\theta(\phi)\Lambda(d\phi) < \infty$ for all $\theta \in \Theta$. Obviously, the loss function $D(\theta, \phi) = F(\theta - \phi)$ where F is a twice continuously differentiable function with $F(0) = 0, F(\phi) > 0$ for $\phi \neq 0, F''(0) > 0$ and $F''(\phi) - F''(\phi_1))| \leq C_\theta|\phi - \phi_1|$ for all $\phi, \phi_1 \in \Gamma_\theta$ where Γ_θ is a neighbourhood of θ and $C_\theta \geq 0$ satisfies the condition (B1), (B3) - (B5), (B7) and (B8). In particular, the quadratic loss function satisfies these conditions.

To prove the main results we start with some preliminary lemmas.

Lemma 5.1 Let conditions (B3), (B4), and (B7) (b) be satisfied. Then for every compact $K \subseteq \Theta$ there exist $d_K \geq 0$ such that

$$\sup_{\theta \in K} \sup_{|\phi - \theta| < d_K} |D_{10}(\theta, \phi)| < \infty.$$

Proof: Similar to Lemma 2 in Strasser (1975). Details are omitted. $\square$

Lemma 5.2 Assume that the conditions (A1) − (A3), (B1) − (B4), (B6) (a), and (B7) (b) be satisfied. Then for every compact $K \subseteq \Theta$ there exist sets $M_{T,\theta}, \theta \in K$ satisfying (5.2) such that on the set $M_{T,\theta}$

$$\left| \int_\Theta D_{10}(\theta_T, \phi)R_T(d\phi) - \int_{V_T(s_K)} D_{10}(\theta_T, \phi)R_T(d\phi) \right| \leq C_K T^{-1}.$$

Proof: Let $\eta > 0$ and $\delta > 0$. Then the inequality

$$\int_{|\phi - \theta| \geq \delta} |D_{10}(\theta_T, \phi)|R_T(d\phi) \geq \exp(-\eta T)$$

yields

$$\frac{\int_{|\phi - \theta| \geq \delta} |D_{10}(\theta_T, \phi)|L_T(\phi)\Lambda(d\phi)}{\int_{|\phi - \theta| \leq \delta_1} L_T(\phi)\Lambda(d\phi)} \geq \exp(-\eta T).$$

for arbitrary $\delta_1 > 0$. Hence

$$\frac{\sup_{|\phi-\theta|\geq\delta} L_T(\phi) \int_\Theta |D_{10}(\theta_T,\phi)|\Lambda(d\phi)}{\inf_{|\phi-\theta|\geq\delta} L_T(\phi)\Lambda\{\phi\in\Theta : |\phi-\theta|<\delta_1\}} \geq \exp(-\eta T).$$

Taking logarithms and dividing by T, we obtain

$$\sup_{|\phi-\theta|<\delta_1}\frac{1}{T}l_T(\phi) - \inf_{|\phi-\theta|>\delta}\frac{1}{T}l_T(\phi) > -\eta + \frac{a}{T}$$

where

$$a := \inf_{\theta\in K}\left[\log\Lambda\{\phi\in\Theta : |\phi-\theta|<\delta_1\} - \log\int_\Theta |D_{10}(\theta_T,\phi)|\Lambda(d\phi)\right].$$

Since there is a compact set $K_1 \subset \Theta$ with $K \subseteq K_1$ and $\theta_T \in K_1$ on the set $M_{T,\theta}$, we may assume that $a \geq a_K > -\infty$ on the set $M_{T,\theta}$. Then completely the same proof as for Theorem 4.3 shows that on the set $M_{T,\theta}$

$$\int_{|\phi-\theta|\geq\delta} |D_{10}(\theta_T,\phi)|R_T(d\phi) \leq \exp(-\eta_K T)$$

for sufficiently small $\eta_K > 0$ and arbitrary $\delta > 0$.

Thus

$$\left|\int_\Theta D_{10}(\theta_T,\phi)R_T(d\phi) - \int_{V_T(s_K)} D_{10}(\theta,\phi)R_T(d\phi)\right|$$
$$\leq \int_{|\phi-\theta|\geq\delta} |D_{10}(\theta_T,\phi)|R_T(d\phi) + \sup_{\theta\in K_1}\sup_{|\phi-\theta|<\delta} |D_{10}(\theta,\phi)|R_T(V_T^c(s_K)).$$

Now Lemma 5.1 and Theorem 4.3 prove the theorem. □

Lemma 5.3 Under (B7), for every compact $K \subseteq \Theta$ and every $\epsilon > 0$ with $\bar{K}_\epsilon \subseteq \Theta$ there exists $C_K > 0$ such that

$$|D_{11}(\phi,\phi_1) - D_{11}(\phi,\phi)| \leq C_K|\phi-\phi_1|$$

Proof: Similar to the proof of Lemma 4.7. Details are omitted. □

Lemma 5.4 Under (B2), for every compact $K \subseteq \Theta$ and every $\delta > 0$, there exists $\epsilon_K > 0$ such that $|\tau-\theta| \geq \delta, \theta \in K, \tau \in \bar{\Theta}$ implies $D(\theta,\theta) < D(\tau,\theta)-\epsilon_K$.
Proof: Similar to Lemma 3 in Strasser (1975). Details are omitted. □

Lemma 5.5 Let (A1)–(A3) be satisfied. Let $g_\theta : \Theta \to \mathbb{R}, \theta \in \Theta$ be a class of continuous Λ integrable functions satisfying the following conditions:
(G1) For every compact $K \subseteq \Theta$

$$\sup_{\theta\in K}\int_\Theta |g_\theta(\phi)|\Lambda(d\phi) < \infty.$$

(G2) For every compact $K \subseteq \Theta$

$$\sup_{\theta \in K} \int_{\Theta} |g_\theta(\theta)| < \infty.$$

(G3) For every compact $K \subseteq \Theta$ and every $\epsilon > 0$, there exists $\delta_K > 0$ such that $|\phi - \theta| < \delta_K, \theta \in K$ implies

$$|g_\theta(\phi) - g_\theta(\theta)| < \epsilon.$$

Then for every $\epsilon > 0$ and every compact $K \subseteq \Theta$

$$\sup_{\theta \in K} P_\theta \left\{ \left| \int_{\Theta} g_\theta(\phi) R_T(d\phi) - g_\theta(\theta) \right| > \epsilon \right\} \leq C_K T^{-1}.$$

Proof: Similar to the proof of Lemma 4 in Strasser (1975) by using Lemma 4.4 and Theorem 4.2. We omit the details. □

Lemma 5.6 Under (A1)–(A3), for every $\epsilon > 0, \tau \in \bar{\Theta}$ there exists $\tilde{U}_\tau \subseteq U_\tau^\lambda$ such that $U \subset \tilde{U}_\tau$ implies

$$\sup_{\theta \in K} P_\theta \left\{ \int_{\Theta} \inf D(U, \phi) R_T(d\phi) \leq \inf D(U, \theta) - \epsilon \right\} \leq C_K T^{-1}.$$

Proof: Similar to Lemma 5 in Strasser (1975). □

We need the following large deviations inequality for the Bayes estimator $\tilde{\theta}_T$ proved in Bishwal (1999a).

Theorem 5.7 Under (C1), (C2), (B1), (B2) for every $\delta > 0$ and every compact $K \subseteq \Theta$

$$\sup_{\theta \in K} P_\theta \left\{ \left| \tilde{\theta}_T - \theta \right| > \delta \right\} \leq C_K \exp(-CT\delta^2).$$

Corollary 5.8 Under (C1)–(C3), (B1)–(B4), (B6)(b), (B7)(a) and (B8), we have for every $\epsilon > 0$,

$$\sup_{\theta \in K} P_\theta \left\{ \left| \int_{\Theta} D_{20}(\theta_T^{**}, \phi) R_T(d\phi) - D_{20}(\theta, \theta) \right| \geq \epsilon \right\} \leq C_K T^{-1}.$$

where θ_T^{**} is defined in (5.1).

Proof: Similar to the proof of Corollary 1 of Strasser (1975). □

Proposition 5.9 Under (A1)–(A3), (A4)$_2$, (C1)–(C3), (B3), (B4), (B6) (a) and (B6) (b), for every compact $K \subseteq \Theta$ there exist sets $M_{T,\theta}, \theta \in K$ satisfying (2.1) such that on the set $M_{T,\theta}$

$$\left| \int_{\Theta} D_{10}(\theta_T, \phi) R_T(d\phi) \right| \leq C_K T^{-1}.$$

Proof: By Lemma 5.2 on the set $M_{T,\theta}$

$$\left| \int_{\Theta} D_{10}(\theta_T, \phi) R_T(d\phi) - \int_{V_T(s_K)} D_{10}(\theta_T, \phi) R_T(d\phi) \right| \leq C_K T^{-1}.$$

Let $\tilde{G}_T$ be the probability measure on $S_T := \{y \in \mathbb{R} : y^2 \leq s_K \log T\}$ which is induced by R_T and u_T. Since on the set $M_{T,\theta}$ we have $V_T(s_K) \subseteq \Theta$, the substitution $y = u_T(\phi)$ if $\phi \in V_T(s_K)$ yields on the set $M_{T,\theta}$

$$\int_{V_T(s_K)} D_{10}(\theta_T, \phi) R_T(d\phi) = \int_{S_T} D_{10}\left(\theta_T, \theta_T + \left(\frac{1}{TI(\theta)}\right)^{1/2} y\right) \tilde{G}_T(dy).$$

By Taylor expansion on the set $M_{T,\theta}$

$$\int_{V_T(s_K)} D_{10}(\theta_T, \phi) R_T(d\phi) = \left(\frac{1}{TI(\theta)}\right)^{1/2} \int_{S_T} y D_{11}(\theta_T, \theta_T^{**}) \tilde{G}_T(dy).$$

where $|\theta_T^{**} - \theta_T| \leq C_K |y| T^{-1/2}$ if $y^2 \leq \log T$.

The first term of the Taylor expansion is omitted since $D_{10}(\theta_T, \theta_T) = 0$ if $\theta_T \in \Theta$. Let $\epsilon > 0$ be such that an ϵ-neighbourhood K^ϵ of K is still contained in Θ. We may assume that on the set $M_{T,\theta}, \theta_T \in K^\epsilon$ and $\theta_T^{**} \in K^\epsilon$ for $y^2 \leq \log T$.

Hence it follows from Lemma 5.3 that on the set $M_{T,\theta}$

$$\left| \int_{V_T(s_K)} D_{10}(\theta_T, \phi) R_T(d\phi) - \left(\frac{1}{TI(\theta)}\right)^{1/2} D_{11}(\theta_T, \theta_T) \int_{S_T} y \tilde{G}_T(dy) \right|$$

$$\leq \left(\frac{1}{TI(\theta)}\right)^{1/2} \int_{S_T} |y| |D_{11}(\theta_T, \theta_T^{**}) - D_{11}(\theta_T, \theta_T)| \tilde{G}_T(dy)$$

$$\leq C_K \left(\frac{1}{TI(\theta)}\right)^{1/2} \int_{S_T} |y|^2 |\theta_T^{**} - \theta_T| \tilde{G}_T(dy)$$

$$\leq C_K T^{-1} \int_{S_T} y^2 \tilde{G}_T(dy).$$

Thus

$$\left| \int_{V_T(s_K)} D_{10}(\theta_T, \phi) R_T(d\phi) \right|$$

$$\leq \left(\frac{1}{TI(\theta)}\right)^{1/2} D_{11}(\theta_T, \theta_T) \left| \int_{S_T} y \tilde{G}_T(dy) \right| + C_K T^{-1} \int_{S_T} y^2 G_T(dy)$$

$$\leq C_K T^{-1}.$$

since from Theorem 4.9, on the set $M_{T,\theta}$

$$\left| \int_{S_T} y G_T(dy) \right| \leq C_K T^{-1/2} \quad \text{and} \quad \left| \int_{S_T} y^2 G_T(dy) \right| \leq C_K. \qquad \square$$

The next theorem provides bound on the asymptotic equivalence of the maximum likelihood and the Bayes estimators.

Theorem 5.10 Under (A1)–(A3), (A4)$_2$, (B1)–(B8), for every compact $K \subseteq \Theta$ there exist sets $M_{T,\theta}, \theta \in K$ satisfying (2.1) such that on the set $M_{T,\theta}$

$$\left| \tilde{\theta}_T - \theta_T \right| \leq C_K T^{-1}$$

i.e.,

$$\sup_{\theta \in K} P_\theta \left\{ \left| \tilde{\theta}_T - \theta_T \right| > C_K T^{-1} \right\} = o(T^{-1/2}).$$

Proof: From the large deviations inequality for the Bayes estimator $\tilde{\theta}_T$ (see Theorem 5.7) we can assume that $\tilde{\theta}_T \in \Theta$, on the set $M_{T,\theta}$. Hence by Taylor expansion

$$\int_\Theta D_{10}(\tilde{\theta}_T, \phi) R_T(d\phi)$$
$$= \int_\Theta D_{10}(\theta_T, \phi) R_T(d\phi) + (\tilde{\theta}_T - \theta_T) \int_\Theta D_{20}(\theta_T^{***}, \phi) R_T(d\phi) \tag{5.1}$$

where

$$|\theta_T^{***} - \theta_T| \leq \left| \tilde{\theta}_T - \theta_T \right|.$$

But $\int_\Theta D_{10}(\tilde{\theta}_T, \phi) R_T(d\phi) = 0$ from the definition of the Bayes estimator $\tilde{\theta}_T$. Therefore,

$$\tilde{\theta}_T - \theta_T = - \frac{\int_\Theta D_{10}(\theta_T, \phi) R_T(d\phi)}{\int_\Theta D_{20}(\theta_T^{***}, \phi) R_T(d\phi)}. \tag{5.2}$$

From Proposition 5.9

$$\left| \int_\Theta D_{10}(\theta_T, \phi) R_T(d\phi) \right| \leq C_K T^{-1}. \tag{5.3}$$

From Corollary 5.8, we have

$$\left| \int_\Theta D_{20}(\theta_T^{***}, \phi) R_T(d\phi) - D_{20}(\theta, \theta) \right| < \epsilon$$

on the set $M_{T,\theta}$. Therefore by condition (B6)(b), we have on the set $M_{T,\theta}$

$$\left| \int_\Theta D_{20}(\theta_T^{***}, \phi) R_T(d\phi) \right| \geq \epsilon_K \tag{5.4}$$

for suitably chosen $\epsilon_K > 0$.

Thus from (5.3) and (5.4), on the set $M_{T,\theta}$

$$\left| \tilde{\theta}_T - \theta_T \right| \leq C_K T^{-1}.$$

This completes the proof of the theorem.   □

In the next theorem we obtain Berry-Esseen bound for the Bayes estimator $\tilde{\theta}_T$.

Theorem 5.11 Under (A1)–(A3), (A4)$_2$, (B1)–(B8), for every compact set $K \subseteq \Theta$

$$\sup_{\theta \in K} \sup_{x \in \mathbb{R}} \left| P_\theta \left\{ ((TI(\theta))^{1/2} (\tilde{\theta}_T - \theta) \leq x \right\} - \Phi(x) \right| \leq C_K T^{-1/2}.$$

Proof: Using Lemma 1.1 (a) we have

$$\sup_{x \in \mathbb{R}} \left| P_\theta \left\{ (TI(\theta))^{1/2} (\tilde{\theta}_T - \theta) \leq x \right\} - \Phi(x) \right|$$

$$= \sup_{x \in \mathbb{R}} \left| P_\theta \left\{ (TI(\theta))^{1/2} (\theta_T - \theta + \tilde{\theta}_T - \theta_T) \leq x \right\} - \Phi(x) \right|$$

$$\leq \sup_{x \in \mathbb{R}} \left| P_\theta \left\{ (TI(\theta))^{1/2} (\theta_T - \theta) \leq x \right\} - \Phi(x) \right|$$

$$+ P_\theta \left\{ (TI(\theta))^{1/2} |\tilde{\theta}_T - \theta_T| > \epsilon \right\} + \epsilon \tag{5.5}$$

By Theorem 3.3, the first term is $O(T^{-1/2})$. By Theorem 5.10,

$$\sup_{\theta \in K} P_\theta \left\{ (TI(\theta))^{1/2} |\tilde{\theta}_T - \theta_T| > C_K T^{-1/2} \right\} = o(T^{-1/2}).$$

Thus choosing $\epsilon = C_K T^{-1/2}$ in (5.5) it is seen that the r.h.s. of (5.5) is of the order $O(T^{-1/2})$.   □

2.6 Example: Hyperbolic Diffusion Model

The hyperbolic diffusion satisfies the following SDE

$$dX_t = -\theta \frac{X_t}{\sqrt{1 + X_t^2}} dt + \sigma dW_t$$

where $\theta \in (\alpha, \beta)$, $\alpha > 0$. This diffusion process is ergodic with a hyperbolic stationary distribution. It is easy to verify the conditions of Sections 2.3, 2.4 and 2.5.

Remarks:

(1) Here we obtained the rate of convergence of the posterior distribution and the Bayes estimator when the diffusion process is stationary and ergodic. In the nonergodic case, the posterior distributions and the Bayes estimators converge to a mixture of normal distribution (see Bishwal (2004b)). It remains open to obtain rate of convergence in this case.

(2) Extension of the results of this chapter to multidimensional process and multiparameter case is worth investigating.

(3) Nonuniform rates of convergence, which are more useful remains to be investigated.

(4) Based on discrete observations of an ergodic diffusion process, Bishwal (2005b) studied the asymptotic normality of pseudo posterior distributions and pseudo Bayes estimators, for smooth prior and loss functions, based on approximate likelihood functions. Rates of convergence in this case is worth investigating.

(5) The Berry-Esseen bounds obtained in this Chapter use parameter dependent nonranom norming. These are useful for testing hypotheses about the parameter. These bounds do not necessarily give confidence interval. In Bishwal and Bose (1995), we obtained the rate $O(T^{-1/2}(\log T)^{1/2}$ using two different random normings. Using the results of this Chapter, Berry-Esseen bounds of the order $O(T^{-1/2})$ for the maximum likelihood estimator and the Bayes estimators are obtained in Bishwal (1999h) using two different random normings which are useful for computation of confidence intervals.

(6) We have only considered the rate of convergence of the MLE, BE and the posterior distribution when the O-U process is stable or positive recurrent, i.e., $\theta > 0$. In the explosive or transient case, i.e., when $\theta < 0$, the MLE, suitably normalized by a nonrandom norming $\frac{e^{\theta T}}{2\theta}$, has asymptotically a Cauchy distribution (see Kutoyants (1994a)) and when normalized by random norming it has asymptotically normal distribution (see Feigin (1976)). The rate of convergence of the MLE in this case remains open. The asymptotic distribution of Bayes estimator and the posterior distribution is Cauchy in this case (see Bishwal (1999i)). Note that in the unstable (null recurrent or critical) case, i.e., when $\theta = 0$, MLE when normalised by T has a distribution of the ratio of a noncentral chi-square to the sum of chi-squares, a distribution concentrated on a half line (see Feigin (1979)). We conjecture that one can obtain faster than $O(T^{-1/2})$ rate of convergence in the explosive and the unstable cases.

(7) When $\theta > 0$, the model (3.1) satisfies the local asymptotic normality (LAN) condition (see Le Cam (1986) for a definition). When $\theta < 0$, the model satisfies the local asymptotic mixed normality (LAMN) condition (see Jeganathan (1982), Le Cam and Yang (1990) for a definition). When $\theta = 0$, the model satisfies the local asymptotic Brownian functional (LABF) condition (see Greenwood and Wefelmeyer (1993) for a definition). The model satisfies the local asymptotic quadraticity (LAQ) condition (see Le Cam and Yang (1990) or Jeganathan (1995) for a definition) for all $\theta \in \mathbb{R}$ and the model is *asymptotically centered* (AC) for all $\theta \in \mathbb{R}$ (see Gushchin (1995)).

(8) Extension of the results of this Chapter to multiparameter case and multidimensional process is worth investigating.

3

Large Deviations of Estimators in Homogeneous Diffusions

3.1 Introduction

In this Chapter we obtain large deviations results for the maximum likelihood estimator and the Bayes estimators in non-linear stochastic differential equations.

The weak consistency, asymptotic normality and asymptotic efficiency of the MLE and the Bayes estimator were obtained by Kutoyants (1977) via studying the local asymptotic normality (LAN) property of the model. Lanska (1979) obtained the strong consistency and asymptotic normality of the more general minimum contrast estimators which includes the MLE. Prakasa Rao and Rubin (1981) obtained the strong consistency and asymptotic normality of the MLE by studying the weak convergence of the least squares random field where the families of stochastic integrals were studied by Fourier analytic methods. Prakasa Rao (1982) studied the weak consistency, asymptotic normality and asymptotic efficiency of the maximum probability estimator. All the above authors assumed stationarity and ergodicity of the model and the parameter was taken as a scalar. For the multidimensional drift parameter, Bose (1986b) obtained the strong consistency and asymptotic normality of the MLE and the Bayes estimator under slightly weaker assumptions.

For the nonlinear nonhomogeneous equation Kutoyants (1978) obtained weak consistency and asymptotic normality of the MLE and the BE of the drift parameter as the intensity of noise decreases to zero. Borkar and Bagchi (1982) obtained weak consistency of the MLE as $T \to \infty$. Mishra and Prakasa Rao (1985) obtained strong consistency and conditional asymptotic normality of the MLE as $T \to \infty$. All the above authors assumed the parameter space to be compact. Levanony et al. (1994) under some stronger regularity conditions, obtained strong consistency and asymptotic normality of the MLE as $T \to \infty$ when the parameter space is open.

In this Chapter, we obtain exponential bounds on the large deviation probability for the MLE (see Theorem 3.1) and for Bayes estimators (see Theorem 4.1) in homogeneous nonlinear SDE. For the large deviation result we follow the method in Ibragimov and Hasminskii (1981). This method was used by Prakasa Rao (1984) to obtain the large deviation probability bound for the least squares estimator in the nonlinear regression model with Gaussian errors.

The Chapter is organised as follows: In Section 3.2 we prepare notations, assumptions and preliminaries. Section 3.3 contains the large deviation inequality for the MLE and Section 3.4 contains the large deviation results for the Bayes estimator. Section 3.5 contains some examples.

This chapter is adapted from Bishwal (1999a).

3.2 Model, Assumptions and Preliminaries

Let $(\Omega, \mathcal{F}, \{\mathcal{F}_t\}_{t \geq 0} P)$ be a stochastic basis satisfying the usual hypotheses on which we define a stationary ergodic diffusion process $\{X_t, t \geq 0\}$ satisfying the Itô SDE

$$dX_t = f(\theta, X_t)dt + dW_t, \ t \geq 0$$
$$X_0 = \xi$$

(2.1)

where $\{W_t, t \geq 0\}$ is a standard Wiener process, $E[\xi^2] < \infty$, $f(\theta, x)$ is a known real valued function continuous on $\Theta \times \mathbb{R}$ where Θ is a closed interval of the real line and the parameter θ is unknown, which is to be estimated on the basis of observation of the process $\{X_t, 0 \leq t \leq T\} \equiv X_0^T$. Let θ_0 be the true value of the parameter which lies inside the parameter space Θ.

Let P_θ^T be the measure generated by the process X_0^T on the space (C_T, B_T) of continuous functions on $[0, T]$ with the associated Borel σ-algebra B_T under the supremum norm when θ is the true value of the parameter. Let E_θ^T be the expectation with respect to the measure P_θ^T. Suppose P_θ^T is absolutely continuous with respect to $P_{\theta_0}^T$. Then it is well known that (see Lipster and Shiryayev (1977))

$$L_T(\theta)$$
$$:= \frac{dP_\theta^T}{dP_{\theta_0}^T}(X_0^T)$$
$$= \exp\{\int_0^T [f(\theta, X_s) - f(\theta_0, X_s)]dW_s - \frac{1}{2}\int_0^T [f(\theta, X_s) - f(\theta_0, X_s)]^2 ds\}$$

(2.2)

is the Radon-Nikodym derivative of P_θ^T with respect to $P_{\theta_0}^T$. The MLE θ_T of θ based on X_0^T is defined as

$$\theta_T := \arg\max_{\theta \epsilon \Theta} L_T(\theta).$$

(2.3)

If $L_T(\theta)$ is continuous in θ, it can be shown that there exists a measurable MLE by using Lemma 3.3 in Schmetterer (1974). Here after, we assume the existence of such a measurable MLE. We will also assume that the following regularity conditions on $f(\theta, x)$ are satisfied. C denotes a generic constant throughout the chapter.

(A1) (i) $f(\theta, x)$ is continuous on $\Theta \times \mathbb{R}$.
   (ii) $|f(\theta, x)| \leq M(\theta)(1 + |x|) \; \forall \theta \in \Theta, x \in \mathbb{R}, \sup\{M(\theta), \theta \in \Theta\} < \infty$.
   (iii) $|f(\theta, x) - f(\theta, y)| \leq M(\theta)|x - y| \; \forall \theta \in \Theta \; \forall x, y \in \mathbb{R}$
   (iv) $|f(\theta, x) - f(\phi, x)| \leq J(x)|\theta - \phi| \; \forall \theta, \phi \in \Theta, \; \forall x \in \mathbb{R}$
      where $J(\cdot)$ is continuous and $E[J^2(\xi)] < \infty$.
   (v) $I(\theta) = E|f(\theta, \xi) - f(\theta_0, \xi)|^2 > 0 \; \forall \theta \neq \theta_0$.

(A2) (i) The first partial derivative of f w.r.t. θ exists and is denoted by $f_\theta^{(1)}(\theta, x)$. The derivative evaluated at θ_0 is denoted by $f_\theta^{(1)}(\theta_0, x)$.
   (ii) $\beta = E[f^{(1)}(\theta, x)]^2 < \infty$.
   (iii) There exists $\alpha > 0$ s.t.

$$|f_\theta^{(1)}(\theta, x) - f_\theta^{(1)}(\theta, x) \leq J(x)|\theta - \phi|^\alpha \; \forall x \in \mathbb{R}$$

$\forall \theta, \phi \in \Theta$ and J is as in (A1) (iv).
   (iv) $|f_\theta^{(1)}(\theta, x)| \leq N(\theta)(1 + |x|) \; \forall \theta \epsilon \Theta, \; \forall x \in \mathbb{R}, \sup\{N(\theta), \theta \in \Theta\} < \infty$

(A3) There exists a positive constant C such that

$$E_\theta[\exp\{-u^2(3T)^{-1} \int_0^T \inf_{\phi \in \Theta} (f_\theta^{(1)}(\phi, X_t))^2 dt\}] \leq C \exp(-u^2 C) \quad \text{for all} \quad u.$$

Under the assumption (A1) and (A2), Prakasa Rao and Rubin (1981) proved the strong consistency and asymptotic normality of θ_T as $T \to \infty$. Assumption (A3) is used to prove our large deviation result.

3.3 Large Deviations for the Maximum Likelihood Estimator

Before we obtain bounds on the probabilities of large deviation for the MLE θ_T we shall give a more general result.

Theorem 3.1 *Under the assumptions (A1) - (A3), for $\rho > 0$, we have*

$$\sup_{\theta \in \Theta} P_\theta^T \{\sqrt{T}|\theta_T - \theta| \geq \rho\} \leq B \exp(-b\rho^2)$$

for some positive constants b and B independent of ρ and T.
   By substituting $\rho = \sqrt{T}\epsilon$ in Theorem 3.1, the following Corollary is obtained.

Corollary 3.1 *Under the conditions of Theorem 3.1, for arbitrary $\epsilon > 0$ and all $T > 0$, we have*

$$\sup_{\theta \in \Theta} P_{\theta}^{T} \{|\theta_T - \theta| > \epsilon\} \leq B \exp(-CT\epsilon^2)$$

where B and C are positive constants independent of ϵ and T.

To prove Theorem 3.1 we shall use the following revised version of Theorem 19 of Ibragimov and Has'minskii (1981, p. 372), (see Kallianpur and Selukar (1993, p. 330)).

Lemma 3.2 *Let $\zeta(t)$ be a real valued random function defined on a closed subset F of the Euclidean space $\mathbb{R}^k$. We shall assume that the random process $\zeta(t)$ is measurable and separable. Assume that the following condition is fulfilled: there exist numbers $m \geq r > k$ and a function $H(x) : \mathbb{R}^k \to \mathbb{R}^1$ bounded on compact sets such that for all $x, h \in F$, $x + h \in F$,*

$$E|\zeta(x)|^m \leq H(x),$$

$$E|\zeta(x + h) - \zeta(x)|^m \leq H(x)|h|^r.$$

Then with probability one the realizations of $\zeta(t)$ are continuous functions on F. Moreover, set

$$w(\delta, \zeta, L) = \sup_{x,y \in F, |x|, |y| \leq L, |x-y| \leq \delta} |\zeta(x) - \zeta(y)|,$$

then

$$E(w(h; \zeta, L)) \leq B_0 (\sup_{|x| < L} H(x))^{1/m} L^k \, h^{(r-k)/m} \log(h^{-1})$$

where $B_0 = 64^k (1 - 2^{-(r-k)m})^{-1} + (2^{(m-r)/m} - 1)^{-1}$.

Let us consider the likelihood ratio process

$$Z_T(u) = \frac{dP_{\theta + uT^{-1/2}}}{dP_{\theta}}(X_0^T).$$

By (2.2) with $g_t(u) = f(\theta + uT^{-1/2}, X_t) - f(\theta, X_t)$, we have

$$Z_T(u) = \exp\{\int_0^T [f(\theta + uT^{-1/2}, X_t) - f(\theta, X_t)]dW_t$$

$$-\frac{1}{2}\int_0^T \left[f(\theta + uT^{-1/2}, X_t) - f(\theta, X_t)\right]^2 dt\}$$

$$= \exp\{\int_0^T g_t(u)dW_t - \frac{1}{2}\int_0^T g_t^2(u)dt\}.$$

Lemma 3.3 *Under the assumptions (A1) - (A3), we have*
(a) $E_\theta^T [Z_T^{1/2}(u_1) - Z_T^{1/2}(u_2)]^2 \leq C(u_2 - u_1)^2$
(b) $E_\theta^T [Z_T^{1/2}(u)] \leq C \exp(-Cu^2)$

Proof. Note that

$$E_\theta^T [Z_T^{1/2}(u_1) - Z_T^{1/2}(u_2)]^2$$
$$= E_\theta^T [Z_T(u_1)] + E_\theta^T [Z_T(u_2)] - 2E_\theta^T [Z_T^{1/2}(u_1) Z_T^{1/2}(u_2)] \qquad (3.1)$$
$$\leq 2 - 2E_\theta^T [Z_T^{1/2}(u_1) Z_T^{1/2}(u_2)].$$

From Gikhman and Skorohod (1972), for all u we have

$$E_\theta^T [Z_T(u)] = E_\theta^T [\exp\{\int_0^T g_t(u)dW_t - \frac{1}{2}\int_0^T g_t^2(u)dt\}] \leq 1. \qquad (3.2)$$

Let

$$\theta_1 = \theta + u_1 T^{-1/2}, \quad \theta_2 = \theta + u_2 T^{-1/2},$$
$$\delta_t = f(\theta_2, X_t) - f(\theta_1, X_t), \qquad (3.3)$$

and

$$V_T = \exp\{\frac{1}{2}\int_0^T \delta_T dW_t - \frac{1}{4}\int_0^T \delta_t^2 \, dt\} = \left(\frac{dP_{\theta_2}^T}{dP_{\theta_1}^T}\right)^{1/2}.$$

By Itô formula, V_T can be represented as

$$V_T = 1 - \frac{1}{8}\int_0^T V_t \delta_t^2 dt + \frac{1}{2}\int_0^T V_t \delta_t dW_t. \qquad (3.5)$$

The random process $\{V_t^2, \mathcal{F}_t, P_\theta^T, 0 \leq t \leq T\}$ is a martingale and from the $\mathcal{F}_t$-measurability of δ_t for each $t \in [0, T]$,

$$E_{\theta_1}^T \int_0^T V_t^2 \delta_t^2 dt = E_{\theta_1}^T \int_0^T E_{\theta_1}^T (V_T^2|\mathcal{F}_t)\delta_t^2 dt$$
$$= E_{\theta_1}^T V_T^2 \int_0^T \delta_t^2 dt$$
$$= \int V_T^2 (\int_0^T \delta_t^2 dt)dP_{\theta_1}^T$$
$$= \int (\int_0^T \delta_t^2 dt)dP_{\theta_2}^T \qquad (3.6)$$
$$= E_{\theta_2}^T (\int_0^T \delta_t^2 dt)$$
$$= E_{\theta_2}^T \int_0^T |f(\theta_2, X_t) - f(\theta_1, X_t)|^2 dt$$

$$\leq E_{\theta_2}^T \int_0^T [J^2(X_t)]|\theta_2 - \theta_1|^2 dt \quad \text{(by A1))}$$
$$\leq (u_2 - u_1)^2 \cdot \frac{1}{T}\int_0^T E_{\theta_2}[J^2(\xi)]dt \qquad (3.7)$$
$$< C(u_2 - u_1)^2 < \infty.$$

Hence $E_{\theta_1}^T \int_0^T V_t \delta_t dW_t = 0$. Therefore, using $|ab| \leq \frac{a^2 + b^2}{2}$, from (3.5) we obtain

$$
\begin{aligned}
E_{\theta_1}^T(V_T) &= 1 - \frac{1}{8} \int_0^T E_{\theta_1}^T(\delta_t V_t . \delta_t) dt \\
&\geq 1 - \frac{1}{16} \int_0^T E_{\theta_1}^T \delta_t^2 dt - \frac{1}{16} \int_0^T E_{\theta_1}^T V_t^2 \delta_t^2 dt \\
&= 1 - \frac{1}{16} E_{\theta_1}^T \int_0^T \delta_t^2 dt - \frac{1}{16} E_{\theta_2}^T \int_0^T \delta_t^2 dt \quad \text{(by (3.6)).}
\end{aligned}
\tag{3.8}
$$

Now

$$
\begin{aligned}
&E_\theta^T[Z_T^{1/2}(u_1) Z_T^{1/2}(u_2)] \\
&= E_\theta^T \left\{ \left[\frac{dP_{\theta + u_1 T^{-1/2}}^T}{dP_\theta^T} \right]^{1/2} \left[\frac{dP_{\theta + u_2 T^{-1/2}}^T}{dP_\theta^T} \right]^{1/2} \right\} \\
&= \int \left[\frac{dP_{\theta_1}^T}{dP_\theta^T} \right]^{1/2} \left[\frac{dP_{\theta_2}^T}{dP_\theta^T} \right]^{1/2} dP_\theta^T \\
&= \int \left[\frac{dP_{\theta_2}^T}{dP_{\theta_1}^T} \right]^{1/2} dP_{\theta_1}^T = E_{\theta_1}^T(V_T).
\end{aligned}
\tag{3.9}
$$

Substituting (3.9) into (3.1) and using (3.8), we obtain

$$
\begin{aligned}
&E_\theta[Z_T^{1/2}(u_1) - Z_T^{1/2}(u_2)]^2 \\
&\leq 2 - 2E_{\theta_1}(V_T) \\
&\leq \frac{1}{8} E_{\theta_1} \int_0^T \delta_t^2 dt + \frac{1}{8} E_{\theta_1} \int_0^T \delta_t^2 dt \\
&\leq C(u_2 - u_1)^2 \text{(using arguments similar to (3.7)).}
\end{aligned}
$$

This completes the proof of (a). Let us now prove (b). By the Hölder inequality,

$$
\begin{aligned}
&E_\theta[Z_T^{1/2}(u)] \\
&= E_\theta \left[\exp \left\{ \frac{1}{2} \int_0^T g_t(u) dW_t - \frac{1}{4} \int_0^T g_t^2(u) dt \right\} \right] \\
&= E_\theta \left[\exp \left\{ \frac{1}{2} \int_0^T g_t(u) dW_t - \frac{1}{6} \int_0^T (g_t(u))^2 dt \right\} \exp \left\{ -\frac{1}{12} \int_0^T (g_t(u))^2 dt \right\} \right] \\
&\leq \left\{ E_\theta \left[\exp \left\{ \frac{1}{2} \int_0^T g_t(u) dW_t - \frac{1}{6} \int_0^T (g_t(u))^2 dt \right\} \right]^{4/3} \right\}^{3/4} \\
&\quad \times \left\{ E_\theta \left[\exp \left\{ -\frac{1}{12} \int_0^T (g_t(u))^2 dt \right\} \right]^4 \right\}^{1/4}
\end{aligned}
$$

$$\leq \left[E_\theta \exp \left\{ \frac{2}{3} \int_0^T g_t(u) dW_t - \frac{2}{9} \int_0^T (g_t^2(u)) dt \right\} \right]^{3/4}$$

$$\times \left[E_\theta \exp \left\{ -\frac{1}{3} \int_0^T (g_t(u))^2 dt \right\} \right]^{1/4} \tag{3.10}$$

Assumptions (A2) and (A3) yield

$$E_\theta \exp \left\{ -\frac{1}{3} \int_0^T (g_t(u))^2 dt \right\}$$

$$= E_\theta \exp \left\{ -\frac{1}{3} \int_0^T \left[f(\theta + uT^{-1/2}, X_t) - f(\theta, X_t) \right]^2 dt \right\}$$

$$= E_\theta \exp \left\{ -\frac{u^2}{3T} \int_0^T (f_\theta^{(1)}(\overline{\theta}, X_t))^2 dt \right\} \quad \text{where } |\theta - \overline{\theta}| \leq uT^{-1/2} \tag{3.11}$$

$$\leq E_\theta \exp \left\{ -\frac{u^2}{3T} \int_0^T \inf_{\phi \in \Theta} (f_\theta^{(1)}(\phi, X_t))^2 dt \right\}$$

$$\leq C \exp(-u^2 C).$$

On the other hand, from Gikhman and Skorohod (1972)

$$E_\theta \left[\exp \left\{ \int_0^T \frac{2}{3} g_t(u) dW_t - \frac{1}{2} \int_0^T \left(\frac{2}{3} g_t(u) \right)^2 dt \right\} \right] \leq 1. \tag{3.12}$$

Combining (3.10) - (3.12) completes the proof of (b). □

Proof of Theorem 3.1 Let $U_T = \{ u : \theta + uT^{-1/2} \in \Theta \}$. Then

$$P_\theta^T \left\{ \sqrt{T} |\theta_T - \theta| > \rho \right\}$$

$$= P_\theta^T \left\{ |\theta_T - \theta| > \rho T^{-1/2} \right\}$$

$$\leq P_\theta^T \left\{ \sup_{|u| \geq \rho, u \in U_T} L_T(\theta + uT^{-1/2}) \geq L_T(\theta) \right\}$$

$$= P_\theta^T \left\{ \sup_{|u| \geq \rho} \frac{L_T(\theta + uT^{-1/2})}{L_T(\theta)} \geq 1 \right\} \tag{3.13}$$

$$= P_\theta^T \left\{ \sup_{|u| \geq \rho} Z_T(u) \geq 1 \right\}$$

$$\leq \sum_{r=0}^{\infty} P_\theta^T \left\{ \sup_{u \in \Gamma_r} Z_T(u) \geq 1 \right\}.$$

where $\Gamma_r = [\rho + r, \rho + r + 1]$.

Applying Lemma 3.2 with $\zeta(u) = Z_T^{1/2}(u)$, we obtain from Lemma 3.3 that there exists a constant $B > 0$ such that

$$\sup_{\theta \in \Theta} E_\theta^T \left\{ \sup_{|u_1-u_2| \le h, |u_1|, |u_2| \le l} \left| Z_T^{1/2}(u_1) - Z_T^{1/2}(u_2) \right| \right\} \le B l^{1/2} h^{1/2} \log h^{-1}.$$

(3.14)

Divide Γ_r into subintervals of length atmost $h > 0$. The number n of subintervals is clearly less than or equal to $[\frac{1}{h}] + 1$. Let $\Gamma_r^{(j)}, 1 \le j \le n$ be the subintervals chosen. Choose $u_j \in \Gamma_r^{(j)}$. Then

$$P_\theta^T \left[\sup_{u \in \Gamma_r} Z_T(u) \ge 1 \right]$$

$$\le \sum_{j=1}^n P_\theta^T \left[Z_T^{1/2}(u_j) \ge \frac{1}{2} \right]$$

$$+ P_\theta^T \left\{ \sup_{|u-v| \le h, |u|, |v| \le \rho+r+1} |Z_T^{1/2}(u) - Z_T^{1/2}(v)| \ge \frac{1}{2} \right\}$$

$$\le 2 \sum_{j=1}^n E_\theta^T [Z_T^{1/2}(u_j)] + 2B(\rho+r+1)^{1/2} h^{1/2} \log(h^{-1})$$

(by Markov inequality and (3.14))

$$\le 2C \sum_{j=1}^n \exp\left(-Cu_j^2\right) + 2B(\rho+r+1)^{1/2} h^{1/2} \log(h^{-1}) \quad \text{(by Lemma 3.3)}$$

$$\le 2C \left(\left[\frac{1}{h}\right] + 1 \right) \exp\left\{-C(\rho+r)^2\right\} + 2B(\rho+r+1)^{1/2} h^{1/2} \log(h^{-1}).$$

Let us now choose $h = \exp\left\{\frac{-C(\rho+r)^2}{2}\right\}$. Then

$$\sup_{\theta \in \Theta} P_\theta^T \left\{ \sup_{u > \rho} Z_T(u) \ge 1 \right\}$$

$$\le B \sum_{r=0}^{\infty} (\rho+r+1)^{1/2} \exp\left\{\frac{-C(\rho+r)^2}{4}\right\}$$

(3.15)

$$\le B \exp(-b\rho^2).$$

where B and b are positive generic constants independent of ρ and T.

Similarly it can be shown that

$$\sup_{\theta \in \Theta} P_\theta^T \left[\sup_{u < -\rho} Z_T(u) \ge 1 \right] \le B \exp(-b\rho^2).$$

(3.16)

Combining (3.15) and (3.16), we obtain

$$\sup_{\theta \in \Theta} P_\theta^T \left[\sup_{|u| > \rho} Z_T(u) \ge 1 \right] \le B \exp(-b\rho^2).$$

(3.17)

The theorem follows from (3.14) and (3.17).

□

3.4 Large Deviations for Bayes Estimators

Let Λ be a prior probability measure on $(\Theta, \mathcal{B})$ where $\mathcal{B}$ is the σ-algebra of Borel subsets of Θ. Suppose that Λ has a density $\lambda(\cdot)$ with respect to the Lebesgue measure on $\mathbb{R}$, which is continuous and positive on Θ and possesses in Θ a polynomial majorant.

Let $p(\theta|X_0^T)$ be the posterior density of θ given X_0^T. By Bayes theorem $p(\theta|X_0^T)$ is given by

$$p(\theta|X_0^T) = \frac{L_T(\theta)\lambda(\theta)}{\int_\Theta L_T(\theta)\lambda(\theta)d\theta}.$$

Let $l(\cdot, \cdot) : \Theta \times \Theta \to \mathbb{R}$ be a loss function which satisfies the following properties
(C1) $l(u, v) = l(u - v)$.
(C2) $l(u)$ is defined and nonnegative on $\mathbb{R}, l(0) = 0$ and $l(u)$ is continuous at $u = 0$ but is not identically equal to 0.
(C3) l is symmetric, i.e., $l(u) = l(-u)$.
(C4) $\{u : l(u) < c\}$ are convex sets and are bounded for all $c > 0$ sufficiently small.
(C5) There exists numbers $\gamma > 0, H_0 \geq 0$ s.t. for $H \geq H_0$

$$\sup\{l(u) : |u| \leq H^\gamma\} \leq \inf\{l(u) : |u| \geq H\}.$$

Clearly all loss functions of the form $|u - v|^r, r > 0$ satisfy the conditions (C1)-(C5). In particular, quadratic loss function $|u - v|^2$ satisfies these conditions.

Then a Bayes estimator $\widetilde{\theta}_T$ of θ with respect to the loss function $l(\theta, \theta')$ and prior density $\lambda(\theta)$ is one which minimizes the posterior risk and is given by

$$\widetilde{\theta}_T = \arg\min_{u \in \Theta} \int_\Theta l(u, \theta)p(\theta|X_0^T)d\theta. \tag{4.1}$$

In particular, for the quadratic loss function $l(u, v) = |u - v|^2$, the Bayes estimator $\widetilde{\theta}_T$ becomes the posterior mean given by

$$\widetilde{\theta}_T = \frac{\int_\Theta u\, p(u|X_0^T)du}{\int_\Theta p(v|X_0^T)dv}.$$

We now state the large deviation inequality for the Bayes estimator $\widetilde{\theta}_T$.

Theorem 4.1 *Suppose (A1) - (A3) and (C1) - (C5) hold. For $\rho > 0$, the Bayes estimator $\widetilde{\theta}_T$ with respect to the prior $\lambda(\cdot)$ and a loss function $l(\cdot, \cdot)$ with $l(u) = l(|u|)$ satisfies*

$$\sup_{\theta \in \Theta} P_\theta^T\{\sqrt{T}|\widetilde{\theta}_T - \theta| \geq \rho\} \leq B \exp(-b\rho^2)$$

for some positive constants B and b independent of ρ and T.

Corollary 4.1 *Under the conditions of Theorem 4.1, for arbitrary $\epsilon > 0$ and all $T > 0$, we have*

$$\sup_{\theta \in \Theta} P_\theta^T \{|\widetilde{\theta}_T - \theta| > \epsilon\} \leq B \exp(-CT\epsilon^2).$$ $\square$

To prove Theorem 4.1 we shall use the following Theorem of Ibragimov and Has'minskii (1981, p. 45).

Theorem 4.2 *Let $Z_{\epsilon,\theta}(u)$ be the likelihood ratio function corresponding to the points $\theta + \phi(\epsilon)u$ and θ where $\phi(\epsilon)$ denotes a normalizing factor such that $|\phi(\epsilon)| \to 0$ as $\epsilon \to 0$. Thus $Z_{\epsilon,\theta}$ is defined on the set $U_\epsilon = (\phi(\epsilon))^{-1}(\Theta - \theta)$. Let $Z_{\epsilon,0}^\theta(u)$ possesses the following properties : given a compact set $K \subset \Theta$ there exist numbers $M_1 > 0$ and $m_1 \geq 0$ and functions $g_\epsilon^K(y) = g_\epsilon(y)$ correspond such that*

(1) For some $\alpha > 0$ and all $\theta \in K$,

$$\sup_{|u_1| \leq R, |u_2| \leq R} |u_2 - u_1|^{-\alpha} E_\theta^{(\epsilon)} |Z_{\epsilon,\theta}^{1/2}(u_2) - Z_{\epsilon,\theta}^{1/2}(u_1)|^2 \leq M_1(1 + R^{m_1}).$$

(2) For all $\theta \in K$ and $u \in U_\epsilon$, $E_\theta^{(\epsilon)} Z_{\epsilon,\theta}^{1/2}(u) \leq e^{-g_\epsilon(u)}$.
(3) $g_\epsilon(u)$ is a monotonically increasing to ∞ function of y

$$\lim_{y \to \infty, \epsilon \to 0} y^N e^{-g_\epsilon(y)} = 0.$$

Let $\{\widetilde{t}_\epsilon\}$ be a family of estimators, Bayesian with respect to the prior density q, which is continuous and positive on K and possesses in Θ a polynomial majorant and a loss function $\omega_\epsilon(u,v) = l((\phi(\epsilon))^{-1}(u - v))$ where l satisfies (C1) - (C5). Then for any N,

$$\lim_{H \to \infty, \epsilon \to 0} H^N \sup_{\theta \in K} P_\theta^{(\epsilon)} \{|(\phi(\epsilon))^{-1}(\widetilde{t}_\epsilon - \theta)| > H\} = 0.$$

If in addition, $l(u) = \tau(|u|)$, then for all ϵ sufficiently small, $0 < \epsilon < \epsilon_0$,

$$\sup_{\theta \in K} P_\theta^{(\epsilon)} \{|(\phi(\epsilon))^{-1}(\widetilde{t}_\epsilon - \theta)| > H\} \leq B_0 e^{-b_0 g_\epsilon(H)}.$$ $\square$

Proof of Theorem 4.1. In view of Lemma 3.3, conditions (1), (2) and (3) of Theorem 4.2 are satisfied with $\alpha = 2$ and $g(u) = u^2$. Hence the result follows from Theorem 4.2. $\square$

As another application of Theorem 4.2, we obtain the following result.

Theorem 4.2. *Under the assumptions (A1) - (A3), for any N, we have for the Bayes estimator $\widetilde{\theta}_T$ w.r.t. the prior $\lambda(\cdot)$ and loss function $l(\cdot, \cdot)$ satisfying the conditions (C1) - (C5),*

$$\lim_{H \to \infty, T \to \infty} H^N \sup_{\theta \in \Theta} P_\theta^T \{\sqrt{T} |\widetilde{\theta}_T - \theta| > H\} = 0.$$

3.5 Examples

(a) Ornstein-Uhlenbeck Process

Consider the stationary Ornstein-Uhlenbeck model

$$dX_t = -\theta X_t \, dt + dW_t$$

where X_0 has $\mathcal{N}(0, -1/2\theta)$ distribution. It satisfies the conditions for the large deviations of MLE. For Bayes estimator choose, squared error loss function.

(b) Kessler-Sørensen Model

Kessler and Sørensen (1999) studied the following model

$$dX_t = -\theta \tan(X_t) \, dt + dW_t$$

$\theta \geq 1/2$. This is an ergodic diffusion with on the interval $(-\pi/2, \pi/2)$. This can be thought as an Ornstein-Uhlenbeck process on a finite interval.

(c) Larsen-Sørensen Model

$$dX_t = -\theta \frac{\sin(\frac{1}{2}\pi(X_t - m)/z) - \rho}{\cos \frac{1}{2}\pi(X_t - m)/z} \, dt + \sigma dW_t$$

where $\theta > 0, \rho \in (-1, 1), z > 0, m \in \mathbb{R}$. This is a generalization of the Kessler-Sørensen model. When $\gamma = 0, m = 0$ and $z = \pi/2$, we obtain the Kessler-Sørensen model. The market volatility is constant, but the central banks intervene very forcefully when the exchange rate comes near the boundaries to try it to keep away from them. When $\theta \geq \frac{1}{2}\sigma^2$ and $-1 + \sigma^2/(2\theta) \leq \rho \leq 1 - \sigma^2/(2\theta)$. This model is used to model exchange rates in a target zone, θ expresses the strength of the intervention of the central bank and ρ measures the asymmetry between two currencies, while σ^2 expresses the volatility of the market.

(d) Jacobi Diffusion Model

Consider the Jacobi diffusion

$$dX_t = -\theta[X_t - (m + \gamma Z)] \, dt + \sigma \sqrt{Z^2 - (X_t - m)^2} dW_t$$

$\theta > 0$ and $\gamma \in (-1, 1)$. This process reverts to the mean $m + \gamma Z$ with a speed that is proportional to the deviation from this model. When $\theta(1 - \gamma) \geq \sigma^2$ and $\theta(1 + \gamma) \geq \sigma^2$ this process is an ergodic diffusion for which the stationary distribution is the Beta distribution on $(m-z, m+z)$ with parameters $\theta(1-\gamma)\sigma^{-2}$ and $\theta(1 + \gamma)\sigma^{-2}$. The parameter γ allows asymmetric stationary distribution which is usually needed to fit observations of exchange rates in a target zone, see Larsen and Sørensen (2003).

Remarks

(1) Following the methods of this Chapter, bounds on the large deviation probability for the MLE and the BE in nonlinear nonhomogeneous SDE is investigated in Bishwal (2003b). Also a Berry-Esseen type bound for the MLE for nonlinear nonhomogeneous diffusions is investigated in Bishwal (2003b).

(2) For the nonlinear SDE it is known that maximum likelihood estimator and the Bayes estimators are asymptotically equivalent as $T \to \infty$ (see Bose (1983b)). It remains open to obtain the bounds on the asymptotic equivalence of these two estimators.

(3) The rate of convergence in the Bernstein-von Mises theorem and the Berry-Esseen bound for the maximum likelihood and the Bayes estimators in the nonlinear (both homogeneous and nonhomogeneous) SDE remain open.

4

Local Asymptotic Mixed Normality for Nonhomogeneous Diffusions

4.1 Introduction

We study the asymptotic properties of various estimators of the parameter appearing nonlinearly in the nonhomogeneous drift coefficient of a functional stochastic differential equation when the corresponding solution process, called the diffusion type process, is observed over a continuous time interval $[0, T]$. We show that the maximum likelihood estimator, maximum probability estimator and regular Bayes estimators are strongly consistent and when suitably normalised, converge to a mixture of normal distribution and are locally asymptotically minimax in the Hajek-Le Cam sense as $T \to \infty$ under some regularity conditions. Also we show that posterior distributions, suitably normalised and centered at the maximum likelihood estimator, converge to a mixture of normal distribution. Further, the maximum likelihood estimator and the regular Bayes estimators are asymptotically equivalent as $T \to \infty$. We illustrate the results through the exponential memory Ornstein-Uhlenbeck process, the nonhomogeneous Ornstein-Uhlenbeck process and the Kalman-Bucy filter model where the limit distribution of the above estimators and the posteriors is shown to be Cauchy.

Time dependent diffusion models are useful for modeling term structure dynamics, see Fan *et al.* (2003). Long time asymptotics and small noise asymptotics of maximum likelihood estimator (MLE) and the Bayes estimators (BEs) of the drift parameter in the linear and nonlinear Markov diffusion processes have been paid a lot of attention during the last two decades, see e.g., Liptser and Shiryayev (1978), Basawa and Prakasa Rao (1980), Kutoyants (1984b, 1994), Linkov (1993) and Prakasa Rao (1999). For the nonlinear homogeneous stationary ergodic diffusion processes, Prakasa Rao and Rubin (1981) and Prakasa Rao (1982) studied the long time asymptotics of the MLE and the maximum probability estimator (MPE) respectively and Bose (1983) studied the long time asymptotics of the posterior distributions and the BEs. For the nonlinear nonhomogeneous nonstationary nonergodic diffusion processes, Borkar and Bagchi (1982), Mishra and Prakasa Rao (1985) and Levanony,

Shwartz and Zeitouni (1994) studied the long time asymptotics of the MLE. Small noise asymptotics of the MLE and the BEs for these processes were studied by Kutoyants (1984b).

For general theory of asymptotic inference for nonergodic models, see Jeganathan (1982) and Basawa and Scott (1983). Asymptotic study of estimation of the drift parameter of nonlinear nonhomogeneous non-Markovian diffusion type processes which are solutions of functional stochastic differential equations has been paid much less attention. Taraskin (1985) showed the local asymptotic conditional normality (LACN) property for diffusion type processes. Dietz (1989) studied the weak consistency and asymptotic mixed normality of maximum likelihood estimator (MLE) as the intensity of noise $\epsilon \to 0$ or the observation time $T \to \infty$. Linkov (1990) also obtained the local asymptotic mixed normality (LAMN) property of such models as observation time $T \to \infty$ under a different set of conditions. Dietz (1992) showed the LAMN property for the concrete example of diffusion type process which is an exponential memory, a non-Markovian relative to the Ornstein-Uhlenbeck (O-U) process and showed the asymptotic properties of the MLE as the observation time $T \to \infty$. Guttorp and Kulperger (1984) studied the asymptotic properties of MLE in Volterra population processes in a random environment as $T \to \infty$. Gushchin and Kuchler (1998) studied the asymptotic behaviour of MLE in a linear SDE with time delay whose solution is a non-Markovian diffusion type process. For a two dimensional parameter they showed that MLE has eleven types of behaviour in eleven parts of the parameter space. Functional stochastic differential equations also include the partially observed Kalman-Bucy filter models for which the asymptotic behaviour of MLE were studied by Campillo and Le Gland (1989), James and Le Gland (1993), Konecky (1991), Kallianpur and Selukar (1991), Kutoyants and Pohlman (1994) and Kutoyants (1994). For asymptotic minimax theory of models satisfying beyond the LAN condition, like the LAID, LAQ, LABF, PLAMN, FLAQ etc. see Taraskin (1985), Spokoiny (1992), Greenwood and Wefelmeyer (1993), Gushchin (1995), Jeganathan (1995) and Shiryayev and Spokoiny (2000).

In this chapter we study the asymptotic properties of the MLE, MPE, BEs and posterior distributions of the parameter appearing in the drift coefficient of nonlinear nonhomogeneous non-Markovian diffusion type process when the length of observation time becomes large under some regularity conditions. The chapter is organised as follows: In Section 4.2 we prepare notations, assumptions and preliminaries. In Section 4.3 we show that the maximum likelihood estimator is strongly consistent. In Section 4.4 we prove the Bernstein-von Mises type theorem showing that a suitably normalised and centered posterior distribution converges to a mixture of normal distribution. As a consequence, we show that the maximum likelihood estimator and a regular class of Bayes estimators are asymptotically equivalent and Bayes estimators converge to a mixture of normal distribution. In Section 4.5 we show that the maximum probability estimator (MPE) is consistent, converge to a mixture of normal distribution and are locally asymptotically minimax (in the Hajek-Le Cam sense). Finally in Section 4.6 we illustrate our results through examples from exponential memory O-U process, nonhomogeneous O-U process and the Kalman-Bucy filter model.

This chapter is adapted from Bishwal (2004b).

4.2 Model, Assumptions and Preliminaries

Let $(\Omega, \mathcal{F}, \{\mathcal{F}_t\}_{t\geq 0}, P)$ be a stochastic basis satisfying the usual hypotheses on which we have a real valued diffusion type process $\{X_t, t \geq 0\}$ satisfying the functional stochastic differential equation (FSDE)

$$dX_t = f(\theta, t, X) \, dt + \, dW_t, \quad t \geq 0,$$
$$X_0 = \xi \tag{2.1}$$

where $\{W_t, t \geq 0\}$ is a standard Wiener process, ξ is a $\mathcal{F}_0$-measurable real valued random variable, $\theta \in \Theta$ a compact subset of $\mathbb{R}$ is the unknown parameter to be estimated on the basis of observation of the process $\{X_t, 0 \leq t \leq T\} =: X_0^T$. The measurable function $f(\theta, t, x), t \geq 0, \theta \in \Theta$ and $x \in C[0, T]$ are assumed to be (for each fixed θ) nonanticipative, that is $\mathcal{B}_t$-measurable for each $t \geq 0$ where $C[0, T]$ is the space of continuous functions from $[0, T]$ to $\mathbb{R}$.

Prime denotes derivative with respect to θ throughout the chapter. We assume the following conditions:

(A1) $\displaystyle\int_0^T f^2(\theta, t, x)dt < \infty, T < \infty, x \in C[0, T], \; \theta \in \mathbb{R},$

(A2) $P\left(\displaystyle\int_0^\infty f'^2(\theta, t, X)dt = \infty\right) = 1, \; \theta \in \mathbb{R}$

(A3) $|f(\theta, t, x) - f(\theta, t, y)| \leq M_1 \int_0^t |x_s - y_s|ds + M_2|x_t - y_t|,$

$$f^2(\theta, t, x) \leq M_1 \int_0^t (1 + |x_s|)ds + M_2(1 + |x_t|),$$

where $x_s, y_s \in C[0, T], \theta \in \mathbb{R}, M_1$ and M_2 are constants.

Under the condition (A3), it is well known that equation (2.1) has a unique solution (see Kutoyants (1984b), Mao (1997)).

Let P_θ^T be the measure generated by the process X_0^T on the space $(C_T, \mathcal{B}_T)$ of continuous functions from $[0, T] \to \mathbb{R}$ with the associated Borel σ-algebra $\mathcal{B}_T$ under the supremum norm and P_W^T be the measure generated by $\xi + W_t$ on the space $(C_T, \mathcal{B}_T)$. Let E_θ^T denote the expectation w.r.t. measure P_θ^T. Under the assumption (A1), $P_\theta^T \ll P_{\theta_0}^T$ and the Radon-Nikodym derivative (likelihood function) is given by (see Liptser and Shiryayev (1977))

$$L_T(\theta) := \frac{dP_\theta^T}{dP_{\theta_0}^T}(X_0^T)$$

$$= \exp\left\{\int_0^T [f(\theta, t, X) - f(\theta_0, t, X)]dX_t - \frac{1}{2}\int_0^T [f^2(\theta, t, X) - f^2(\theta_0, t, X)]dt\right\}.$$

The maximum likelihood estimate (MLE) is defined as

$$\theta_T := \arg\sup_{\theta \in \Theta} L_T(\theta).$$

One can show that there exists a $\mathcal{F}_T$ measurable MLE by using Lemma 3.3 in Schemetterer (1974) since $L_T(\theta)$ is continuous in θ and Θ is compact. Hereafter we assume the existence of such a measurable MLE.

For the MLE θ_T, Kutoyants (1984a) proved the weak consistency, asymptotic normality and convergence of moments to that of a normal variable when the model satisfies the local asymptotic normality (LAN) condition. He also showed similar properties for a regular class of Bayes estimators.

Let $l_T(\theta) := \log L_T(\theta)$. We shall further assume the following conditions:
(**A4**) $P_{\theta_1} \neq P_{\theta_2}$ for $\theta_1 \neq \theta_2$ in Θ.
(**A5**) For any $\theta \in \Theta$, there exists a neighbourhood V_θ of θ in Θ such that

$$P_\theta \left\{ \int_0^\infty [f(\theta_1, t, X) - f(\theta, t, X)]^2 dt = \infty \right\} = 1 \quad \text{for} \quad \theta_1 \in V_\theta \backslash \{\theta\}.$$

There exists a nonrandom sequence $m_T = m_T(\theta) \uparrow \infty$ as $T \uparrow \infty$ satisfying the following conditions:

(**A6**) $\theta \to f(\theta, t, x) \in C_2(\Theta)$ and for $g = f$ or f' or f''

$$P_{\theta_0} \left\{ \frac{1}{m_T} \int_0^T \sup_{\theta \in \Theta} g^2(\theta, s, X) ds < \infty \right\} = 1.$$

(**A7**) There exists a random variable $\zeta(\theta_0)$ with $P(\zeta(\theta_0) > 0) > 0$ such that

$$\frac{1}{m_T} \int_0^T |f'(\theta_0, s, X)|^2 ds \to \zeta(\theta_0) \quad P_{\theta_0} - \text{a.s. as } T \to \infty \text{ for all } \theta \in \Theta.$$

(**A8**) For all $\theta \in \Theta$ and $u \in \mathbb{R}$,

$$P_\theta - \limsup_{v \in [0,1]} \int_0^T |f'(\theta + uvm_T^{-1/2}, s, X) - f'(\theta, s, X)|^2 ds = 0.$$

Suppose that Λ is a prior probability on (Θ, Ξ) where Ξ is the σ-algebra of Borel subsets of Θ. Assume that Λ has a density $\lambda(.)$ with respect to the Lebesgue measure and the density is continuous and positive in an open neighbourhood of θ_0.

The posterior density of θ given X_0^T is given by

$$p(\theta|X_0^T) = \frac{L_T(\theta)\lambda(\theta)}{\int_\Theta L_T(\theta)\lambda(\theta)d\theta}.$$

Let K be a nonnegative measurable function satisfying

(K1) There exists a number ϵ, $P_{\theta_0}(0 < \epsilon < \zeta(\theta_0)) = 1$, for which

$$\int_{-\infty}^{\infty} K(t) \exp(-\frac{1}{2}(\zeta(\theta_0) - \epsilon)u^2) du < \infty \quad P_{\theta_0} - \text{a.s.}$$

(K2) For every $d > 0$ and every random variable η with $P_{\theta_0}(0 < \eta < \infty) = 1$

$$\exp(-m_T\eta) \int_{|u| \geq d} K(m_T^{1/2}u)\lambda(\theta_T + u)du \to 0 \quad P_{\theta_0} - \text{a.s. as } T \to \infty.$$

Suppose that $\omega(\theta_1, \theta_2)$ is a loss function defined on $\Theta \times \Theta$. Assume that $\omega(\theta_1, \theta_2) = \omega(|\theta_1 - \theta_2|) \geq 0$ and $\omega(t)$ is nondecreasing. Suppose that R is a nonnegative function and K and G are functions satisfying

(B1) $R(m_T)\omega(um_T^{-1/2}) \leq G(u)$ for all $T \geq 0$.

(B2) $R(m_T)\omega(um_T^{-1/2}) \to K(u)$ uniformly on bounded intervals of u as $T \to \infty$.

(B3) $\int_{-\infty}^{\infty} K(t+r) \exp(-\frac{1}{2}\zeta(\theta_0)u^2)du$ has a strict minimum at $r = 0$ P_{θ_0}-a.s.

(B4) The function G satisfies **(K1)** and **(K2)**.

A regular Bayes estimator $\tilde{\theta}_T$ based on X_0^T is defined as

$$\tilde{\theta}_T := \arg\inf_{\psi \in \Theta} B_T(\psi)$$

where $B_T(\psi) = \int_\Theta \omega(\theta, \psi)p(\theta|X_0^T)d\theta$. Assume that such an estimator exists.

Define $Z_T(\theta) := \int_{\theta-m_T^{-1/2}}^{\theta+m_T^{-1/2}} l_T(\phi)d\phi$. The maximum probability estimator is defined as

$$\hat{\theta}_T := \arg\sup_{\theta \in \Theta} Z_T(\theta).$$

Assume that such an estimator exists.

First we show the LAMN property of the model (2.1). Note that LAMN property was shown for more general semimartingale models under more restricted assumptions (see Luschgy (1992)). This result would be used to prove some of our main results.

Theorem 2.1 *Let* $\theta = \theta_0 + um_T^{-1/2}, u \in \mathbb{R}$. *Under the assumptions (A1)– (A8), for all* $u \in \mathbb{R}$,

$$\log \frac{dP_\theta^T}{dP_{\theta_0}^T} = u\Delta_T(\theta_0) - \frac{1}{2}u^2\zeta_T(\theta_0) + \psi(u, \theta_0)$$

where $(\Delta_T(\theta_0), \zeta_T(\theta_0)) \overset{\mathcal{D}[P_{\theta_0}]}{\to} (\zeta^{1/2}(\theta_0)\eta, \zeta(\theta_0))$ *as* $T \to \infty$ *and* $\lim_{T\to\infty} \psi_T(\theta_0, u) = 0$ *in* P_{θ_0}-*probability as* $T \to \infty$ *with* $\eta \sim \mathcal{N}(0,1)$ *independent of* $\zeta(\theta_0)$.

Proof. For $\theta = \theta_0 + um_T^{-1/2}$, we have

$$\log \frac{dP_\theta^T}{dP_{\theta_0}^T}(X_0^T)$$

$$= \int_0^T [f(\theta, t, X) - f(\theta_0, t, X)]dX_t - \frac{1}{2}\int_0^T [f^2(\theta, t, X) - f^2(\theta_0, t, X)]dt$$

$$= \int_0^T [f(\theta, t, X) - f(\theta_0, t, X)]dW_t - \frac{1}{2}\int_0^T [f(\theta, t, X) - f(\theta_0, t, X)]^2 dt$$

$$= um_T^{-1/2}\int_0^T f'(\theta_0, t, X)dW_t$$

$$+ um_T^{-1/2}\int_0^T [f'(\theta_0 + uvm_T^{-1/2}, t, X) - f'(\theta_0, t, X)]dW_t$$

$$- \frac{1}{2}u^2 m_T^{-1}\int_0^T f'^2(\theta_0, t, X)dt$$

$$- \frac{1}{2}u^2 m_T^{-1}\int_0^T [f'^2(\theta_0 + uvm_T^{-1/2}, t, X) - f'^2(\theta_0, t, X)]dt$$

(where $v \in [0, 1]$)

$$=: u\Delta_T(\theta_0) - \frac{1}{2}u^2\zeta_T(\theta_0) + \psi_T(\theta_0, u)$$

where $\Delta_T(\theta_0) = m_T^{-1/2}\int_0^T f'(\theta_0, t, X)dW_t$,

$$\zeta_T(\theta_0) = m_T^{-1}\int_0^T f'^2(\theta_0, t, X)dt,$$

$$\psi_T(\theta_0, u) = um_T^{-1/2}\int_0^T [f'(\theta_0 + uvm_T^{-1/2}, t, X) - f'(\theta_0, t, X)]dW_t$$

$$- \frac{1}{2}u^2 m_T^{-1}\int_0^T [f'^2(\theta_0 + uvm_T^{-1/2}, t, X) - f'^2(\theta_0, t, X)]dt.$$

By (A8), we have $\psi_T(\theta_0, u) \to 0$ in P_{θ_0}- probability as $T \to \infty$.
Notice that $m_T^{1/2}\Delta_T(\theta_0)$ is a zero mean square integrable martingale with quadratic variation process $m_T\zeta_T(\theta_0)$. By the martingale central limit theorem and stability of weak convergence (see Jacod and Shiryayev (1987)) and (A7),

$$(\Delta_T(\theta_0), \zeta_T(\theta_0)) \overset{\mathcal{D}[P_{\theta_0}]}{\to} (\mathcal{N}(0, \zeta(\theta_0)), \zeta(\theta_0)) \quad \mathcal{G}-\text{stably as } T \to \infty. \qquad \square$$

4.3 Asymptotics of the Maximum Likelihood Estimator

Dietz (1989) showed that $\theta_T \rightarrow \theta_0$ in P_{θ_0}- probability as $T \rightarrow \infty$. We strengthen the result by proving strong consistency of the MLE.

Theorem 3.1 *Under the assumptions (A1)–(A5), there exists a root of the likelihood equation which is strongly consistent, i.e., $\theta_T \rightarrow \theta_0$ P_{θ_0}-a.s. as $T \rightarrow \infty$.*

Proof. Observe that, for $\delta > 0$

$$
l_T(\theta_0 \pm \delta) - l_T(\theta_0)
$$
$$
= \log \frac{dP^T_{\theta_0 \pm \delta}}{dP^T_{\theta_0}}
$$
$$
= \int_0^T [f(\theta_0 \pm \delta, t, X) - f(\theta_0, t, X)]dX_t
$$
$$
- \frac{1}{2} \int_0^T [f^2(\theta_0 \pm \delta, t, X) - f^2(\theta_0, t, X)]dt
$$
$$
= \int_0^T [f(\theta_0 \pm \delta, t, X) - f(\theta_0, t, X)]dW_t
$$
$$
- \frac{1}{2} \int_0^T [f(\theta_0 \pm \delta, t, X) - f(\theta_0, t, X)]^2 dt
$$
$$
= \int_0^T A_t^{\theta_0} dW_t - \frac{1}{2} \int_0^T (A_t^{\theta_0})^2 dt.
$$

where $A_t^\theta := f(\theta \pm \delta, t, X) - f(\theta, t, X)$.
Let $N_T := \int_0^T (A_t^{\theta_0})^2 dt$. Then

$$
\frac{l_T(\theta_0 \pm \delta) - l_T(\theta_0)}{N_T} = \frac{\int_0^T A_t^{\theta_0} dW_t}{\int_0^T (A_t^{\theta_0})^2 dt} - \frac{1}{2}
$$
$$
= \frac{W^*(\int_0^T (A_t^\theta)^2 dt)}{\int_0^T (A_t^\theta)^2 dt} - \frac{1}{2} \tag{3.1}
$$
$$
= \frac{W^*(N_T)}{N_T} - \frac{1}{2}
$$

by (A5) and the Skorohod embedding of the continuous martingale $\int_0^T A_t^\theta dW_t$, where W^* is some other Brownian motion with respect to the enlarged filtration $(\mathcal{G}_t)_{t \geq 0}$ where $\mathcal{G}_t = \mathcal{F}_{\tau_t}$ with $\tau_t = \inf\{s : N_s > t\}$ and W^* independent of N_T.

Using the assumption (A5) and the strong law of large numbers for Brownian motion (see Lemma 17.4 in Liptser and Shiryayev (1978, p. 210)), the first term on the r.h.s. of (3.1) converges to zero P_{θ_0}-a.s. as $T \rightarrow \infty$. Hence,

$$
\frac{l_T(\theta_0 \pm \delta) - l_T(\theta_0)}{N_T} \rightarrow -\frac{1}{2} \quad P_{\theta_0} - \text{a.s. as } T \rightarrow \infty.
$$

Furthermore, $N_T > 0$ P_{θ_0}–a.s. by (A4). Therefore, for almost every $w \in \Omega$, δ and θ, there exist some T_0 such that for $T \geq T_0$

$$l_T(\theta_0 \pm \delta) < l_T(\theta_0). \tag{3.2}$$

Since $l_T(\theta)$ is continuous on the compact set $[\theta - \delta, \theta + \delta]$, it has a local maximum and it is attained at a measurable θ_T in $[\theta - \delta, \theta + \delta]$. In view of (3.2), $\theta_T \in (\theta_0 - \delta, \theta_0 + \delta)$ for $T > T_0$. Since $l_T(\theta)$ is differentiable with respect to θ, it follows that $l'_T(\theta_T) = 0$ for $T \geq T_0$ and $\theta_T \to \theta_0$ P_{θ_0}-a.s. as $T \to \infty$. $\square$

The LAMN property (Theorem 2.1) together with Theorem 7 in Jeganathan (1982) gives

Theorem 3.2 Under the assumptions (A1)–(A8),

$$m_T^{1/2}(\hat{\theta}_T - \theta_0) \overset{\mathcal{D}[P_{\theta_0}]}{\to} \mathcal{N}(0, \zeta^{-1}(\theta_0)) \text{ as } T \to \infty$$

and the maximum likelihood estimator is *locally asymptotically minimax*.

4.4 The Bernstein-von Mises Type Theorem and Asymptotics of Bayes Estimators

In this section we prove the Bernstein-von Mises type theorem concerning the convergence of suitably normalized and centered posterior distribution to a mixture of normal distribution.

Let $u = m_T^{1/2}(\theta - \theta_T)$. Then the posterior density of $m_T^{1/2}(\theta - \theta_T)$ given X_0^T is given by

$$p^*(u|X_0^T) = m_T^{-1/2} p(\theta_T + u m_T^{-1/2}|X_0^T).$$

Let

$$\gamma_T(u) = \frac{dP_{\theta_T + u m_T^{-1/2}}^T}{dP_{\theta_0}^T}(X_0^T) \Big/ \frac{dP_{\theta_T}^T}{dP_{\theta_0}^T}(X_0^T),$$

$$C_T = \int_{-\infty}^{\infty} \gamma_T(u) \lambda(\theta_T + u m_T^{-1/2}) du.$$

Clearly, $p^*(u|X_0^T) = C_T^{-1} \gamma_T(u) \lambda(\theta_T + u m_T^{-1/2}).$

Remarks Under the assumptions (A1)–(A3), all the stochastic integrals occurring henceforth can be defined pathwise. Further it is possible to differentiate (with respect to θ) within the stochastic integral (indeed outside a fixed null set of the basic probability space). See Karandikar (1983) and Hutton and Nelson (1984) for details.

Lemma 4.1

(a) For all $\theta \in \Theta$, for $g = f$ or f' or f'',

$$\frac{1}{m_T} \int_0^T g(\theta, s, X) dW_s \to 0 \quad P_{\theta_0} - \text{a.s. as } T \to \infty.$$

(b) For all $\delta > 0$,

$$\inf_{|\theta - \theta_0| > \frac{\delta}{2}} \frac{1}{m_T} \int_0^T |f(\theta, s, X) - f(\theta_0, s, X)|^2 ds \to h(\delta)\zeta(\theta_0) \quad P_{\theta_0}\text{-a.s. as } T \to \infty$$

for some $h(.) > 0$.

Proof (a) By the Skorohod embedding of a continuous martingale in a Brownian motion,

$$\frac{1}{m_T} \int_0^T \sup_{\theta \in \Theta} g(\theta, s, X) dW_s = \frac{W^*(\tau_T)}{m_T}$$

where $\{W_t^*, t \geq 0\}$ is a $\mathcal{F}_T$-Wiener process and $\tau_T = \frac{1}{m_T}\int_0^T \sup_{\theta \in \Theta} g^2(\theta, s, X) ds$. Then (A6) and the fact that

$$\lim_{T \to \infty} \frac{W_T^*}{T} = 0 \text{ a.s. as } T \to \infty$$

imply

$$\frac{1}{m_T} \int_0^T \sup_{\theta \in \Theta} g(\theta, s, X) dW_s \to 0 \quad P_{\theta_0} - \text{a.s. as } T \to \infty.$$

(b) Using (A7) we have the result. $\square$

Lemma 4.2

(a) For each fixed u, $\log \gamma_1(u) \to \frac{1}{2}\zeta(\theta_0)u^2$ P_{θ_0}-a.s. as $T \to \infty$.
(b) For every ϵ, $P_{\theta_0}(0 < \epsilon < \zeta(\theta_0)) = 1$, there exists δ_0 and T_0 such that
$\gamma_T(u) \leq \exp(-\frac{1}{2}(\zeta(\theta_0) - \epsilon)u^2$ P_{θ_0}-a.s. for $|u| \leq \delta_0 m_T^{1/2}$ and $T \geq T_0$.
(c) For every $\delta > 0$, there exists a T_0 such that

$$\sup_{|u| \geq \delta m_T^{1/2}} \gamma_T(u) \leq \exp(-\frac{1}{2}m_T \zeta(\theta_0)h(\frac{\delta}{2})) \quad P_{\theta_0}\text{-a.s. for all } T \geq T_0.$$

Proof. Note that

$$\log \gamma_T(u)$$
$$= \int_0^T [f(\theta_T + um_T^{-1/2}, t, X) - f(\theta_T, t, X)] dW_t$$
$$- \frac{1}{2} \int_0^T [f(\theta_T + um_T^{-1/2}, t, X) - f(\theta_0, t, X)]^2 dt$$
$$+ \frac{1}{2} \int_0^T [f(\theta_T, t, X) - f(\theta_0, t, X)]^2 dt.$$

Applying the mean-value theorem and the likelihood equation it follows that

$$\log \gamma_T(u) = I_1 + I_2 + I_3 + I_4$$

with

$$I_1 := -\frac{u^2}{2m_T} \int_0^T f'^2(\theta_0, t, X)dt,$$

$$I_2 := \frac{u^2}{2m_T} \int_0^T [f'^2(\theta_0, t, X) - f'^2(\theta_T^{**}, t, X)]dt,$$

$$I_3 := \frac{u^2}{2m_T} \int_0^T f''(\theta_T^*, t, X)dW_t,$$

$$I_4 := \int_0^T [f(\theta_T, t, X) - f(\theta_0, t, X)][f(\theta_T + um_T^{-1/2}, t, X) - f(\theta_T, t, X)$$

$$- um_T^{-1/2} f'(\theta_T, t, X)]dt$$

where $\max(|\theta_T^* - \theta_T|, |\theta_T^{**} - \theta_T|) \le |u|m_T^{-1/2}$.

Let us first prove (a). By (A7), $I_1 \to -\frac{1}{2}\zeta(\theta_0)u^2$ P_{θ_0}-a.s. as $T \to \infty$.
By the mean-value theorem and Cauchy-Schwarz inequality we obtain

$$|I_2| \le \frac{|u|^3}{m_T^{1/2}} \left\{ \frac{\int_0^T \sup_{\theta \in \Theta} f'^2(\theta, t, X)dt}{m_T} \right\}^{1/2} \left\{ \frac{\int_0^T \sup_{\theta \in \Theta} f''^2(\theta, t, X)dt}{m_T} \right\}^{1/2} \to 0$$

P_{θ_0}-a.s. as $T \to \infty$ by Lemma 4.1 and (A7).
Again by the mean-value theorem and Cauchy-Schwarz inequality we obtain

$$|I_4|$$
$$\le |u|^2|\theta_T - \theta_0| \left\{ \frac{\int_0^T \sup_{\theta \in \Theta} f'^2(\theta, t, X)dt}{m_T} \right\}^{1/2} \left\{ \frac{\int_0^T \sup_{\theta \in \Theta} f''^2(\theta, t, X)dt}{m_T} \right\}^{1/2}$$
$$\to 0 \qquad P_{\theta_0}\text{-a.s. as } T \to \infty$$

by the strong consistency of θ_T, Lemma 4.1 and (A7).
Clearly, $I_3 \to 0$ P_{θ_0}-a.s. as $T \to \infty$ by Lemma 4.1.
Thus $\log \gamma_T(u) \to \frac{1}{2}\zeta(\theta_0)u^2$ P_{θ_0}-a.s. as $T \to \infty$. This completes the proof of
(a).
Let us prove (b) next. Fix $\epsilon > 0$. Clearly there exists a T_1 such that for $T \ge T_1$
$-\frac{1}{2}\frac{|u|^2}{m_T} \int_0^T f'^2(\theta_0, t, X)dt \le -\frac{1}{2}(\zeta(\theta_0) - \epsilon)u^2$ P_{θ_0}-a.s.
By Lemma 4.1 there exists a T_2 such that for all $T \ge T_2$,
$\sup_{\theta \in \Theta} \frac{1}{m_T} \int_0^T f''(\theta_0, t, X)dW_t \le \frac{\epsilon}{3}$ P_{θ_0}-a.s. Next if $\frac{|u|}{m_T^{1/2}} \le \delta_0$

$$|I_2|$$

$$\leq \left(\frac{|u|}{m_T^{1/2}} + |\theta_T - \theta| \right) u^2 \left\{ \frac{\int_0^T \sup_{\theta \in \Theta} f'^2(\theta, t, X) dt}{m_T} \right\}^{1/2}$$

$$\times \left\{ \frac{\int_0^T \sup_{\theta \in \Theta} f''^2(\theta, t, X) dt}{m_T} \right\}^{1/2}$$

$$\leq (\delta_0 + |\theta_T - \theta|) u^2 \left\{ \frac{\int_0^T \sup_{\theta \in \Theta} f'^2(\theta, t, X) dt}{m_T} \right\}^{1/2}$$

$$\times \left\{ \frac{\int_0^T \sup_{\theta \in \Theta} f''^2(\theta, t, X) dt}{m_T} \right\}^{1/2} .$$

Choosing δ_0 suitably and using Theorem 3.1 it follows that there exists a δ_0 and T_3 such that $\frac{|u|}{m_T^{1/2}} \leq \delta_0$ and $T \geq T_3$ implies $I_2 \leq \frac{\epsilon}{3} P_{\theta_0}$-a.s. Similarly using the mean-value theorem and arguing as above there exists a δ_1 and T_4 such that $\frac{|u|}{m_T^{1/2}} \leq \delta_1$ and $T \geq T_4$ implies $I_4 \leq \frac{\epsilon}{3} P_{\theta_0}$-a.s. Now it suffices to combine the results to obtain (b). Next we prove (c). Note that

$$\frac{1}{m_T} \log \gamma_T(u)$$

$$= \frac{1}{m_T} \int_0^T [f(\theta_T + u m_T^{-1/2}, t, X) - f(\theta_0, t, X)] dW_t$$

$$- \frac{1}{2m_T} \int_0^T [f(\theta_T + u m_T^{-1/2}, t, X) - f(\theta_T, t, X)]^2 dt$$

$$+ \frac{1}{2m_T} \int_0^T [f(\theta_T, t, X) - f(\theta_0, t, X)]^2 dt$$

$$=: J_1(u, T) + J_2(u, T) + J_3(T).$$

By the mean-value theorem, we obtain

$$J_3(T) = \frac{1}{2m_T} |\theta_T - \theta_0|^2 \int_0^T f'^2(\bar{\theta}_T, t, X) dt$$

where $\bar{\theta}_T = \theta_0 + \alpha_T(\theta_T - \theta_0), 0 < \alpha_T < 1$ and $J_3(T) \to 0$ P_{θ_0}-a.s. as $T \to \infty$ by Theorem 3.1 and Lemma 4.1.
Next by Lemma 4.1

$$\sup_{u \in \mathbb{R}} |J_1(u, T)| \leq 2 \sup_{\theta \in \Theta} \frac{1}{m_T} \left| \int_0^T f(\theta, t, X) dW_t \right| \to 0 \quad P_{\theta_0} - \text{a.s. as } T \to \infty$$

Finally by the strong consistency of θ_T, for all $\delta > 0$ there exists a T_0 such that for all $T \geq T_0$, $|\theta_T - \theta_0| \leq \delta$ P_{θ_0}-a.s. Hence if $|u| m_T^{-1/2} \geq \delta$ and $T \geq T_0$, then

$$|\theta_T + |u| m_T^{-1/2} - \theta_0| > \frac{\delta}{2}.$$

Thus by Lemma 4.1

$$|J_2(u, T)| \leq -\frac{1}{2} \inf_{|\theta - \theta_0| \geq \frac{\delta}{2}} \frac{1}{m_T} \int_0^T [f(\theta, t, X) - f(\theta_0, t, X)]^2 dt \to -\frac{1}{2} h(\frac{\delta}{2})\zeta(\theta_0)$$

P_{θ_0}-a.s. as $T \to \infty$. Combining these estimates of J_1, J_2 and J_3, (c) is proved. $\square$

Lemma 4.3 Under the assumptions (A1)–(A8) and (K1)–(K2),
(a) there exists δ_0 such that
$\int_{|u| \leq \delta_0 m_T^{1/2}} K(u)|\gamma_T(u)\lambda(\theta_T + um_T^{-1/2}) - \lambda(\theta_0)\exp(-\frac{1}{2}\zeta(\theta_0)u^2)|du \to 0$ P_{θ_0}-a.s.
as $T \to \infty$,
(b) for every $\delta > 0$
$\int_{|u| \geq \delta m_T^{1/2}} K(u)|\gamma_T(u)\lambda(\theta_T + um_T^{-1/2}) - \lambda(\theta_0)\exp(-\frac{1}{2}\zeta(\theta_0)u^2)|du \to 0$ P_{θ_0}-a.s.
as $T \to \infty$.

Proof We omit the proof since the arguments are already available in Borwanker *et al.* (1971) (Lemmas 3.2 and 3.3). $\square$

With the aid of Lemma 4.1 and Lemma 4.3 we are now ready to prove a generalized version of the Bernstein-von Mises theorem.

Theorem 4.1 Under the assumptions (A1)–(A8) and (K1)–(K2),

$$\lim_{T \to \infty} \int_{-\infty}^{\infty} K(u)|p^*(u|X_0^T) - \left(\frac{\zeta(\theta_0)}{2\pi}\right)^{1/2} \exp(-\frac{1}{2}\zeta(\theta_0)u^2)|du = 0 \quad P_{\theta_0} - \text{a.s.}$$

Proof. By Lemma 4.3, we have
$$\lim_{T \to \infty} \int_{-\infty}^{\infty} K(u)|\gamma_T(u)\lambda(\theta_T + um_T^{-1/2}) - \lambda(\theta_0)\exp(-\frac{1}{2}\zeta(\theta_0)u^2)|du = 0$$
$P_{\theta_0} - \text{a.s.}$ $\qquad (4.1)$
Putting $K(u) = 1$ which trivially satisfies (K1)-(K2), we obtain
$C_T = \int_{-\infty}^{\infty} \gamma_T(u)\lambda(\theta_T + um_T^{-1/2})du \to \lambda(\theta_0)\int_{-\infty}^{\infty} \exp(-\frac{1}{2}\zeta(\theta_0)u^2)du$
$P_{\theta_0} - \text{a.s. as} T \to \infty.$ $\qquad (4.2)$
    Therefore,

$$\int_{-\infty}^{\infty} K(u)|p^*(m_T^{1/2}(\theta - \theta_T)|X_0^T) - \left(\frac{\zeta(\theta_0)}{2\pi}\right)^{1/2} \exp(-\frac{1}{2}\zeta(\theta_0)u^2)|du$$

$$\leq \int_{-\infty}^{\infty} K(u)|C_T^{-1}\gamma_T(u)\lambda(\theta_T + um_T^{-1/2}) - C_T^{-1}\lambda(\theta_0)\exp(-\frac{1}{2}\zeta(\theta_0)u^2)|du$$

$$+ \int_{-\infty}^{\infty} K(u)|C_T^{-1}\lambda(\theta_0)\exp(-\frac{1}{2}\zeta(\theta_0)u^2) - \left(\frac{\zeta(\theta_0)}{2\pi}\right)^{1/2} \exp(-\frac{1}{2}\zeta(\theta_0)u^2)|du$$

$$\to 0 \ P_{\theta_0}\text{-a.s. as } T \to \infty$$

by (4.1) and (4.2).

$\square$

Corollary 4.1 In additions to assumptions of Theorem 2.1, if further $\int_{-\infty}^{\infty} |\theta|^n \lambda(\theta) d\theta < \infty$ for $n \in \mathbb{N}$, then

$$\lim_{T \to \infty} \int_{-\infty}^{\infty} |u|^n |p^*(u|X_0^T) - \left(\frac{\zeta(\theta_0)}{2\pi}\right)^{1/2} \exp(-\frac{1}{2}\zeta(\theta_0)u^2)| du = 0 \quad P_{\theta_0} - \text{a.s.}$$

Remarks

(1) The case $n = 0$ and conditionally on $\zeta(\theta_0)$, Corollary 2.1 gives a generalized version of the classical form of Bernstein-von Mises type theorem.
Thus posterior distributions, suitably normalised and centered, converges to a mixture of normal distribution in total variation norm a.s. as $T \to \infty$.
(2) Note that if one verifies the condition

$$\lim_{\alpha \to \infty} \limsup_{T \to \infty} P_{\theta_0}[\int_{|u| \geq \alpha} |u|^a L_T(\theta_0 + um_T^{-1/2}) du > \epsilon] = 0$$

for every $\epsilon > 0$ and some $a \geq 0$, then by Theorem 5 in Jeganathan (1982) along with Theorem 2.1 one obtains an in-probability version of the Bernstein-von Mises theorem, i.e., for every $0 \leq b \leq a$

$$\int_{-\infty}^{\infty} |u|^b |p^*(u|X_0^T) - \left(\frac{\zeta(\theta_0)}{2\pi}\right)^{1/2} \exp(-\frac{1}{2}\zeta(\theta_0)u^2)| du = 0$$

in P_{θ_0} – probability as $T \to \infty$.

We have the following asymptotic properties of the Bayes estimators.

Theorem 4.2 Under the assumptions (A1)–(A8), (K1)–(K2) and (B1)–(B4),
(a) $m_T^{1/2}(\theta_T - \tilde{\theta}_T) \to 0$ P_{θ_0}-a.s. as $T \to \infty$.
(b) $R(m_T)B(\theta_T) = R(m_T)B(\tilde{\theta}_T) = \left(\frac{\zeta(\theta_0)}{2\pi}\right)^{1/2} \int_{-\infty}^{\infty} K(u) \exp(-\frac{1}{2}\zeta(\theta_0)u^2) du$
P_{θ_0}-a.s.

Proof. The proof is omitted as it follows the arguments in Borwanker *et al.* (1973). □

Thus maximum likelihood estimator and a regular class of Bayes estimators are asymptotically equivalent as $T \to \infty$.

Corollary 4.2 Under the assumptions (A1)–(A8), (K1)–(K2) and (B1)–(B4),
(a) $\tilde{\theta}_T \to \theta_0$ P_{θ_0}-a.s. as $T \to \infty$.
(b) $m_T^{1/2}(\tilde{\theta}_T - \theta_0) \overset{\mathcal{D}[P_{\theta_0}]}{\to} \mathcal{N}(0, \zeta^{-1}(\theta_0))$ as $T \to \infty$.
Thus Bayes estimators are strongly consistent and converge to a mixture of normal distribution as $T \to \infty$.

4.5 Asymptotics of Maximum Probability Estimator

We need the following Lemma in the sequel.

Lemma 5.1 Let $\Gamma_T(\theta) = \frac{d^2 l_T(\theta)}{d\theta^2}$. Then

$$\frac{1}{m_T} \int_0^1 l_T''(\theta + \lambda \tau_T) d\lambda \to -\zeta(\theta_0) \; P_{\theta_0} - \text{a.s. as } T \to \infty$$

where $\tau_T \to 0$ as $T \to \infty$.

Proof. Martingale Convergence Theorem (see Dietz (1989)) and (A8) prove the result.   □

We first prove the strong consistency of the MPE.

Theorem 5.1 Under the assumptions (A1)–(A5), for sufficiently large T with P_{θ_0}-probability approaching one, there exists an MPE which is $m_T^{1/2}$-consistent estimate of θ,
i.e., $m_T^{1/2}(\hat\theta_T - \theta_0) = O_{P_{\theta_0}}(1)$.

Proof. Consider the maximum probability equation

$$Z_T'(\theta) = l_T(\theta + m_T^{-1/2}) - l_T(\theta - m_T^{-1/2}) = 0. \tag{5.1}$$

For d in a neighborhood of θ,

$$l_T(d \pm m_T^{-1/2})$$
$$= l_T(\theta) + (d - \theta \pm m_T^{-1/2})l_T'(\theta)$$
$$+ \frac{1}{2}(d - \theta \pm m_T^{-1/2})^2 \int_0^1 l_T''(\theta + \lambda(d - \theta \pm m_T^{-1/2}))d\lambda$$

Let

$$D_T^\pm := \frac{1}{m_T} \int_0^1 l_T''(\theta + \lambda(d - \theta \pm m_T^{-1/2}))d\lambda$$
$$=: \quad F_T((d - \theta \pm m_T^{-1/2}) + \theta)$$

Replacing θ by d in (5.1), we have

$$(d - \theta + m_T^{-1/2})l_T'(\theta) + \frac{1}{2}(d - \theta + m_T^{-1/2})^2 m_T D_T^+$$
$$= (d - \theta - m_T^{-1/2})l_T'(\theta) + \frac{1}{2}(d - \theta - m_T^{-1/2})^2 m_T D_T^-,$$

from which we obtain,

$$\gamma_T(d) := 4l_T'(\theta) + \{m_T^{1/2}(d - \theta)\}^2(D_T^+ - D_T^-) + 2m_T^{1/2}(d - \theta)(D_T^+ - D_T^-)$$
$$+ (D_T^+ - D_T^-) = 0.$$

Let $N > 0$ be fixed. Thus
$$\gamma_T(\theta_0 + Nm_T^{-1/2}) \quad = \quad 4l_T'(\theta_0) + N^2\{F_T(\theta_0 + (N + 1)m_T^{-1/2}) - F_T(\theta_0 +$$

$(N-1)m_T^{-1/2})\} + 2N\{F_T(\theta_0 + (N+1)m_T^{-1/2}) + F_T(\theta_0 + (N-1)m_T^{-1/2})\} +$
$\{F_T(\theta_0 + (N+1)m_T^{-1/2}) + F_T(\theta_0 + (N-1)m_T^{-1/2})\}$
and $\gamma_T(\theta_0 - Nm_T^{-1/2}) = 4l_T'(\theta_0) + N^2\{F_T(\theta_0 - (N-1)m_T^{-1/2}) - F_T(\theta_0 -$
$(N+1)m_T^{-1/2})\} - 2N\{F_T(\theta_0 - (N-1)m_T^{-1/2}) + F_T(\theta_0 - (N+1)m_T^{-1/2})\} +$
$\{F_T(\theta_0 + (N+1)m_T^{-1/2}) + F_T(\theta_0 + (N-1)m_T^{-1/2})\}.$
We have

$$l_T'(\theta_0) \overset{\mathcal{D}[P_{\theta_0}]}{\to} \mathcal{N}(0, \zeta(\theta_0)) \text{ as } T \to \infty \text{ conditionally on } \zeta(\theta_0).$$

On the other hand by Lemma 5.1, F_T above converges P_{θ_0}-a.s. to $-\zeta(\theta_0)$ as $T \to \infty$. Hence for any $N > 0$,

$$\gamma_T(\theta_0 + Nm_T^{-1/2}) \overset{\mathcal{D}[P_{\theta_0}]}{\to} \mathcal{N}(-4N\zeta(\theta), 16\zeta(\theta_0)) \text{ conditionally on } \zeta(\theta_0)$$

and

$$\gamma_T(\theta_0 - Nm_T^{-1/2}) \overset{\mathcal{D}[P_{\theta_0}]}{\to} \mathcal{N}(4N\zeta(\theta), 16\zeta(\theta_0)) \text{ conditionally on } \zeta(\theta_0).$$

Since $\zeta(\theta_0) > 0$ with P_{θ_0}-probability 1 for $\epsilon > 0$ and large T, $P_{\theta_0}[\gamma_T(\theta_0 + Nm_T^{-1/2}) < 0] > 1 - \frac{\epsilon}{2}$ and $P_{\theta_0}[\gamma_T(\theta_0 + Nm_T^{-1/2}) > 0] > 1 - \frac{\epsilon}{2}$ for sufficiently large N. However, $\gamma_T(d)$ is continuous in d. Hence, there exists $\hat{\theta}_T = \hat{\theta}_T(X_0^T)$ such that $\hat{\theta}_T \in [\theta_0 - Nm_T^{-1/2}, \theta_0 + Nm_T^{-1/2}]$ with P_{θ_0}-probability greater than $1 - \epsilon$ and $\gamma_T(\hat{\theta}_T) = 0$. Hence the theorem. $\qquad\square$

Theorem 2.1 and Theorem 5.1 with the aid of Theorem 4 in Jeganathan (1982) yield the following result.

Theorem 5.2 Under the assumptions (A1)–(A8),

$$m_T^{1/2}(\hat{\theta}_T - \theta_0) \overset{\mathcal{D}[P_{\theta_0}]}{\to} \mathcal{N}(0, \zeta^{-1}(\theta_0)) \text{ as } T \to \infty$$

and the maximum probability estimator is *locally asymptotically minimax*.

4.6 Examples

(a) Exponential Memory Ornstein-Uhlenbeck Process

Consider the process $\{X_t, t \geq 0\}$ satisfying the functional SDE

$$dX_t = \theta\alpha \int_0^t e^{-\alpha(t-s)}X_s ds dt + dW_t, t \geq 0, X_0 = 0$$

Suppose $\alpha > 0$ is known and the problem is to estimate $\theta \in (0, \infty)$. Here $f(\theta, t, X) = \theta\alpha \int_0^t e^{-\alpha(t-s)}X_s ds$. The SDE here is a nonhomogeneous Volterra integro-differential equation with a white noise input. The solution $\{X_t\}$ is a non-Markovian process, see Dietz (1992). Note that the range of the memory process is governed by the parameter α which is assumed to be known. As $\alpha \to \infty$, X_t converges a.s. to the classical Ornstein-Uhlenbeck process, i.e., the process losses memory.

Note that in this case the LAMN property holds with $m_T(\theta) = \frac{e^{\theta T}}{\sqrt{2\theta}}$.

$$\theta_T = \hat{\theta}_T = \frac{\int_0^T \int_0^t e^{-\alpha(t-s)} X_s ds dX_t}{\alpha \int_0^T \left(\int_0^t e^{-\alpha(t-s)} X_s ds\right)^2 dt},$$

$$\zeta(\theta) = \left(e^{\frac{1-\sqrt{\alpha^2+4\theta}}{2} s} dW_s\right)^2.$$

(b) Nonhomogeneous Ornstein-Uhlenbeck Process

Consider the nonhomogeneous Ornstein-Uhlenbeck process satisfying the SDE

$$dX_t = \theta g(t) X_t dt + dW_t, \quad t \geq 0, \quad X_0 = 0$$

where $g : \mathbb{R}_+ \to \mathbb{R}$ is measurable with $\int_0^t g^2(s) ds < \infty$ for every t and $\Theta = \{\theta \in \mathbb{R} : \theta \int_0^T g(s) ds \to \infty \text{ as } T \to \infty \text{ and } \int_0^\infty \exp(-2\theta \int_0^t g(s) ds) dt < \infty\}$. Note that $\{X_t\}$ is a nonstationary nonhomogeneous Markov process.

Here $m_T(\theta_0) = \int_0^T g^2(s) e^{2a(s,\theta_0)} ds$ where $a(s, \theta) = \theta \int_0^s g(u) du$. Note that $m_T(\theta_0) \uparrow \infty$ as $T \uparrow \infty$. Here

$$\theta_T = \hat{\theta}_T = \frac{\int_0^T g(s) X_s dX_s}{\int_0^T g^2(s) X_s^2 ds},$$

$$\zeta(\theta) = \left(\int_0^\infty e^{-a(s,\theta)} dW_s\right)^2,$$

which is a chi-square random variable. Thus $\mathcal{N}(0, \zeta(\theta_0))$ is standard Cauchy. Hence Theorem 3.1 becomes

$$\lim_{T \to \infty} \int_{-\infty}^\infty K(u) \left| p^*(u|X_0^T) - \frac{1}{\pi(1+u^2)} \right| du = 0 \quad P_{\theta_0} - \text{a.s. as } T \to \infty.$$

Thus the normalized and centered (at the MLE) posterior distribution converges to the Cauchy distribution as $T \to \infty$. In this case the MPE and the MLE coincide. As a consequence of Corollary 3.1, Bayes estimators have asymptotically Cauchy distribution as $T \to \infty$. As a consequence of Theorem 5.2, MPE and hence MLE also converge to Cauchy distribution as $T \to \infty$.

Let us obtain the normings and the parameter spaces in some special cases. When $g(s) = 1$ (homogeneous case, the classical O-U process), we have

$m_T(\theta) = \frac{e^{\theta T}}{\sqrt{2\theta}}$ and $\Theta = (0,\infty)$. When $g(s) = s$, we have $m_T(\theta) = \frac{e^{\theta T^2/2}}{4\theta}$ and $\Theta = (0,\infty)$.

When $g(s) = \frac{1}{1+s}$, we have $m_T(\theta) = T^{\theta-\frac{1}{2}}$ and $\Theta = (\frac{1}{2},\infty)$.

(c) Kalman-Bucy Filter Model

Let $\{X_t, Y_t, 0 \le t \le T\}$ be the two dimensional diffusion process satisfying the SDEs

$$dX_t = \theta Y_t dt + dW_t, \quad X_0 = 0,$$

$$dY_t = \beta Y_t dt + dV_t, \quad Y_0 = y_0 \ne 0$$

where $\{W_t\}$ and $\{V_t\}$ are independent Wiener processes, $\beta > 0$ is a known parameter and $\theta \in (a,b)$, $a > 0$ is a unknown parameter. Here $\{X_t\}$ is observable while $\{Y_t\}$ is unobservable and it is required to estimate θ by the realization $\{X_t, 0 \le t \le T\}$. This type of model in the LAN situation was studied by Kallianpur and Selukar (1991).

This model is a particular case of the model (2.1) with $f(\theta, t, X) = E(\theta Y_t | X_s, 0 \le s \le t)$.

Let $q_t := E(Y_t | X_s, 0 \le s \le t)$. It is known that q_t satisfies the filtering equation (see Liptser and Shiryayev (1978))

$$dq_t = \beta q_t dt + \gamma_t(\theta)\theta dW_t, \quad q_0 = 0$$

where the function $\gamma_t(\theta) = E_\theta(Y_t - q_t)^2$ satisfies the Riccati equation

$$\frac{\partial \gamma_t(\theta)}{\partial t} = 2\beta\gamma_t(\theta) + \gamma_t^2(\theta)\theta^2 + 1, \quad \gamma_0(\theta) = 0.$$

It is known that $\gamma_t(\theta) \to \gamma(\theta) = \frac{(\beta + \sqrt{\theta^2 + \beta^2})}{\theta^2}$ as $t \to \infty$ (see Liptser and Shiryayev (1978)). We assume that the system has reached the steady state, i.e., we assume that q_t satisfies the SDE

$$dq_t = \beta q_t dt + \theta\gamma(\theta)dW_t, \quad q_0 = 0.$$

It is easy to verify that

$$q_t = \theta\gamma(\theta) \int_0^t e^{-\sqrt{\theta^2 + \beta^2}(t-s)} dX_s.$$

Here the process $\{X_t\}$ is a nonstationary process and the likelihood is given by

$$L_T(\theta, X) = \exp\{\theta \int_0^T q_t dX_t - \frac{1}{2}\theta^2 \int_0^T q_t^2 dt\}$$

and the MLE is given by

$$\theta_T = \frac{\int_0^T q_t dX_t}{\int_0^T q_t^2 dt}.$$

In this case the LAMN property holds with

$$m_T(\theta) = e^{\theta T}, \quad \zeta(\theta) = \left(\int_0^\infty e^{-\sqrt{\theta^2 + \beta^2}(t-s)} dW_s \right)^2.$$

(d) Null Recurrent Diffusions

Consider the null recurrent diffusion

$$dX_t = \theta \frac{X_t}{1 + X_t^2} dt + dW_t, \quad t \ge 0, \quad X_0 = 0.$$

Let the parameter space be $\Theta = \left(-\frac{1}{2}, \frac{1}{2}\right)$, the maximal open interval on which the above diffusion is null recurrent.

Let $\gamma(\theta) := \frac{1}{2} - \theta \in (0, 1), \theta \in \Theta$. The limit distribution of the estimator a variance mixture of normals where the mixing variable is strictly positive and has a Mittag Leffler law of index $\gamma(\theta)$, i.e., the mixing variable is a level crossing time for the stable increasing process of index $\gamma(\theta)$. This model satisfies LAMN propery with local scale $T^{-\gamma(\theta)/2}$.

The MLE here is

$$\theta_T = \frac{\int_0^T \frac{X_t}{1+X_t^2} dX_t}{\int_0^T \left(\frac{X_t}{1+X_t^2}\right)^2 dt}$$

which converges to a mixed normal limit, i.e.,

$$T^{-\gamma(\theta)/2}(\theta_T - \theta) \to^{\mathcal{D}} K(\theta, f)^{-1/2} \frac{B(W_1^{\gamma(\theta)})}{W_1^{\gamma(\theta)}}$$

where

$$K(\theta, f) = \frac{\Gamma(1+\gamma)}{\Gamma(1-\gamma)\gamma^{2\gamma}} \frac{(1-2\theta)^{-\frac{4\theta}{1-2\theta}}}{4} \int (f(x))^2 \mu^\theta(dx)$$

and the invariant measure of the process X in the recurrent case is

$$\mu^\theta(dx) = 2\sqrt{1+x^2}^{2\theta} dx.$$

See Hopfner and Kutoyants (2003) for details.

Remarks

(1) Rates of convergence (both large deviations and Berry-Esseen bounds) for the MLE, MPE and BEs remain to be investigated. Also the rate of convergence in the Bernstein-von Mises type theorem remains to be investigated.

(2) The extension of the results of this chapter to vector parameter can be done without any difficulty.

5

Bayes and Sequential Estimation in Stochastic PDEs

5.1 Long Time Asymptotics

5.1.1 Introduction

In Chapters 2-4 we were concerned with the study of asymptotic properties of several estimators of real valued drift parameter in linear and nonlinear Itô stochastic differential equations (SDEs) whose solutions are real valued diffusions. In many cases the results can be generalized to multidimensional stochastic differential equations. Parameter estimation in finite dimensional stochastic differential equations (SDEs) has been paid a lot of attention during the last three decades. See e.g., Liptser and Shiryayev (1978), Basawa and Prakasa Rao (1980), Kutoyants (1984a, 1994a) and Prakasa Rao (1999). On the other hand, this problem for infinite dimensional SDEs has received very little attention. Loges (1984) initiated the study of asymptotic properties of maximum likelihood estimator in Hilbert-space valued SDEs. Koski and Loges (1985) applied the above theory to a stochastic heat flow problem. Koski and Loges (1986) studied the consistency and asymptotic normality of minimum contrast estimators in Hilbert space valued SDEs which include the MLE. Kim (1996) also proved the consistency and asymptotic normality of MLE in a Hilbert space valued SDE using Fourier expansion of the solution as the observation time $T \to \infty$. Mohapl (1994) obtained strong consistency and asymptotic normality of MLE of a vector parameter in nuclear space valued SDEs from both time continuous observations as well as spatial observations. Huebner, Khasminskii and Rozovskii (1992) studied the asymptotic properties properties of the MLE of a parameter in the drift coefficient of parabolic stochastic partial differential equations (SPDEs) as the amplitude of the noise goes to zero. Huebner and Rozovskii (1995) introduced the spectral method and studied the asymptotics of the MLE of the drift parameter in parabolic SPDEs when the number of Fourier coefficients of the solutions of the SPDEs becomes large, both the observation time and the intensity of the noise remaining fixed. In the same asymptotic framework of Huebner and Rozovskii (1995), Bishwal

(2002a) studied the Bernstein-von Mises theorem concerning the convergence of posterior distribution to normal distribution and the strong consistency and the asumptotic normality of regular Bayes estimators for parabolic SPDEs. Piterbarg and Rozovskii (1995) used the last approach and studied the asymptotic properties of the MLE of a parameter in the SPDE which is used to model the upper ocean variability in physical oceanography using both continuous and discrete time sampling. Piterbarg and Rozovskii (1996) gave necessary and sufficient conditions for the consistency, asymptotic normality and asymptotic efficiency of the MLE based on discrete time sampling when the number of observable Fourier coefficients of the random field governed by the SPDEs becomes large. Mohapl (1996) compared the least squares, optimal estimating function and maximum likelihood estimators of a planar (spatial and temporal) Ornstein-Uhlenbeck process satisfying an SPDE based on lattice sampling.

The MLEs were studied in a similar setting but with space dependent parameter by Bagchi and Borkar (1984), Aihara and Bagchi (1988, 1989, 1991), Aihara (1992, 1994, 1995) when continuous observation are available but since the observations were assumed to be corrupted by additional noise, i.e., only partial observations were available, only consistency of a specified function of the MLE under suitable conditions could be proved. All the above authors assumed linear equations whose solutions are infinite dimensional stationary Markov processes. So far as we know, no results are known about the properties of Bayes estimators and sequential estimators in Hilbert space vlaued SDEs. Our aim in the Chapter is to bridge this gap. Also we prove the Bernstein-von Mises theorem concerning the posterior asymptotic normality for Hilbert space valued diffusion processes.

For the previous work on Bernstein-von Mises theorem, concerning the convergence of posterior density to normal density, and the asymptotic properties of Bayes estimators in finite dimensional SDEs, see the references in the Chapter 1. For previous work on sequential maximum likelihood estimation in finite dimensional SDEs, see Novikov (1972) and Liptser and Shiryayev (1978).

This section is organized as follows: Subsection 5.1.2 contains the notations, assumptions and preliminaries, Subsection 5.1.3 contains the Bernstein-von Mises theorem. In Subsection 5.1.4 asymptotic properties of Bayes estimators are studied and in Section 5.2 we study the properties of sequential MLE. In Section 5.2 we study spectral asymptotics of a stochastic partial differential equation.

This Chapter is adapted from Bishwal (1999b) and Bishwal (2002a).

5.1.2 Model, Assumptions and Preliminaries

Let $(\Omega, \mathcal{F}, \{\mathcal{F}_t\}_{t \geq 0}, P)$ be a stochastic basis satisfying the usual hypotheses on which we define the infinite dimensional SDE

$$dX(t) = \theta AX(t)dt + dW(t), \ X(0) = X_0, t \geq 0 \qquad (2.1)$$

where A is the infinitesimal generator of a strongly continuous semigroup acting on a real separable Hilbert space H with scalar product $< \cdot, \cdot >$ and norm $\| \cdot \|$, $\{W(t), t \geq 0\}$ is a H-valued Wiener process (see Curtain and Pritchard (1978)) and $\theta \in \Theta \subset \mathbb{R}$, which is to be estimated on the basis of H-valued process $\{X(t), 0 \leq t \leq T\}$. Let θ_0 be the true value of the unknown parameter θ.

It is well known that the covariance operator Q of $\{W(t), t \geq 0\}$ is nuclear and $W(t)$ has zero mean (see Itô (1984)). Let also $W(0) = 0$ and assume that the eigen values $\lambda_i, i \geq 1$ of Q are positive. One can write $W(t) = \sum_{i=1}^{\infty} \beta_i(t) e_i$ P-a.s. where $\{e_i, i \geq 1\}$ is a complete orthonormal basis for H consisting of eigen vectors $\{\lambda_i\}$ of Q and $\{\beta_i(t)\}_{t \geq 0} \equiv \{< W(t), e_i >\}_{t \geq 0}$ are mutually independent real valued Wiener processes with incremental covariances λ_i (see Curtain and Pritchard (1978)). For existence, uniqueness and other properties of solutions of Hilbert space valued SDEs, and other applications see Itô (1984) and Kallianpur and Xiong (1995).

Let us assume that there exists a unique strong solution of (2.1). Sufficient conditions for this in terms of the operators in the SDE can be found in Curtain and Pritchard (1978).

Let P_{θ}^{T} be the measure in the space $(\mathcal{C}([0, T], H), \mathcal{B})$ induced by the solution $X(t) = X^{\theta}(t)$ of (2.1) and P_0^{T} corresponds to $(W(t) + X_0)$, $t \in [0, T]$. By $\mathcal{C}([0, T], H)$ we mean the Banach space of continuous functions $f : [0, T] \to H$, which is equipped with the sup-norm. By $\mathcal{B}$, we denote the associated Borel σ-algebra. Let $X_0^{T} \equiv \{X(t), 0 \leq t \leq T\}$.

We assume the following conditions:

(A1) There exist two positive constants C and D such that

$$E \exp(D(\sum_{i=1}^{\infty} \frac{1}{\lambda_i} < AX(t), e_i >^2)) \leq C \text{ for all } t \geq 0.$$

(A2) $\sum_{i=1}^{n_0} \frac{1}{\lambda_i} \int_0^{\infty} < AX(t), e_i >^2 dt = \infty$ P - a.s. for some $n_0 \in \mathbb{N}$.

(A3) $E(\sum_{i=1}^{\infty} \frac{1}{\lambda_i} \int_0^{l} < AX(t), e_i >^2 dt) < \infty$ for all $l \in (0, \infty)$.

(A4) $\lim_{T \to \infty} \frac{1}{T} \sum_{i=1}^{\infty} \frac{1}{\lambda_i} \int_0^{T} < AX(t), e_i >^2 dt = \Gamma$ in probability where Γ is a positive constant.

Under the condition (A1), Loges (1984) showed that the measures P_{θ}^{T} and P_0^{T} are equivalent and the Radon-Nikodym derivative (likelihood) of P_{θ}^{T} with respect to P_0^{T} is given by

$$L_T(\theta) = \frac{dP_\theta^T}{dP_0^T}(X_0^T)$$

$$= \exp\left\{-\frac{1}{2}\theta^2 \sum_{i=1}^{\infty}\frac{1}{\lambda_i}\int_0^T <AX(t), e_i>^2 dt \right.$$

$$\left. +\theta\, L_2 - \lim_{n\to\infty}\sum_{i=1}^{n}\frac{1}{\lambda_i}\int_0^T <AX(t), e_i> d(<X(t), e_i>)\right\}. \qquad (2.2)$$

The maximum likelihood estimate (MLE) is given by

$$\theta_T = \frac{L_2 - \lim_{n\to\infty}\sum_{i=1}^{n}\frac{1}{\lambda_i}\int_0^T <AX(t), e_i> d(<X(t), e_i>)}{\sum_{i=1}^{\infty}\frac{1}{\lambda_i}\int_0^T <AX(t), e_i>^2 dt} \qquad (2.3)$$

if the denominator is greater than zero. Loges (1984) proved the following theorem on the asymptotic behaviour of MLE.

Theorem 1.2.1 *Under the assumptions (A2) - (A3), $\theta_T \to \theta_0$ a.s. $[P_{\theta_0}]$ and under the condition (A2) - (A4), θ_T is asymptotically normally distributed, i.e., $\sqrt{T}(\theta_T - \theta_0) \xrightarrow{D} N(0, \Gamma^{-1})$ as $T \to \infty$.*

5.1.3 Bernstein-von Mises Theorem

Suppose that Λ is a prior probability on (Θ, Ξ), where Ξ is the Borel σ-algebra of Θ. Assume that Λ has a density $\lambda(\cdot)$ with respect to the Lebesgue measure and the density is continuous and positive in an open neighbourhood of θ_0. The posterior density of θ given X_0^T is given by

$$p(\theta|X_0^T) = \frac{\frac{dP_\theta^T}{dP_{\theta_0}^T}(X_0^T)\lambda(\theta)}{\int_\Theta \frac{dP_\theta^T}{dP_{\theta_0}^T}(X_0^T)\lambda(\theta)d\theta} \qquad (3.1)$$

Let $u = T^{1/2}(\theta - \theta_T)$.

Then the posterior density of $\sqrt{T}(\theta - \theta_T)$ is given by

$$p^*(u|X_0^T) = T^{-1/2}p(\theta_T + uT^{-1/2}|X_0^T). \qquad (3.2)$$

Let

$$\gamma_T(u) = \frac{dP_{\theta_T+uT^{-1/2}}^T}{dP_{\theta_0}^T}(X_0^T)\Big/\frac{dP_{\theta_T}^T}{dP_{\theta_0}^T}(X_0^T)$$

$$= \frac{dP_{\theta_T+uT^{-1/2}}^T}{dP_{\theta_T}^T}(X_0^T), \qquad (3.3)$$

$$C_T = \int_{-\infty}^{\infty} \gamma_T(u)\lambda(\theta_T + uT^{-1/2})du. \tag{3.4}$$

Then

$$p^*(u|X_0^T) = C_T^{-1}\gamma_T(u)\lambda(\theta_T + uT^{-1/2}). \tag{3.5}$$

One can reduce equation (2.1) to

$$d(<X(t), e_i>) = \theta < AX(t), e_i > dt + d(<W(t), e_i>). \tag{3.6}$$

Hence we can write

$$\sqrt{T}(\theta_T - \theta) = \frac{L_2 - \lim_{n \to \infty} \sum_{i=1}^{n} \frac{1}{\lambda_i} T^{-1/2} \int_0^T < AX(t), e_i > d(<W(t), e_i>)}{\sum_{i=1}^{\infty} \frac{1}{\lambda_i} T^{-1} \int_0^T < AX(t), e_i >^2 dt}. \tag{3.7}$$

From (3.7) and (2.2), it is easy to check that

$$\log \gamma_T(u) = -\frac{1}{2}u^2 \Gamma_T. \tag{3.8}$$

where $\Gamma_T = \frac{1}{T} \sum_{i=1}^{\infty} \frac{1}{\lambda_i} \int_0^T < AX(t), e_i >^2 dt$.

Let $K(\cdot)$ be a measurable function satisfying the following conditions.

(B1) There exists a number $\epsilon, 0 < \epsilon < \Gamma$ for which

$$\int_{-\infty}^{\infty} K(u) \exp\left\{-\frac{1}{2}u^2(\Gamma - \epsilon)\right\} du < \infty.$$

(B2) For every $h > 0$ and every $\delta > 0$

$$\exp(-T\delta) \int_{|u|>h} K(T^{1/2}u)\lambda(\theta_T + u)du \to 0 \quad \text{a.s. } [P_{\theta_0}] \text{ as } T \to \infty.$$

Theorem 1.3.1. *Under the assumptions (A1) - (A3) and (B1) - (B2), we have*

$$\lim_{T \to \infty} \int_{-\infty}^{\infty} K(u) \left| p^*(u|X_0^T) - \left(\frac{\Gamma}{2\pi}\right)^{1/2} \exp(-\frac{1}{2}\Gamma u^2) \right| du = 0 \quad a.s. \ [P_{\theta_0}].$$

Proof: The proof of this theorem is analogous to that of Theorem 3.1 in Borwanker, Kallianpur and Prakasa Rao (1971) once one reduces the infinite dimensional equation (2.1) to a finite dimensional equation (3.6). Analogous proof for one dimensional diffusion processes is given in Prakasa Rao (1980), Bose (1983b) and Mishra (1989). We omit the details. $\square$

As a consequence of this theorem we obtain the following corollary.

Corollary 1.3.1 *If further $\int_{-\infty}^{\infty} |\theta|^m \lambda(\theta)d\theta < \infty$ for some positive integer m, then*

$$\lim_{T \to \infty} \int_{-\infty}^{\infty} |u|^m \left| p^*(u|X_0^T) - \left(\frac{\Gamma}{2\pi}\right)^{1/2} \exp(-\frac{1}{2}\Gamma u^2) \right| du = 0 \quad a.s. \ [P_{\theta_0}].$$

Remark: The case $m = 0$ gives is the classical *Bernstein-von Mises theorem* for Hilbert space valued diffusions in its simplest form.

5.1.4 Asymptotics of Bayes Estimators

We will study the asymptotic properties of the Bayes estimators in this Section.

Suppose that $l(\theta, \phi)$ is a loss function defined on $\Theta \times \Theta$. Assume that $l(\theta, \phi) = l(|\theta - \phi|) \geq 0$ and $l(t)$ is non decreasing for $t \geq 0$. Suppose that $J(\cdot)$ is a non negative function and $K(\cdot)$ and $G(\cdot)$ are functions such that

(C1) $J(T)l(uT^{-1/2}) \leq G(u)$ for all $T \geq 0$.

(C2) $J(T)l(uT^{-1/2}) \to K(u)$ uniformly on bounded intervals of u as $T \to \infty$.

(C3) $\int_{-\infty}^{\infty} K(u + v) \exp(-\frac{1}{2}u^2\Gamma)du$ has a strict minimum at $v = 0$.

(C4) G satisfies (B1) and (B2).

A regular Bayes estimator $\tilde{\theta}_T$ of θ based on X_0^T is one which minimizes

$$B_T(\phi) = \int_{\Theta} l(\theta, \phi)p(\theta|X_0^T)d\theta.$$

Assume that a measurable Bayes estimator exists. We have the following theorem.

Theorem 1.4.1 *Under the condition (A1) - (A4) and (C1) - (C4) we have*

(i) $\sqrt{T}(\theta_T - \tilde{\theta}_T) \to 0$ a.s. $[P_{\theta_0}]$ as $T \to \infty$.

(ii) $\lim_{T \to \infty} J(T)B_T(\theta_T) = \lim_{T \to \infty} J(T)B_T(\tilde{\theta}_T)$

$$= \left(\frac{\Gamma}{2\pi}\right)^{1/2} \int_{-\infty}^{\infty} K(u) \exp(-\frac{1}{2}\Gamma u^2)du.$$

Proof: The proof is similar to that of Theorem 3.1 in Borwanker, Kallianpur and Prakasa Rao (1971). ☐

Theorem 1.4.1 says that asymptotically Bayes and maximum likelihood estimators are asymptotically equivalent as $T \to \infty$. To distinguish between these two estimators one has to investigate the second order efficiency of these two estimators.

Combining Theorems 1.2.1 and 1.4.1 we obtain the following theorem.

Theorem 1.4.2 *Under the conditions (A1) - (A4) and (C1) - (C4), we have*

(i) $\tilde{\theta}_T \to \theta_0$ a.s. $[P_{\theta_0}]$ as $T \to \infty$.

(ii) $\sqrt{T}(\tilde{\theta}_T - \theta_0) \overset{\mathcal{D}[P_{\theta_0}]}{\to} \mathcal{N}(0, \Gamma^{-1})$ as $T \to \infty$.

In otherwords, Bayes estimators $\tilde{\theta}_T$ are strongly consistent and asymptotically normally distributed as $T \to \infty$. □

5.2 Sequential Estimation

5.2.1 Sequential Maximum Likelihood Estimation

We know that the MLE and the Bayes estimators have good asymptotic properties under stationarity of the process. However, in addition to asymptotic theory which certainly play a predominant role in statistical theory, sequential estimation has got certain advantages. In the finite dimensional linear SDEs, Novikov (1972) (see also Liptser and Shiryayev (1978), Sørensen (1983)) studied the properties of sequential maximum likelihood estimate (SMLE) of the drift parameter which is the MLE based on observation on a random time interval. He showed that SMLE is better than the ordinary MLE in the sense that the former is unbiased, uniformly normally distributed and efficient (in the sense of having the least variance). His plan is to observe the process until the observed Fisher information exceeds a predetermined level of precision. Of course, this type of sampling plan dates back to Anscombe (1952) which has been used in many other situations, e.g., in autoregressive parameter estimation see, e.g., Lai and Siegmund (1985), Greenwood and Shiryayev (1992). Under the assumption that the mean duration of observation in the sequential plan and the ordinary (fixed time) plan are the same, Novikov (1972) showed that the SMLE is more efficient than the ordinary MLE. In this Section, our aim is to extend the problem to Hilbert space valued SDE (2.1).

We assume that the process $\{X(t)\}$ is observed until the observed Fisher information of the process exceeds a predetermined level of precision H, i.e., we observe $\{X(t)\}$ over the random time interval $[0, \tau]$ where the stop time τ is defined as

$$\tau \equiv \tau_H := \inf \left\{ t \geq 0 : \sum_{i=1}^{\infty} \frac{1}{\lambda_i} \int_0^t < AX(s), e_i >^2 ds = H \right\}, 0 < H < \infty.$$
(5.1)

Under the condition (A1) it is well known that the measures P_θ^τ and P_0^τ are equivalent (See Loges (1984), Liptser and Shiryayev (1977)) and the Random-Nikodym derivative of P_θ^τ with respect to P_0^τ is given by

$$\frac{dP_\theta^\tau}{dP_0^\tau}(X_0^\tau) = \exp\left\{-\frac{1}{2}\theta^2 \sum_{i=1}^\infty \frac{1}{\lambda_i} \int_0^\tau < AX(t), e_i >^2 dt\right.$$

$$\left. + \theta L_2 - \lim_{n\to\infty} \sum_{i=1}^n \frac{1}{\lambda_i} \int_0^\tau < AX(t), e_i > d(< X(t), e_i >)\right\}.$$

$$(5.2)$$

Maximizing (5.2) with respect to θ provides the Sequential MLE

$$\theta_\tau = \frac{L_2 - \lim_{n\to\infty} \sum_{i=1}^n \frac{1}{\lambda_i} \int_0^\tau < AX(t), e_i > d < X(t), e_i >}{\sum_{i=1}^\infty \frac{1}{\lambda_i} \int_0^\tau < AX(t), e_i >^2 dt.} \qquad (5.3)$$

$$= \frac{1}{H}L_2 - \lim_{n\to\infty} \sum_{i=1}^n \frac{1}{\lambda_i} \int_0^\tau < AX(t), e_i > d(< X(t), e_i >).$$

We obtain the following properties of the SMLE θ_τ.

Theorem 2.1.1 *Under the conditions (A1) - (A4), we have*
(i) the sequential plan is closed, i.e., $\tau < \infty$ with prob. one.
(ii) $E_{\theta_0}(\theta_\tau) = \theta_0$, i.e., the SMLE is unbiased.
(iii) θ_τ is distributed normally with parameters θ and $\frac{1}{H}$, for all $\theta \in \Theta$,
i.e., $\sqrt{H}(\theta_\tau - \theta_0) \sim N(0, 1)$.
(iv) In the class of unbiased sequential plans $(\gamma, \hat{\theta}_\gamma)$, the plan (τ, θ_τ) is optimal
in the mean square sense, i.e.,

$$E_{\theta_0}(\theta_\tau - \theta_0)^2 \le E_{\theta_0}(\hat{\theta}_\gamma - \theta)^2.$$

Proof: From (5.1), we have

$$P_{\theta_0}(\tau > t) = P_{\theta_0}\left\{\sum_{i=1}^\infty \frac{1}{\lambda_i} \int_0^t < AX(s), e_i >^2 ds < H\right\}$$

from which, due to (A2), it follows that

$$P_{\theta_0}(\tau = \infty) = P_{\theta_0}\left\{\sum_{i=1}^\infty \frac{1}{\lambda_i} \int_0^\infty < AX(s), e_i >^2 ds < H\right\} = 0.$$

Hence $P_{\theta_0}(\tau < \infty) = 1$.
    Because of

$$d(< X(t), e_i >) = \theta_0 < AX(t), e_i > + d(< W(t), e_i >)$$

we can write

$$\theta_\tau = \theta_0 + \frac{1}{H}\left[L_2 - \lim_{n\to\infty}\sum_{i=1}^{n}\frac{1}{\lambda_i}\int_0^\tau <AX(t), e_i> d(W(t), e_i>)\right].$$

Hence $E_{\theta_0}(\theta_\tau) = \theta$ and $\sqrt{H}(\theta_\tau - \theta)$ has standard normal distribution for all $\theta \in \Theta$ since $L_2 - \lim_{n\to\infty}\sum_{i=1}^{n}\frac{1}{\lambda_i}\int_0^\tau <AX(t), e_i> d(<W(t), e_i>)$ is a standard Wiener process (see Curtain and Pritchard (1978)).

Thus

$$E_{\theta_0}(\theta_\tau - \theta_0)^2 = \frac{1}{H} \quad \text{for all } \theta \in \Theta. \tag{5.4}$$

Let $(\gamma, \hat{\theta}_\gamma)$ be any other sequential plan satisfying

$$P_{\theta_0}\left\{\sum_{i=1}^{\infty}\frac{1}{\lambda_i}\int_0^\gamma <AX(t), e_i>^2 dt < \infty\right\} = 1,$$

$$E_{\theta_0}(\hat{\theta}_\gamma)^2 < \infty,$$

$$\text{and } E\left(\sum_{i=1}^{\infty}\frac{1}{\lambda_i}\int_0^\gamma <AX(t), e_i>^2 dt\right) \leq H.$$

Then from the Cramer-Rao-Wolfowitz inequality (see Liptser and Shiryayev (1977)) it follows that

$$E_{\theta_0}(\hat{\theta}_\gamma - \theta_0)^2 \geq \left[E_{\theta_0}\left(\sum_{i=1}^{\infty}\frac{1}{\lambda_i}\int_0^\gamma <AX(t), e_i>^2 dt\right)\right]^{-1} \geq \frac{1}{H}, \tag{5.5}$$

for all $\theta \in \Theta$.

From (5.4) and (5.5), we obtain

$$E(\theta_\tau - \theta_0)^2 \leq E_{\theta_0}(\hat{\theta}_\gamma - \theta_0)^2.$$

Hence SMLE is efficient and the sequential plan (τ, θ_τ) is optimal in the mean square sense. □

5.2.2 Example

Consider the stochastic partial differential equation (SPDE) (see Walsh (1986))

$$dX(t, u) = \theta\frac{\partial^2}{\partial u^2}X(t, u)dt + dW(t, u)$$

where $\theta \in \mathbb{R}$, $W(t, u)$ is a cylindrical Brownian motion and the Laplacian $\frac{\partial^2}{\partial u^2}$ acts on the Hilbert space $L^2[0, T]$ endowed with the usual scalar product with the domain of definition

$$\mathcal{D}(A) = \{h \in H : h'_u, h''_{uu} \in H; h(0) = h(1) = 0\}.$$

Selecting $\{-k_i\}_{i \in \mathbb{N}}$ where $k_i = (\pi i)^2, i = 1, 2, \ldots$ as the set of eigenvalues, we obtain the stability of the corresponding semigroup. For this model, assumptions (A1)-(A4) hold. Hence the MLE θ_T is strongly consistent and asymptotically normally distributed as $T \to \infty$. Also the posterior distributions converge to standard normal distribution and the Bayes estimators are strongly consistent and asymptotically normally distributed as $T \to \infty$. Further the SMLE is unbiased and normally distributed.

Remarks

(1) It would be interesting to prove the asymptotic properties of the Bayes and maximum likelihood estimators by showing the local asymptotic normality property of the model as in Kutoyants (1977), (1984a).

(2) The Berry-Esseen bound and large deviation results for the Bayes and maximum likelihood estimators and the rate of convergence of the posterior distribution remains to be investigated.

(3) Asymptotic properties of MLE and Bayes estimators, the Bernstein-von Mises theorem in Hilbert space valued nonlinear SDE remains to be investigated.

(4) Here we have studied the asymptotics based on continuous observation $\{X(t), 0 \leq t \leq T\}$ of the solution of the SDE. The study of asymptotic properties based on discrete observations of $X(t)$ at time points, $\{t_0, t_1, \ldots, t_n\} \subset [0, T]$ from infinite dimensional SDE has been paid least amount of attention and it is a very fertile field to work with in our opinion. The ideas in Chapters 5 through 8 should be helpful for this work.

(5) The SMLE definitely has certain advantages : the SMLE is unbiased but the ordinary MLE is not so. Ordinary MLE is not asymptotic normally distributed for all $\theta \in \Theta$, but the SMLE is uniformly normally distributed for all $\theta \in \Theta$. Hence the variance of SMLE is a constant. Since $\sqrt{H}(\theta_T - \theta)$ is Gaussian $\mathcal{N}(0, 1)$ one can construct confidence interval for θ.

(6) Here we have studied the properties of sequential maximum likelihood estimator. It would be interesting to study the properties of sequential Bayes estimator.

5.3 Spectral Asymptotics

5.3.1 Introduction

In this section the Bernstein-von Mises theorem, concerning the convergence of suitably normalized and centered posterior density to normal density, is proved for a certain class of linearly parametrized parabolic stochastic partial differential equations (SPDEs) as the number of Fourier coefficients in the expansion of the solution increases to infinity. As a consequence, the Bayes estimators of the drift parameter, for smooth loss functions and priors, are shown to be strongly consistent, asymptotically normal and locally asymptotically minimax (in the Hajek-Le Cam sense), and asymptotically equivalent to the maximum likelihood estimator as the number of Fourier coefficients become large. Unlike in the classical finite dimensional SDEs, here the total observation time and the intensity of noise remain fixed.

Recently stochastic partial differential equations (SPDEs) are being paid a lot of attention in view of their modeling applications in neurophysiology, turbulence, oceanography and finance, see Itô (1984), Walsh (1986) and Kallianpur and Xiong (1995), Holden et al. (1996) and Carmona and Rozovskii (1999). In view of this it becomes necessary to estimate the unknown parameters in SPDEs.

Various methods of estimation in finite dimensional SDEs has been extensively studied during the last three decades as the observation time tends to infinity (see, Liptser and Shiryayev (1978), Basawa and Prakasa Rao (1980), Prakasa Rao (1999) and Kutoyants (1999)) or as the intensity of noise tends to zero (see, Ibragimov and Has'minskii (1981), Kutoyants (1984, 1994)). On the other hand, this problem for infinite dimensional SDEs is young. Loges (1984) initiated the study of parameter estimation in such models. When the length of the observation time becomes large, he obtained consistency and asymptotic normality of the maximum likelihood estimator (MLE) of a real valued drift parameter in a Hilbert space valued SDE. Koski and Loges (1986) extended the work of Loges (1984) to minimum contrast estimators. Koski and Loges (1985) applied the work to a stochastic heat flow problem. Mohapl (1992) studied the asymptotics of MLE in a in a nuclear space valued SDE. Kim (1996) also studied the properties of MLE in a similar set up.

Huebner, Khasminskii and Rozovskii (1992) introduced spectral method to study consistency, asymptotic normality and asymptotic efficiency of MLE of a parameter in the drift coefficient of an SPDE. This approach allows one to obtain asymptotics of estimators under conditions which guarantee the singularity of the measures generated by the corresponding diffusion field for different parameters. Unlike in the finite dimensional cases, where the total observation time was assumed to be long ($T \to \infty$) or intensity of the noise was assumed to be small ($\epsilon \to 0$), here both are kept fixed. Here the asymptotics are obtained when the number of Fourier coefficients (n) of the solution of SPDE becomes large.

The spectral asymptotics or Fourier asymptotics for MLE was extended by Huebner and Rozovskii (1995) to more general SPDEs where the partial differential operators commute and satisfy some order conditions. Piterberg and Rozovskii (1995) studied the properties MLE of a parameter in SPDE which are used to model the upper ocean variability in physical oceonogoaphy. Piterbarg and Rozovskii (1996) studied the properties of MLE based on discrete observations of the corresponding diffusion field. Huebner (1997) extended the problem to the ML estimation of multidimensional parameter. Lototsky and Rozovskii (1999) studied the same problem without the commutativity condition.

The Bernstein-von Mises theorem, concerning the convergence of suitably normalised and centered posterior distribution to normal distribution, plays a fundamental role in asymptotic Bayesian inference, see Le Cam and Yang (1990). In the i.i.d. case, the theorem was first proved Le Cam (1953). Since then the theorem has been extended to many depended cases. Borwanker *et al.* (1972) obtained the theorem for discrete time Markov processes. For the linear homogeneous diffusion processes, the Bernstein - von Mises theorem was proved by Prakasa Rao (1980). Prakasa Rao (1981) extended the theorem to a two parameter diffusion field. Bose (1983) extended the theorem to the homogeneous nonlinear diffusions and Mishra (1989) to the nonhomogeneous diffusions. As a further refinement in Bernstein-von Mises theorem, Bishwal (2000b) obtained sharp rates of convergence to normality of the posterior distribution and the Bayes estimators.

All these above work on Bernstein-von Mises theorem are concerned with finite dimensional SDEs. Recently Bishwal (1999b) proved the Bernstein-von Mises theorem and obtained asymptotic properties of regular Bayes estimator of the drift parameter in a Hilbert space valued SDE when the corresponding diffusion process is observed continuously over a time interval $[0, T]$. The asymptotics are studied as $T \to \infty$ under the condition of absolute continuity of measures generated by the process. Results are illustrated for the example of an SPDE. The situation is analogous to the finite dimensional SDEs, where the measures are absolutely continuous.

Our aim here is to use the spectral approach to study Bernstein-von Mises theorem and Bayes estimation in parabolic SPDE.

5.3.2 Model and Preliminaries

Let $(\Omega, \mathcal{F}, P)$ be a complete probability space on which is defined the parabolic SPDE

$$du^\theta(t, x) = A^\theta u^\theta(t, x)dt + dW(t, x), \ 0 \leq t \leq T, \ x \in G \qquad (2.1)$$

with Dirichlet boundary conditions

$$u(0, x) = u_0(x) \qquad (2.2)$$

$$D^\gamma u(t, x)|_{\partial G} = 0 \qquad (2.3)$$

for all indices γ with $|\gamma| \leq m - 1$.

where $A^\theta = \theta A_1 + A_0$, A_1 and A_0 are partial differential operators of orders m_1 and m_2 respectively, A^θ has order $2m = \max(m_1, m_0)$, $W(t, x)$ is a cylindrical Brownian motion in $L^2([0, T] \times G)$ where G is a bounded domain in $\mathbb{R}^d$ and $u_0 \in L_2(G)$. Here $\theta \in \Theta \subseteq \mathbb{R}$ is the unknown parameter to be estimated on the basis of the observations of the field $u^\theta(t, x), t \in [0, T], x \in G$. Let θ_0 be the true value of the unknown parameter.

Here $u^\theta(t, x)$ is the observation at time t at point x. In practice, it is impossible to observe the field $u^\theta(t, x)$ at all points t and x. Hence, it is assumed that only a finite dimensional projection $u^n := u^{n,\theta} = (u_1^\theta(t), \ldots, u_n^\theta(t)), t \in [0, T]$ of the solution of the equation (2.1) are available. In other words, we observe the first n highest nodes in the Fourier expansion

$$u^\theta(t, x) = \sum_{t=1}^{\infty} u_i^\theta(t)\phi_i(x)$$

corresponding to some orthogonal basis $\{\phi_i(x)\}_{i=1}^{\infty}$. We consider observation continuous in time $t \in [0, T]$. Note that $u_i^\theta(t), i \geq 1$ are independent one dimensional Ornstein-Uhlenbeck processes (see Huebner and Rozovskii (1995)).

Since here the basic set up is same as in HR, for different terminology the reader is referred to HR.

The following conditions are assumed:

(H1) $m_1 \geq m - d/2$.

(H2) The operators A_1 and A_0 are formally self-adjoint, i.e., for $i = 0, 1$

$$\int_G A_i uv dx = \int_G u A_i v dx \quad \text{for all } u, v \in C_0^\infty(G).$$

(H3) There is a compact neighbourhood Θ of θ_0 so that $\{A^\theta, \theta \in \Theta\}$ is a family of uniformly strongly elliptic operators.

(H4) There exists a complete orthonormal system $\{h_i\}_{i=1}^{\infty}$ in $L_2(G))$ such that for every $i = 1, 2, \ldots, h_i \in W_0^{m,2}(G) \cap C^\infty(\overline{G})$ and

$$\Lambda_\theta h_i = \lambda_i(\theta) h_i, \text{ and } \mathcal{L}_\theta h_i = \mu_i(\theta) h_i \text{ for all } \theta \in \Theta$$

where $\mathcal{L}_\theta$ is a closed self adjoint extension of A^θ, $\Lambda_\theta := (k(\theta)I - \mathcal{L}_\theta)^{1/2m}, k(\theta)$ is a constant and and the spectrum of the operator Λ_θ consists of eigen values $\{\lambda_i(\theta)\}_{i=1}^{\infty}$ of finite multiplicities and $\mu_i = -\lambda_i^{2m} + k(\theta)$.

(H5) The operator A_1 is uniformly strongly elliptic and has the same system of eigen functions $\{h_i\}_{i=1}^{\infty}$ as $\mathcal{L}_\theta$.

For $\alpha > d/2$, define the Hilbert space $H^{-\alpha}$ as in Huebner and Rozovskii (1995). Let P_θ^T the measure generated by the solution $\{u^\theta(t, x), t \in [0, T], x \in G\}$ to the problem (2.1)-(2.3) on the space $\mathcal{C}([0, T]; H^{-\alpha})$ with the associated Borel σ-algebra $\mathcal{B}_T$. Note that, under (H1), for different θ the measures P_θ^T are singular.

Consider the projection of $H^{-\alpha}$ on to the subspace $\mathbb{R}^n$. Let $P_\theta^{T,n}$ be the measure generated by $u^{n,\theta}$ on $\mathcal{C}[(0, T]; \mathbb{R}^n)$ with the associated Borel σ-algebra $\mathcal{B}_T^n$.

It is a classical fact (see Liptser-Shiryayev (1977)) that for any $\theta \in \Theta$, the measures $P_\theta^{T,n}$ and $P_{\theta_0}^{T,n}$ are mutually absolutely continuous with Radon-Nikodym derivative (likelihood ratio) given by

$$Z_n^\theta(u) := \frac{dP_\theta^{T,n}}{dP_{\theta_0}^{T,n}}(u^n) = \exp\left\{(\theta - \theta_0)\int_0^T (A_1 u^n(s), du^n(s))_0 \right.$$
$$-\frac{1}{2}(\theta^2 - \theta_0^2)\int_0^T \|A_1 u^n(s)\|_0^2 ds \qquad (2.4)$$
$$\left. -(\theta - \theta_0)\int_0^T (A_1 u^n(s), A_0 u^n(s))_0 ds\right\}.$$

Maximizing $Z_n^\theta(u)$ w.r.t. θ provides the MLE given by

$$\hat{\theta}^n := \frac{\int_0^T (A_1 u^n(s), du^n(s) - A_0 u^n(s)ds)_0}{\int_0^T \|A_1 u^n(s)\|_0^2 ds}. \qquad (2.5)$$

The Fisher information I_n related to $\frac{dP_\theta^n}{dP_{\theta_0}^n}$ is given by

$$I_n := E_{\theta_0} \int_0^T \|A_1 u^n(s)\|_0^2 ds.$$

Define

$$\psi_n := \begin{cases} \dfrac{\zeta}{4\beta} T n^{2\beta} & \text{if } m_1 > m - d/2 \\ \dfrac{\zeta}{2} T \log n & \text{if } m_1 = m - d/2 \end{cases}$$

where $\beta = \frac{1}{d}(m_1 - m) + \frac{1}{2}$,

$$\zeta := (2\pi)^{2(m_1 - m)} \frac{\left(\int_{P_{A^\theta(x,v)}<1} dx dv\right)^{2m/d}}{\left(\int_{P_{A_1(x,v)}<1} dx dv\right)^{2m_1/d}}.$$

Note that $\frac{I_n}{\psi_n} \to 0$ as $n \to \infty$. Let ω be a real valued, non-negative loss function of polynomial majorant defined on $\mathbb{R}$, which are symmetric $\omega(0) = 0$ and monotone on the positive real line.

Under the conditions (H1) - (H5), Huebner and Rozovskii (1995) showed that $\hat{\theta}_n$ is strongly consistent, asymptotically normally distributed with normalization $\psi_n^{1/2}$ and asymptotically efficient with respect to the loss function ω.

Suppose that Π is a prior probability measure on $(\Theta, \mathcal{D})$, where $\mathcal{D}$ is the σ-algebra of Borel subsets of Θ. Assume that Π has a density $\pi(\cdot)$ w.r.t. the Lebesgue measure and the density is continuous and positive in an open neighbourhood of θ_0.

The posterior density of θ given in u^n is given by

$$p(\theta|u^n) := \frac{Z_n^\theta(u)\pi(\theta)}{\int_\Theta Z_n^\theta(u)\pi(\theta)d\theta}. \qquad (2.6)$$

Let $\tau = \psi_n^{1/2}(\theta - \hat{\theta}^n)$. Then the posterior density of $\psi_n^{1/2}(\theta - \hat{\theta}^n)$ is given by

$$p^*(\tau|u^n) := \psi_n^{-1/2} p(\hat{\theta}^n + \psi_n^{-1/2}\tau|u^n).$$

Let

$$\nu_n(\tau) := \frac{dP_{\hat{\theta}^n + \psi_n^{-1/2}\tau}^n / dP_{\theta_0}^n}{dP_{\hat{\theta}^n}^n / dP_{\theta_0}^n} = \frac{dP_{\hat{\theta}^n + \psi_n^{-1/2}\tau}^n}{dP_{\hat{\theta}^n}^n},$$

$$C_n := \int_{-\infty}^{\infty} \nu_n(\tau) \pi(\hat{\theta}^n + \psi_n^{-1/2}\tau) d\tau.$$

Clearly

$$p^*(\tau|u^n) = C_n^{-1} \nu_n(\tau) \pi(\hat{\theta}^n + \psi_n^{-1/2}\tau).$$

5.3.3 The Bernstein-von Mises Theorem

Let $K(\cdot)$ be a non-negative measurable function satisfying the following two conditions:

(K1) There exists a number η, $0 < \eta < 1$, for which

$$\int_{-\infty}^{\infty} K(\tau) \exp\{-\frac{1}{2}\tau^2(1 - \eta)\} d\tau < \infty.$$

(K2) For every $\epsilon > 0$ and $\delta > 0$

$$e^{-\epsilon\psi_n} \int_{|\tau|>\delta} K(\tau\psi_n^{1/2})\pi(\hat{\theta}^n + \tau) d\tau \to 0 \quad \text{a.s. } [P_{\theta_0}] \text{ as } n \to \infty.$$

We need the following Lemma to prove the Bernstein-von Mises theorem.
Lemma 3.3.1 Under the assumptions (H1) - (H5) and (K1) - (K2),
(i) There exists a $\delta_0 > 0$ such that

$$\lim_{n\to\infty} \int_{|\tau|\leq\delta_0\psi_n^{1/2}} K(\tau) \left| \nu_n(\tau)\pi(\hat{\theta}^n + \psi_n^{-1/2}\tau) - \pi(\theta_0)\exp(-\frac{1}{2}\tau^2) \right| d\tau = 0 \quad \text{a.s. } [P_{\theta_0}].$$

(ii) For every $\delta > 0$,

$$\lim_{n\to\infty} \int_{|\tau|\geq\delta\psi_n^{1/2}} K(\tau) \left| \nu_n(\tau)\pi(\hat{\theta}^n + \psi_n^{-1/2}\tau) - \pi(\theta_0)\exp(-\frac{1}{2}\tau^2) \right| d\tau = 0 \quad \text{a.s. } [P_{\theta_0}].$$

Proof. From (2.7) and (2.8), it is easy to check that

$$\log \nu_n(\tau) = -\frac{1}{2}\tau^2\psi_n^{-1} \int_0^T \|A_1 u^n(s)\|_0^2 ds$$

Now (i) follows by an application of dominated convergence theorem.
For every $\delta > 0$, there exists $\epsilon > 0$ depending on δ and β such that

$$\int_{|\tau| \geq \delta \psi_n^{1/2}} K(\tau) \left| \nu_n(\tau) \pi(\hat{\theta}^n + \psi_n^{-1/2} \tau) - \pi(\theta_0) \exp(-\frac{1}{2} \tau^2) \right| d\tau$$

$$\leq \int_{|\tau| \geq \delta \psi_n^{1/2}} K(\tau) \nu_n(\tau) \pi(\hat{\theta}^n + \psi_n^{-1/2} \tau) d\tau + \int_{|\tau| \geq \delta \psi_n^{1/2}} \pi(\theta_0) \exp(-\frac{1}{2} \tau^2) d\tau$$

$$\leq e^{-\epsilon \psi_n} \int_{|\tau| \geq \delta \psi_n^{1/2}} K(\tau) \pi(\hat{\theta}^n + \psi_n^{-1/2} \tau) d\tau + \pi(\theta_0) \int_{|\tau| \geq \delta \psi_n^{1/2}} \exp(-\frac{1}{2} \tau^2) d\tau$$

$$=: F_n + G_n$$

By condition (K2), it follows that $F_n \to 0$ a.s. $[P_{\theta_0}]$ as $n \to \infty$ for every $\delta > 0$. Condition K(1) implies that $G_n \to 0$ as $n \to \infty$. This completes the proof of the Lemma. $\qquad\square$

Now we are ready to prove the generalized version of the Bernstein-von Mises theorem for parabolic SPDEs.

Theorem 3.3.1 Under the assumptions (H1) - (H5) and (K1) - (K2), we have

$$\lim_{n \to \infty} \int_{-\infty}^{\infty} K(\tau) \left| p^*(\tau | u^n) - (\frac{1}{2\pi})^{1/2} \exp(-\frac{1}{2} \tau^2) \right| d\tau = 0 \quad \text{a.s. } [P_{\theta_0}].$$

Proof. From Lemma 3.3.1, we have

$$\lim_{n \to \infty} \int_{-\infty}^{\infty} K(\tau) \left| \nu_n(\tau) \pi(\hat{\theta}^n + \psi_n^{-1/2} \tau) - \pi(\theta_0) \exp(-\frac{1}{2} \tau^2) \right| d\tau = 0 \quad \text{a.s. } [P_{\theta_0}]. \tag{3.1}$$

Putting $K(\tau) = 1$ which trivially satisfies (K1) and (K2), we have

$$C_n = \int_{-\infty}^{\infty} \nu_n(\tau) \pi(\hat{\theta}^n + \psi_n^{-1/2} \tau) d\tau \to \pi(\theta_0) \int_{-\infty}^{\infty} \exp(-\frac{1}{2} \tau^2) d\tau \quad \text{a.s. } [P_{\theta_0}]. \tag{3.2}$$

Therefore, by (3.1) and (3.2), we have

$$\int_{-\infty}^{\infty} K(\tau) \left| p^*(\tau | u^{n,\theta}) - (\frac{1}{2\pi})^{1/2} \exp(-\frac{1}{2} \tau^2) \right| d\tau$$

$$\leq \int_{-\infty}^{\infty} K(\tau) \left| C_n^{-1} \nu_n(\tau) \pi(\hat{\theta}^n + \psi_n^{-1/2} \tau) - C_n^{-1} \pi(\theta_0) \exp(-\frac{1}{2} \tau^2) \right| d\tau$$

$$+ \int_{-\infty}^{\infty} K(\tau) \left| C_n^{-1} \pi(\theta_0) \exp(-\frac{1}{2} \tau^2) - (\frac{1}{2\pi})^{1/2} \exp(-\frac{1}{2} \tau^2) \right| d\tau$$

$$\longrightarrow 0 \quad \text{a.s. } [P_{\theta_0}] \text{ as } n \to \infty. \quad \square$$

Theorem 3.3.2 Suppose (H1)-(H5) and $\int_{-\infty}^{\infty} |\theta|^r \pi(\theta) d\theta < \infty$ for some nonnegative integer r hold. Then

$$\lim_{n \to \infty} \int_{-\infty}^{\infty} |\tau|^r \left| p^*(\tau | u^n) - (\frac{1}{2\pi})^{1/2} \exp(-\frac{1}{2} \tau^2) \right| d\tau = 0 \quad \text{a.s. } [P_{\theta_0}].$$

Proof. For $r = 0$, the verification of (K1) and (K2) is easy and the theorem follows from Theorem 3.1. Suppose $r \geq 1$. Let $K(\tau) = |\tau|^r, \delta > 0$ and $\epsilon > 0$. Using $|a + b|^r \leq 2^{r-1}(|a|^r + |b|^r)$, we have

$$e^{-\epsilon\psi_n} \int_{|\tau|>\delta} K(\tau\psi_n^{1/2})\pi(\hat{\theta}^n + \tau)d\tau$$

$$\leq \psi_n^{r/2} e^{-\epsilon\psi_n} \int_{|\tau-\hat{\theta}^n|>\delta} \pi(\tau)|\tau - \hat{\theta}^n|^r d\tau$$

$$\leq 2^{r-1}\psi_n^{r/2} e^{-\epsilon\psi_n} \left[\int_{|\tau-\hat{\theta}^n|>\delta} \pi(\tau)|\tau|^r d\tau + \int_{|\tau-\hat{\theta}^n|>\delta} \pi(\tau)|\hat{\theta}^n|^r d\tau \right]$$

$$\leq 2^{r-1}\psi_n^{r/2} e^{-\epsilon\psi_n} \left[\int_{-\infty}^{\infty} \pi(\tau)|\tau|^r d\tau + |\hat{\theta}^n|^r \right]$$

$$\longrightarrow 0 \text{ a.s. } [P_{\theta_0}] \text{ as } n \to \infty$$

from the strong consistency of $\hat{\theta}^n$ and hypothesis of the theorem. Thus the theorem follows from Theorem 3.1.1 □

Remark For $r = 0$ in Theorem 3.3.2, we have

$$\lim_{n\to\infty} \int_{-\infty}^{\infty} \left| p^*(\tau|u^n) - \left(\frac{1}{2\pi}\right)^{1/2} \exp\left(-\frac{1}{2}\tau^2\right) \right| d\tau = 0 \text{ a.s. } [P_{\theta_0}].$$

This is the classical form of Bernstein-von Mises theorem for parabolic SPDEs in its simplest form.

As a special case of Theorem 3.3.2, we obtain

$$E_{\theta_0}[\psi_n^{1/2}(\hat{\theta}^n - \theta_0)]^r \to E[\xi^r]$$

as $n \to \infty$ where $\xi \sim \mathcal{N}(0, 1)$.

5.3.4 Bayes Estimation

As an application of Theorem 3.3.1, we obtain the asymptotic properties of a regular Bayes estimator of θ. Suppose $l(\theta, \phi)$ is a loss function defined on $\Theta \times \Theta$. Assume that $l(\theta, \phi) = l(|\theta - \phi|) \geq 0$ and $l(\cdot)$ is non decreasing. Suppose that J is a non-negative function on $\mathbb{N}$ and $K(\cdot)$ and $G(\cdot)$ are functions on $\mathbb{R}$ such that

(**B1**) $J(n)l(\tau\psi_n^{-1/2}) \leq G(\tau)$ for all n,
(**B2**) $J(n)l(\tau\psi_n^{-1/2}) \to K(\tau)$ as $n \to \infty$ uniformly on bounded subsets of $\mathbb{R}$.
(**B3**) $\int_{-\infty}^{\infty} K(\tau + s) \exp\{-\frac{1}{2}\tau^2\}d\tau$ has a strict minimum at $s = 0$.
(**B4**) $G(\cdot)$ satisfies (K1) and (K2).

Let

$$B_n(\phi) = \int_\theta l(\theta, \phi)p(\theta|u^n)d\theta.$$

A regular Bayes estimator $\tilde{\theta}^n$ based on u^n is defined as

$$\tilde{\theta}^n := \arg \inf_{\phi \in \Theta} B_n(\phi).$$

Assume that such an estimator exists.

The following Theorem shows that MLE and Bayes estimators are asymptotically equivalent as $n \to \infty$.

Theorem 3.4.1 Assume that (H1) - (H5), (K1) - (K2) and (B1) - (B4) hold. Then we have

(i) $\psi_n^{1/2}(\tilde{\theta}^n - \hat{\theta}^n) \to 0$ a.s.-$[P_{\theta_0}]$ as $n \to \infty$,

(ii) $\lim_{n\to\infty} J(n)B_n(\tilde{\theta}^n) = \lim_{n\to\infty} J(n)B_n(\hat{\theta}^n)$

$$= (\frac{1}{2\pi})^{1/2} \int_{-\infty}^{\infty} K(\tau) \exp(-\frac{1}{2}\tau^2)d\tau \quad \text{a.s. } [P_{\theta_0}].$$

Proof. The proof is analogous to Theorem 4.1 in Borwanker *et al.* (1972). We omit the details. □

Corollary 3.4.2 Under the assumptions of Theorem 4.1, we have
(i) $\tilde{\theta}^n \to \theta_0$ a.s. $[P_{\theta_0}]$ as $n \to \infty$.
(ii) $\psi_n^{1/2}(\tilde{\theta}^n - \theta_0) \overset{\mathcal{L}}{\to} \mathcal{N}(0,1)$ as $n \to \infty$.

Proof. (i) and (ii) follow easily by combining Theorem 3.4.1 and the strong consistency and asymptotic normality results of the MLE in Huebner and Rozovskii (1995). □

The following theorem shows that Bayes estimators are locally asymptotically minimax in the Hajek-Le Cam sense, i.e., equality is achieved in Hajek-Le Cam inequality.

Theorem 3.4.3 Under the assumptions of Theorem 3.4.1, we have

$$\lim_{\delta \to \infty} \lim_{n \to \infty} \sup_{|\theta - \theta_0| < \delta} E\omega\left(I_n^{1/2}(\tilde{\theta}^n - \theta_0)\right) = E\omega(\xi), \quad \mathcal{L}(\xi) = \mathcal{N}(0,1),$$

where $\omega(\cdot)$ is a loss function as defined in Theorem earlier.

Proof. The Theorem follows from Theorem III.2.1 in Ibragimov-Has'minskii (1981) since here conditions (N1) - (N4) of the said theorem are satisfied using Lemma 3.1-3.3 and local asymptotic normality (LAN) property obtained in Huebner and Rozovskii (1995). □

5.3.5 Example: Stochastic Heat Equation

We illustrate the results of the previous sections through the following heat equation

$$du(t, x) - \theta \Delta u(t, x) = dW(t, x), \ t \in [0, T], x \in (0, 1)$$
$$u(0, x) = u_0(x), x \in (0, 1)$$
$$u(t, 0) = u(t, 1) = 0, t \in [0, T]$$

where $u_0 \in L_2(0, 1), W(t, x)$ is a cylindrical Brownian motion in $L_2(0, 1)$ and $\theta > 0$.

Here $m_1 = ord(\Delta) = 2, m_0 = 0, m = \frac{1}{2}\max(m_1, m_0) = 1, d = 1$. So $m - d/2 = 1/2 < m_1$. Thus (H1) is satisfied. By standard arguments, the operator $-\theta\Delta$ with zero boundary conditions extends to a self adjoint operator on $L_2(0, 1)$ which we denote by $-\theta\Delta$. The domain $\mathcal{D}(-\theta\Delta) = W^{2,2}(0, 1) \cap W_0^{1,2}(0, 1)$. It is readily checked that $-\theta\Delta$ is positive, so we can take $k(\theta) = 0$ and set $\Lambda = \sqrt{-\theta\Delta}$. It is a standard fact that $\mathcal{D}(\sqrt{-\theta\Delta}) = W_0^{1,2}(0, 1)$. Write $h_i := \sqrt{2}\sin(i\pi x)$. Obviously the sequence $h_i, i = 1, 2, \cdots$ forms a complete orthonormal system in $L_2(0, 1)$ and $\sqrt{-\theta\Delta}h_i = \lambda_i(\theta)h_i$ where $\lambda_i(\theta) = \sqrt{\theta}\pi i$. It is readily checked that for $s \in \mathbb{R}$, the norm

$$\| u \|_s := \left(2 \sum_{j=1}^{\infty} \theta^s (\pi i)^{2s} | \int_0^1 u(s) \sin(j\pi x) dx |^2 \right)^{1/2}$$

is equivalent to the norm of the Sobolev space $W_0^{s,2}(0, 1)$. Let us choose $\alpha = 1$. Obviously the system $h_{i,\theta}^{-1} := \lambda_i h_i = \sqrt{\theta}\pi i \sqrt{2}\sin(\pi i x), i = 1, 2, \cdots$ is an orthonormal basis in H^{-1}. Hence assumptions (H2)-(H5) are satisfied.

Take squared error loss function $l(\theta, \phi) = |\theta - \phi|^2$. Now for this heat equation example all the results of subsections 5.3.3 and 5.3.4 on posterior convergence and asymptotics of Bayes estimators, which are posterior mean for squared error loss, hold.

Remarks

(1) General set of conditions of posterior convergence through the LAN property was given in Ghosal et al. (1994) extending methods in Ibragimov and Khasminskii (1981). For the model here, using the LAN property of the model along with Lemma 3.1-3.3 in Huebner and Rozovskii (1995), one verifies the conditions in Ghosal et al. (1994) trivially and obtains the in probability version of the Bernstein-von Mises theorem and asymptotic equivalence in probability of the MLE and the Bayes estimators. However, we obtained the almost sure versions of these results.

(2) Large deviations and Berry-Esseen inequality for the MLE through the spectral approach needs to be investigated. Extension of this problem to Bayes estimators remains to be investigated. Also to obtain rates of convergence of the posterior distributions to normal distribution and bounds on the asymptotic equivalence of the MLE and the Bayes estimators remains to be investigated.

(3) Nonparametric estimation of the coefficients of SPDEs is studied in Ibragimov and Khasminskii (1998).

6

Maximum Likelihood Estimation in Fractional Diffusions

6.1 Introduction

In view of recent empirical findings of long memory in finance, it becomes necessary to extend the diffusion models to processes having long-range dependence. One way is to use stochastic differential equations with fractional Brownian motion (fBm) driving term, with Hurst index greater than $1/2$, the solution of which is called fractional diffusion. The fBm being not a Markov process and not a semimartingale (except for the case where Hurst index is $1/2$), the classical Itô calculus is not applicable to develop its theory.

First we review the recent developments in fractional stochastic calculus: stochastic integral with respect to fBm, fractional version of Itô formula and fractional version of Girsanov formula. Then we study the asymptotic behavior of the maximum likelihood estimator of the drift parameter in the SDE driven by fBm with Hurst index greater than $1/2$ when the corresponding fractional diffusion is observed continuously in a long time interval. We study strong consistency and asymptotic normality of maximum likelihood estimator for both directly observed process and partially observed process. The results are illustrated for fractional Ornstein-Uhlenbeck model.

Recently long memory processes or stochastic models having long range dependence phenomena have been paid a lot of attention in view of their applications in finance, hydrology and computer networks (see Beran (1994), Mandelbrot (1997), Shiryaev (1999), Rogers (1997), Dhehiche and Eddahbi (1999)). While parameter estimation in discrete time models having long-range dependence like the autoregressive fractionally integrated moving average (ARFIMA) models have already been paid a lot of attention, this problem for continuous time models is not well settled. Here we study estimation problem for continuous time long memory processes.

It has been empirically found recently that log share prices allow long-range dependence between returns on different days. So modeling with fBm is an alternative. But on the other hand, it allows arbitrage opportunities due to the degeneracy of the fBm covariance kernel (see Rogers (1997)). A remedy was suggested by choosing some homogeneous covariance kernels that are non-singular in zero and generate Gaussian processes with the same long range dependence property as the fBm. Shiryaev (1998) considered fractional version of Bachelier and the Black-Scholes model.

Tyurin and Phillips (1999) developed approximation of the local time of a fBm by local times of continuous mixtures of diffusions and studied the rate of convergence of spatial density estimator based on discrete sample of observations for a wide class of long-memory processes.

Statistical inference for semimartingales is now well developed, see Prakasa Rao (1999b) for a recent survey. First we will review recent development on fractional stochastic calculus and then focus on asymptotic behavior of maximum likelihood estimator of the drift parameter of diffusions driven by fBm. We shall illustrate our results for fractional Ornstein-Uhlenbeck process.

This chapter is adapted from Bishwal (2003a).

6.2 Fractional Stochastic Calculus

Fractional Brownian Motion

The fractional Brownian motion (fBm, in short), which provides a suitable generalization of the Brownian motion, is one of the simplest stochastic processes exhibiting long range-dependence. It was introduced by Kolmogorov (1940) in a Hilbert space framework and later on studied by Levy (1948) and in detail by Mandelbrot and Van Ness (1968).

Consider a probability space $(\Omega, \mathcal{F}, \mathbb{P})$ on which all random variables and processes below are defined.

A fractional Brownian motion $\{W_t^H, t \geq 0\}$ with Hurst parameter $H \in (0, 1)$ is a centered Gaussian process with continuous sample paths whose covariance kernel is given by

$$E(W_t^H W_s^H) = \frac{V_H}{2}(s^{2H} + t^{2H} - |t - s|^{2H}), \quad s, t \geq 0.$$

where

$$V_H := var(W_1^H) = \frac{1}{[\Gamma(H + \frac{1}{2})]^2} \left\{ \frac{1}{2H} + \int_1^\infty \left[u^{H-\frac{1}{2}} - (u-1)^{H-\frac{1}{2}} \right]^2 du \right\}.$$

With $V_H = 1$, fBm is called a *normalized fBm*.

Fractional Brownian motion has the following properties.
(1) It has stationary increments: $E(W_t^H - W_s^H)^2 = |t - s|^{2H}, t, s \geq 0$.
(2) $W_0^H = 0, E(W_t^H) = 0, E(W_t)^2 = |t|^{2H}, t \geq 0$.
(3) When $H = \frac{1}{2}, W_t^{\frac{1}{2}}$ is the standard Brownian motion. The increments are independent.

(4) The process is self similar or scale invariant, i.e.,

$$(W_{\alpha t}^H, t \geq 0) =^d (\alpha^H W_t^H, t \geq 0), \alpha > 0.$$

H is also called the self similarity parameter.

(5) The increments of the fBm are negatively correlated for $H < \frac{1}{2}$ and positively correlated for for $H > \frac{1}{2}$.

(6) For $H > \frac{1}{2}$, fBm is a long memory process since the covariance between far apart increments decrease to zero as a power law:

$$r(n) := E[W_1^H(W_{1+n}^H - W_n^H)] \sim C_H n^{2H-2}.$$

$$\sum_{n=1}^{\infty} r(n) = \infty.$$

This property is also called long range dependence. The parameter H, measures the intensity of the long range dependence.

For $H = \frac{1}{2}$, the process is a standard Brownian motion.

(7) The sample paths of W^H are almost surely Holder continuous of any order less than H, but not Holder continuous of any order greater than H, hence continuous but nowhere differentiable.

(8) For any H, it has a a a finite $\frac{1}{H}$ variation, i.e.,

$$0 < \sup_{\Pi} E \sum_{t_i \in \Pi} \left[\left| W_{t_{i+1}}^H - W_{t_i}^H \right|^{\frac{1}{H}} \right] < \infty.$$

(9) Law of the Iterated Logarithm (Arcones (1995)):

$$P\left(\overline{\lim}_{t \to 0+} \frac{W_t^H}{t^H (\log \log t^{-1})^{\frac{1}{2}}} = \sqrt{V_H} \right) = 1.$$

Self similarity of fBm leads to

$$P\left(\overline{\lim}_{t \to 0+} \frac{W_{\frac{1}{t}}^H}{(\log \log t^{-1})^{\frac{1}{2}}} = \sqrt{V_H} \right) = 1.$$

Setting $u = \frac{1}{t}$,

$$P\left(\overline{\lim}_{u \to \infty} \frac{W_u^H}{u^H (\log \log u^{-1})^{\frac{1}{2}}} = \sqrt{V_H} \right) = 1.$$

Strong Law of Large Numbers for fBm follows from this.

(10) fBm can be represented as a stochastic integral with respect to standard Brownian motion B (Mandelbrot and van Ness (1968)). For $H > \frac{1}{2}$,

$$W_t^H = \frac{1}{\Gamma(H + \frac{1}{2})} \left\{ \int_{-\infty}^0 [(t-s)^{H-\frac{1}{2}} - (-s)^{H-\frac{1}{2}}] dB_s + \int_0^t (t-s)^{H-\frac{1}{2}} dB_s \right\}.$$

Standard Brownian motion can be written as a stochastic integral with respect to W_t^H (see, Igloi and Terdik (1999)):

$$B_t = \frac{1}{\Gamma(\frac{3}{2}-H)} \left\{ \int_{-\infty}^0 [(t-s)^{-H+\frac{1}{2}} - (-s)^{-H+\frac{1}{2}}]dW_s^H + \int_0^t (t-s)^{-H+\frac{1}{2}}dW_s^H \right\}.$$

(11) With topological dimension n, the fractal dimension of fBm is $n+1-H$. Hausdorff dimension of one dimensional fBm is $2-H$.

(12) *Existence of fBm*:

(i) It can be defined by a stochastic integral w.r.t. Brownian motion.

(ii) It can be constructed by Kolmogorov extension theorem (see, Samorodnitsky and Taqqu (1994)).

(iii) It can be defined as the weak limit of some random walks with strong correlations (see, Taqqu (1975)).

(13) For $H \neq \frac{1}{2}$, the fBm is not a semimartingale and not a Markov process, but a Dirichlet process.

(14) *Dirichlet Process*: A process is called a Dirichlet process if it can be decomposed as the sum of a local martingale and an adapted process of zero quadratic variation (zero energy). Obviously is a larger class of processes than semimartingales.

For $H < \frac{1}{2}$, the quadratic variation of W^H is infinite. For $H > \frac{1}{2}$, the quadratic variation of W^H is zero. Hence for $H > \frac{1}{2}, W^H$ is a Dirichlet process.

Note that the estimation of the parameter H based on observation of fractional Brownian motion has already been paid some attention, see, e.g., Peltier and Levy Vehel (1994) and the references there in. However we assume H to be known.

Generalised Fractional Brownian Motion

As a further generalization of fractional Brownian motion we get the Hermite process of order k with Hurst parameter $H \in (\frac{1}{2}, 1)$ which is defined as a multiple Wiener-Itô integral of order k with respect to standard Brownian motion $(B(t))_{t \in \mathbb{R}}$

$$Z_t^{H,k} := c(H,k) \int_{\mathbb{R}}^k \int_0^t \Pi_{j=1}^k (s - y_i)_+^{-(\frac{1}{2}+\frac{H-1}{2})} ds \, dB(y_1)dB(y_2)\cdots dB(y_k)$$

where $x_+ = \max(x, 0)$.

For $k = 1$ the process is fractional Brownian motion W_t^H with Hurst parameter $H \in (0,1)$. For $k = 2$ the process is Rosenblatt process. For $k \geq 2$ the process is non-Gaussian.

The covariance kernel $R(t,s)$ is given by

$$R(t,s) := E[Z_t^{H,k} Z_s^{H,k}]$$

$$= c(H,k)^2 \int_0^t \int_0^s \left[(u-y)_+^{-(\frac{1}{2}+\frac{H-1}{2})} ds (v-y)_+^{-(\frac{1}{2}+\frac{H-1}{2})} dy \right]^k du dv.$$

Let

$$\beta(p, q) = \int_0^1 z^{p-1}(1 - z)^{q-1}dz, \ p, q > 0$$

be the beta function.

Using the identity

$$\int_{\mathbb{R}} \left[(u - y)_+^{a-1}ds(v - y)_+^{a-1}dy\right] = \beta(a, 2a - 1)|u - v|^{2a-1}$$

we have

$$R(t, s) = c(H, k)^2 \beta\left(\frac{1}{2} - \frac{1 - H}{k}, \frac{2H - 2}{k}\right)^k \int_0^t \int_0^s \left(|u - v|^{\frac{2H-2}{k}}\right)^k dvdu$$

$$= c(H, k)^2 \frac{\beta(\frac{1}{2} - \frac{1-H}{k}, \frac{2H-2}{k})^k}{H(2H - 1)} \frac{1}{2}(t^{2H} + s^{2H} - |t - s|^{2H})$$

In order to obtain $E(Z_t^{(H,k)})^2 = 1$,
choose

$$c(H, k)^2 = \left(\frac{\beta(\frac{1}{2} - \frac{1-H}{k}, \frac{2H-2}{k})^k}{H(2H - 1)}\right)^{-1}$$

and we have

$$R(t, s) = \frac{1}{2}(t^{2H} + s^{2H} - |t - s|^{2H}).$$

Thus the covariance structure of the Hermite process and fractional Brownian motion are the same.

The process $Z_t^{(H,k)}$ is H-self similar with stationary increments and all moments are finite.

For any $p \geq 1$,

$$E|Z_t^{(H,k)} - Z_s^{(H,k)}|^p \leq c(p, H, k)|t - s|^{pH}.$$

Thus the Hermite process has Holder continuous paths of order $\delta < H$.

Fractional Poisson Process

A fractional Poisson process $\{P_H(t), t > 0\}$ with Hurst parameter $H \in (1/2, 1)$ is defined as

$$P_H(t) = \frac{1}{\Gamma(H - \frac{1}{2})} \int_0^t u^{\frac{1}{2}-H}\left(\tau^{H-\frac{1}{2}}(\tau - u)^{H-\frac{3}{2}}d\tau\right)dq(u)$$

where $q(u) = \frac{N(u)}{\sqrt{\lambda}} - \sqrt{\lambda}u$ where $N(u)$ is a Poisson process.

The process is self-similar in the wide sense, has wide sense stationary increments, has fat-tailed non-Gaussian distribution, and exhibits long range dependence. The process converges to fractional Brownian motion in distribution. The process is self similar in the asymptotic sense. Strict sense, wide sense and asymptotic sense self-similarity are equivalent for fractional Brownian motion. Stock returns are far from being self-similar in strict sense.

Fractional Stochastic Integral

For $H \neq \frac{1}{2}$, the classical theory of stochastic integration with respect to semimartingales is not applicable to stochastic integration with respect to fBm. Since fBm is a Gaussian process, stochastic integration with respect to Gaussian process is applicable.

For integration questions related to fractional Brownian motion, see Pipiras and Taqqu (2000). Now there exists several approaches to stochastic integration with respect to fBm.

(i) Classical Riemann sum approach: Lin (1995), Dai and Heyde (1996), Kleptsyna, Kloeden and Anh (1998c)

(ii) Malliavin calculus approach: Decreusefond and Ustunel (1998, 1999), Coutin and Decreusefond (1999a), Alos, Mazet and Nualart (1999, 2000)

(iii) Wick calculus approach: Duncan, Hu and Pasik-Duncan (1999)

(iv) Pathwise calculus: Zahle (1998, 1999), Ruzmaikina (2000)

(v) Dirichlet calculus: Lyons and Zhang (1994)

(vi) Rough path analysis: Lyons (1994)

Lin (1996) introduced the stochastic integral as follows: Let ϕ be a left continuous bounded Lebesgue measurable function with right limits, called *sure processes*. Then

$$\int_0^1 \psi(t)dW_t^H = \text{l.i.m.}_{|\pi|\to\infty} \sum_{t_i \in \pi} \psi(t_{i-1})(W_{t_i}^H - W_{t_{i-1}}^H).$$

The indefinite integral is defined as

$$\int_0^t \psi(s)dW_s^H = \int_0^1 \psi(t)I_{[0,t]}dW_t^H.$$

This integral has a continuous version and a Gaussian process. However,

$$E\left(\int_0^t \psi(s)dW_s^H\right) \neq 0.$$

To overcome this situation, Duncan, Hu and Pasik-Duncan (2000) introduced an integral using *Wick calculus* for which

$$E\left(\int_0^t f(s)dW_s^H\right) = 0.$$

They defined integrals of both Itô and Stratonovich type.

We shall discuss the Wick calculus approach here.

Wiener integral for deterministic kernel was defined by Gripenberg and Norros (1996).

Let $\phi : \mathbb{R}_+ \times \mathbb{R} \to \mathbb{R}$ be a Borel measurable deterministic function. Let

$$L^2_\phi(\mathbb{R}_+) := \left\{ f : |f|^2_\phi : \int_0^\infty \int_0^\infty f(s)f(t)\phi(s,t)dsdt < \infty \right\}.$$

The inner product in the Hilbert space L^2_ϕ is denoted by $\langle \cdot, \cdot \rangle_\phi$.

If $f, g \in L^2_\phi$, then $\int_0^\infty f_s dW^H_s$ and $\int_0^\infty g_s dW^H_s$ are well defined zero mean, Gaussian random variables with variances $|f|^2_\phi$ and $|g|^2_\phi$ respectively and covariance

$$E\left(\int_0^\infty f_s dW^H_s \int_0^\infty g_s dW^H_s \right) = \int_0^\infty \int_0^\infty f_s g_s \phi(s,t)dsdt =: \langle f, g \rangle_\phi.$$

Let $(\Omega, \mathcal{F}, P)$ be the probability space on which W^H is defined. For $f \in L^2_\phi$, define $\epsilon : L^2_\phi \to L^1(\Omega, \mathcal{F}, P)$ as

$$\epsilon(f) := \exp\left\{ \int_0^\infty f_t dW^H_t - \frac{1}{2} \int_0^\infty \int_0^\infty f_s f_t \phi(s,t)dsdt \right\}$$

$$= \exp\left\{ \int_0^\infty f_t dW^H_t - \frac{1}{2} \int_0^\infty |f|^2_\phi \right\}$$

which is called an exponential function.

Let $\mathcal{E}$ be the linear span of exponentials, i.e.,

$$\mathcal{E} = \{ \sum_{k=1}^n a_k \epsilon(f_k) : n \in \mathbb{N}, a_k \in \mathbb{R}, f_k \in L^2_\phi(\mathbb{R}_+), k = \overline{1,n} \}.$$

The Wick product of two exponentials is defined as

$$\epsilon(f) \diamond \epsilon(g) = \epsilon(f + g).$$

For distinct $f_1, f_2, \cdots, f_n \in L^2_\phi, \epsilon(f_1), \epsilon(f_2), \cdots, \epsilon(f_n)$ are independent. It can be extended to define the Wick product of two functionals F and G in $\mathcal{E}$.

An analogue of Malliavin Derivative: Wick Derivative

The ϕ-derivative of a random variable $F \in L^p$ in the direction of Φg where $g \in L^2_\phi$ is defined as

$$D_{\Phi g} F(\omega) = \lim_{\delta \to 0} \frac{1}{\delta} \left[F\left(\omega + \delta \int_0^\cdot (\Phi g)(u)du \right) - F(\omega) \right]$$

if the limit exists in $L^p(\Omega, \mathcal{F}, P)$.

If there is a process $(D^\phi F_s, s \geq 0)$ such that

$$D_{\Phi g}F = \int_0^\infty D^\phi F_s g_s ds \quad a.s.$$

for all $g \in L_\phi^2$, then F is said to be ϕ-differentiable. Let $F : [0,T] \times \Omega \to \mathbb{R}$ be a stochastic process. The process is said to be ϕ-differentiable if for each $t \in [0,T]$, $F(t,\cdot)$ is $\phi-$ differentiable and $D_s^\phi F_t$ is jointly measurable.

Chain Rule: If $f : \mathbb{R} \to \mathbb{R}$ is smooth and $F : \Omega \to \mathbb{R}$ is $\phi-$ differentiable, then $f(F)$ is also $\phi-$ differentiable and

$$D_{\Phi g}f(F) = f'(F)D_{\phi g}F$$

and

$$D_s^\phi f(F) = f'(F)D_s^\phi(F).$$

(1) If $g \in L_\phi^2, F \in L^2(\Omega, \mathcal{F}, P)$ and $D_{\Phi g}F \in L^2(\Omega, \mathcal{F}, P)$, then

$$F \diamond \int_0^\infty g_s dW_s^H = F \int_0^\infty g_s dW_s^H - D_{\Phi g}F.$$

(2) If $g, h \in L_\phi^2$ and $F, G \in \mathcal{E}$, then

$$E\left(F \diamond \int_0^\infty g_s dW_s^H \quad G \diamond \int_0^\infty h_s dW_s^H\right) = E\left[D_{\Phi g}F D_{\Phi h}G + FG\langle g, h\rangle_\phi\right].$$

Let $\pi_n : 0 < t_1 < t_2 < \cdots < t_n = T$.

Let $\mathcal{L}[0,T]$ be the family of stochastic processes on F on $[0,T]$ such that $E|F|_\phi^2 < \infty$, F is $\phi-$ differentiable, the trace of $(D_s^\phi F_t, 0 \leq s \leq T, 0 \leq t \leq T)$ exists and $E\int_0^T (D_s^\phi F_s)^2 ds < \infty$ and for each sequence of partitions $\{\pi_n, n \in \mathbb{N}\}$ such that $|\pi_n| \to 0$

$$\sum_{i=0}^{n-1} E\left\{\int_{t_i^{(n)}}^{t_{i+1}^{(n)}} |D_s^\phi F_{t_i^{(n)}}^\pi - D_s^\phi F_s| ds\right\}^2 \to 0$$

and $E|F^\pi - F|_\phi^2 \to 0$ as $n \to \infty$.

For $F \in \mathcal{L}[0,T]$, define

$$\int_0^T F_s dW_s^H = l.i.m._{|\pi|\to 0} \sum_{i=0}^{n-1} F_{t_i} \diamond (W_{t_{i+1}}^H - W_{t_i}^H).$$

Proposition 2.1 Let $F, G, \in \mathcal{L}[0,T]$. Then
(i) $E\left(\int_0^T F_s dW_s^H\right) = 0$.

(ii) $E\left(\int_0^T F_s dW_s^H\right)^2 = E\left\{\left(D_s^\phi F_s ds\right)^2 + |I_{[0,T]}F|_\phi^2\right\}$

(iii) $\int_o^t (aF_s + bG_s)dW_s^H = a\int_0^t F_s dW_s^H + b\int_0^t G_s dW_s^H$ a.s.

(iv) If $E\left[\sup_{0\le s\le T} F_s\right]^2 < \infty$ and $\sup_{0\le s\le T} E|D_s^\phi F_s|^2 < \infty$, then $\{\int_0^t F_s dW_s^H, 0\le t\le T\}$ has a continuous version.

Here it is not assumed that $(F_s, s\in[0,T])$ is adapted to the fBm. Assume that $D_s^\phi F_s = 0, s\in[0,T]$. Then

(v) $E\left(\int_0^T F_s dW_s^H\right)^2 = |I_{[0,T]}F|_\phi^2 = E\int_0^T\int_0^T F_u F_v \phi(u,v)dudv$.

Fractional version of *Stratonovich Integral* is defined as

$$\int_0^t F_s \delta W_s^H := \int_0^t F_s dW_s^H + \int_0^t D_s^\phi F_s ds \quad a.s.$$

Fractional Itô Formula

If $f : \mathbb{R} \to \mathbb{R}$ is a twice continuously differentiable function with bounded derivatives of order two, then

$$f(W_T^H) - f(W_0^H) = \int_0^T f'(W_s^H)dW_s^H + H\int_0^T s^{2H-1}f''(W_s^H)ds \quad a.s.$$

For $H = \frac{1}{2}$, it gives the classical Itô formula for standard Brownian motion.

General Itô Formula

Let $\{F_u, 0\le u\le T\}$ and $\{G_u, 0\le u\le T\}$ be stochastic processes in $\mathcal{L}[0,T]$. Assume that there exists an $\alpha > 1 - H$ such that

$$E|F_u - F_v|^2 \le C|u-v|^{2\alpha},$$

$$\lim_{|u-v|\to 0} E\{|D_u^\phi(F_u - F_v)|^2\} = 0$$

and

$$E\sup_{0\le s\le T}|G_s| < \infty.$$

Let

$$dX_t = G_t dt + F_t dW_t^H, X_0 = \xi \in \mathbb{R}, 0\le t\le T,$$

i.e.,

$$X_t = \xi + \int_0^t G_s ds + \int_0^t F_s dW_s^H.$$

Let $f : \mathbb{R} \to \mathbb{R}$ be C_b^1 in the first variable and C_b^2 in the second variable and let $\left(\frac{\partial f}{\partial x}(s, X_s), s\in[0,T]\right) \in \mathcal{L}[0,T]$. Then

$$f(t, X_t)$$

$$= f(0, \xi) + \int_0^t \frac{\partial f}{\partial s}(s, X_s)ds + \int_0^t \frac{\partial f}{\partial x}(s, X_s)G_s ds + \int_0^t \frac{\partial f}{\partial x}(s, X_s)F_s dW_s^H$$

$$+ \int_0^t \frac{\partial^2 f}{\partial x^2}(s, X_s)F_s D_s^\phi X_s ds.$$

Itô formula for Stratonovich Type integral:

Let $\{F_t, 0 \le t \le T\}$ satisfy the above assumptions. Let

$$\eta_t = \int_0^t F_s \delta W_s^H.$$

Let $g \in C_b^2$ and $\left(\frac{\partial g}{\partial x}(s, \eta_s)F_s, s \in [0, T] \right) \in \mathcal{L}[0, T]$.
 Then for $t \in [0, T]$,

$$g(t, \eta_t) = g(0, 0) + \int_0^t \frac{\partial g}{\partial s}(s, \eta_s)ds + \int_0^t \frac{\partial g}{\partial x}(s, \eta_s)F_s dW_s^H$$

i.e.,

$$\delta g(t, \eta_t) = g_t(t, \eta_t)dt + g_x(t, \eta_t)d\eta_t.$$

Fractional Girsanov Formula

Decreusefond and Ustunel (1999) gave a Girsanov formula using stochastic calculus of variation. Kleptsyna, Le Breton and Roubaud (1999) obtained the following theorem.

Theorem 2.1 *Let h be a continuous function from $[0, T]$ to $\mathbb{R}$. Define for $0 < t \le T$, the function $k_h^t = (k_h^t(s), 0 < s < t)$ by*

$$k_h^t(s) := -\rho_H^{-1} s^{\frac{1}{2} - H} \frac{d}{ds} \int_s^t d\omega \omega^{2H-1}(\omega - s)^{\frac{1}{2} - H} \frac{d}{d\omega} \int_0^\omega dz z^{\frac{1}{2} - H}(\omega - z)^{\frac{1}{2} - H} h(z)$$

  where

$$\rho_H = \Gamma^2(\frac{3}{2} - H)\Gamma(2H + 1)\sin \pi H.$$

Define for $0 \le t \le T$,

$$N_t^h := \int_0^t k_h^t(s)dW_s^H, \quad \langle N^h \rangle_t := \int_0^t h(s)k_h^t(s)ds.$$

Then the process $\{N_t^h, 0 \le t \le T\}$ is a Gaussian martingale with variance function $\{\langle N^h \rangle_t, 0 \le t \le T\}$.

For $h = 1$, the function k_h^t is

$$k_*^t(s) := \tau_H^{-1}(s(t-s))^{\frac{1}{2}-H}$$

where $\tau_H := 2H\Gamma(\frac{3}{2} - H)\Gamma(H + \frac{1}{2})$.
Then the corresponding Gaussian martingale is

$$N_t^* = \int_0^t k_*^t(s)dW_s^H$$

and

$$\langle N^*\rangle_t = \int_0^t k_*^t(s)ds$$
$$= \lambda_H^{-1}t^{2-2H}$$

where

$$\lambda_H = \frac{2H\Gamma(3-2H)\Gamma(H + \frac{1}{2})}{\Gamma(\frac{3}{2} - H)}.$$

6.3 Maximum Likelihood Estimation in Directly Observed Fractional Diffusions

For directly observed diffusions, the study of asymptotic properties of several estimators of a parameter in the drift coefficient of the Itô stochastic differential equation driven by a standard Brownian motion based on observations of the corresponding diffusion process continuously inside a time interval or at discrete time points is now well developed: see Kutoyants (1984, 1994), Ait-Sahalia (2002), Bishwal (1999a) and Prakasa Rao (1999a).

As far as estimation of unknown parameter in fractional diffusions is concerned, maximum likelihood estimator (MLE) of the constant drift parameter of a fractional Brownian motion was obtained by Decreusefond and Ustunel (1999) who developed stochastic analysis of the fBm in a Malliavin calculus framework. Norros, Valkeila and Virtamo (1999) studied the properties of the MLE of the constant drift parameter of fBm using martingale tools. They showed that the MLE is unbiased and normally distributed. They also showed that the MLE is strongly consistent and proved a law of the iterated logarithm as $T \to \infty$. The problem was generalized by Le Breton (1998) to a stochastic differential equation driven by fBm with drift and the diffusion coefficient being nonrandom functions and the unknown parameter in the drift coefficient. He obtained the best linear unbiased estimator (BLUE) of the drift parameter which coincides with the MLE. He also obtained the least squares estimator (LSE) and compared the relative efficiency of the LSE and the BLUE.

We shall study the maximum likelihood estimation of the parameter appearing in more general drift coefficient of a fractional diffusion process.

Consider the stochastic differential equation driven by the fBm W^H

$$dX_t = \theta f(t, X_t)\, dt + dW_t^H \tag{3.1}$$
$$X_0 = \xi$$

where ξ is a P-a.s finite random variable, $H \in (\frac{1}{2}, 1)$ and $\theta \in \Theta$ open in $\mathbb{R}$ is the unknown parameter to be estimated on the basis of observation of the process $\{X_t, 0 \leq t \leq T\}$.

We assume that the function $f : \mathbb{R} \to \mathbb{R}$ is known and it satisfies

(A1) $|f(t, x)| \leq K(1 + |x|)$ for all $t \in [0, T]$;

(A2) $|f(t, x) - f(t, y)| \leq K(|x - y|)$ for all $t \in [0, T]$.

for some constant $K > 0$.

Under the conditions (A1) and (A2), it is known that there exists a unique solution of the SDE (1) (see Lin (1996)).

In order to obtain the MLE, we proceed as follows. Let τ_H, $k(t, s)$ and N^* be as defined in Theorem 2.1. Then by Theorem 2.1, N^* is a Gaussian martingale whose variance function is

$$\langle N^* \rangle_t = \frac{t^{2-2H}}{\lambda_H}$$

where $\lambda_H = \frac{2H\Gamma(3-2H)\Gamma(H+\frac{1}{2})}{\Gamma(\frac{3}{2}-H)}$. N_t^* can be represented as

$$N_t^* = \int_0^t h(s)\, dB_s$$

where $h(s) = \sqrt{\frac{2(1-H)}{\lambda_H}} s^{\frac{1}{2}-H}$ and B_s is a standard Brownian motion. Hence from (3.1), we have

$$Y_t := \int_0^t k(t, s) dX_s = \theta \int_0^t k(t, s) f(s, X_s)\, ds + \int_0^t k(t, s) dW_s^H$$

$$= \theta \int_0^t k(t, s) f(s, X_s)\, ds + \int_0^t h(s)\, dB_s \tag{3.2}$$

Let consider the probability $\tilde{P}$ such that

$$\frac{d\tilde{P}}{dP} = \exp\left(-\theta \int_0^T \frac{k(T, s) f(s, X_s)}{h(s)} dB_s - \frac{\theta^2}{2} \int_0^T \left(\frac{k(T, s) f(s, X_s)}{h(s)}\right)^2 ds\right)$$

Under $\tilde{P}$ the process

$$\tilde{B}_t = B_t + \int_0^t \theta \frac{k(T, s) f(s, X_s)}{h(s)} ds$$

is a Brownian motion for $t \in [0, T]$ by the classical Girsanov's theorem. Therefore, under $\tilde{P}$,

$$Y_T = \int_0^T h(s) \, d\tilde{B}_s, \text{ and } Y_T \sim N\left(0, \frac{T^{2-2H}}{\lambda_H}\right).$$

Hence we know the distribution of Y_T under $\tilde{P}$ and the distribution of Y_T under P is absolutely continuous with respect to it and it has the density

$$L_T(\theta) = \exp\left(\theta \int_0^T \frac{k(T,s)f(s, X_s)}{h(s)} dB_s + \frac{\theta^2}{2} \int_0^t \left(\frac{k(T,s)f(s, X_s)}{h(s)}\right)^2 ds\right)$$

$$= \exp\left(\theta \int_0^T \frac{k(T,s)f(X_s)}{(h(s))^2} dY_s - \frac{\theta^2}{2} \int_0^T \left(\frac{k(T,s)f(X_s)}{h(s)}\right)^2 ds\right).$$

Let $l_T(\theta) := \log L_T(\theta)$. Then the maximum likelihood estimator based on the observation $\{X_t, 0 \le t \le T\}$ is given by

$$\hat{\theta}_T := \arg\max_{\theta \in \Theta} l_T(\theta)$$

$$= \frac{\int_0^T \frac{k(T,s)f(s, X_s)}{(h(s))^2} dY_s}{\int_0^T \left(\frac{k(T,s)f(X_s)}{h(s)}\right)^2 ds}.$$

Note that the numerator of the MLE is a stochastic integral with respect to an integral functional of the observation process $\{X_t\}$.

We shall prove the consistency and asymptotic normality of the MLE with suitable normalizations.

We need the following exponential inequality for a type of stochastic integrals those are not martingales.

Theorem 3.1 (*Nualart and Rovira (2000)*): *Let $M : [0, T] \times [0, T] \times \Omega \to \mathbb{R} \times \mathbb{R}$ be a $\mathcal{B}([0, T]) \otimes \mathcal{B}([0, T]) \otimes \mathcal{F}$ measurable process satisfying the following conditions:*
(i) $M(t, s) = 0$ if $s > t$.
(ii) $M(t, s)$ is $\mathcal{F}_s$-adapted.
(iii) There exists a positive random variable η and $\alpha \in (0, 2]$ such that for all $t, r \in [0, T]$

$$\int_0^{r \wedge t} |M(t, s) - M(r, s)|^2 \le \eta |t - r|^\alpha.$$

Then for any $\beta, 0 < \beta \le 1 \wedge \alpha$, there exist positive constants K_1 (depending only on β), K_2, K_3, such that

$$P\left\{\sup_{0 \le t \le T} \left|\int_0^t M(t, s) dW_s\right| > \lambda, \| M \|_\infty \le K_M, \eta \le C_M\right\}$$

$$\le \exp\left(-\frac{\lambda^2}{(TK_M^2 + T^\alpha C_M)} K_3\right)$$

for any $\lambda > 0, C_M \geq 0$ and $K_M \geq 0$ such that $\lambda(T^{\alpha-\beta}C_M + T^{1-\beta}K_M^2)^{-1/2} \geq K_1 \vee K_2(1+T)T^{\beta/2})$.

Theorem 3.2 *Under the conditions (A1)-(A2), $\hat{\theta}_T \longrightarrow \theta$ a.s. as $T \to \infty$, i.e., $\hat{\theta}_T$ is a strongly consistent estimator of θ.*

Proof. Note that by (3) we have

$$\hat{\theta}_T = \frac{\int_0^T \frac{k(T,s)f(s,X_s)}{(h(s))^2}dY_s}{\int_0^T \left(\frac{k(T,s)f(s,X_s)}{h(s)}\right)^2 ds}$$

$$= \theta + \frac{\int_0^T \frac{k(T,s)f(s,X_s)}{h(s)}dB_s}{\int_0^T \left(\frac{k(T,s)f(s,X_s)}{h(s)}\right)^2 ds}$$

Now by using Theorem 3.1, we have

$$P\{|\hat{\theta}_T - \theta| > \lambda\}$$

$$= P\left\{\left|\frac{\int_0^T \frac{k(T,s)f(s,X_s)}{h(s)}dB_s}{\int_0^T \left(\frac{k(T,s)f(s,X_s)}{h(s)}\right)^2 ds}\right| > \lambda\right\}$$

$$\leq P\left\{\left|\int_0^T \frac{k(T,s)f(s,X_s)}{h(s)}dB_s\right| > \lambda_1, \left|\int_0^T \left(\frac{k(T,s)f(s,X_s)}{h(s)}\right)^2 ds\right| < \lambda_2\right\}$$

(where $\lambda = \frac{\lambda_1}{\lambda_2}$)

$$\leq \exp\left(-K_3 \frac{\lambda_1^2}{TK_M^2 + T^\alpha C_M}\right)$$

Now using Borel-Cantelli argument completes the proof.   □

Theorem 3.3 *Under the conditions (A1)-(A2),*

$$\sqrt{I_T^H}(\hat{\theta}_T - \theta) \longrightarrow \mathcal{N}(0,1) \text{ in distribution as } T \to \infty$$

where

$$I_T^H := \int_0^T \left(\frac{k(T,s)f(s,X_s)}{h(s)}\right)^2 ds$$

Proof. Note that

$$\sqrt{I_T^H}(\hat{\theta}_T - \theta) = \frac{\int_0^T \frac{k(T,s)f(s,X_s)}{h(s)}dB_s}{\sqrt{\int_0^T \left(\frac{k(T,s)f(s,X_s)}{h(s)}\right)^2 ds}}.$$

It is known that the Volterra stochastic integral $\int_0^t g(t, s, X_s) dB_s$ is a continuous semimartingale with the following decomposition:

$$\int_0^t g(t, s, X_s) dB_s = \int_0^t g(s, s, X_s) dB_s + \int_0^t \left(\int_0^r \frac{\partial g}{\partial r}(r, u, X_u) dB_u \right) dr.$$

(see Tudor and Tudor (1995))

Now using Skorohod embedding for continuous semimartinagales (see Monroe (1972)),

$$\int_0^T g(T, s, X_s) dB_s = B^* \left(\int_0^T g^2(T, s, X_s) ds \right)$$

where B^* is some other Brownian motion on an enlarged probability space.

Setting $g(T, s, x) = \frac{k(T,s)f(s,x)}{h(s)}$, we have

$$\frac{\int_0^T \frac{k(T,s)f(s,X_s)}{h(s)} dB_s}{\sqrt{\int_0^T \left(\frac{k(T,s)f(s,X_s)}{h(s)} \right)^2 ds}} = \frac{B^*(I_T^H)}{\sqrt{I_T^H}}$$

By standard argument, the last term converges in distribution to $\mathcal{N}(0, 1)$ as $T \to \infty$. $\qquad\qquad\square$

6.4 Maximum Likelihood Estimation in Partially Observed Fractional Diffusions

In Chapters 2-5, we studied directly observed process. In practice, there could be hidden unobserved process in the model and the process is not fully observable. Examples of such system are a stochastic filtering model and a stochastic volatility model. In the latter case, price is observed while volatility process is unobserved. Asymptotic behaviour of estimators in partially observed stochastic differential system was first studied by Kutoyants (1984) when the state and the observation equations are driven by ordinary Brownian motions when the observation time becomes large. The problem was generalized by Kallianpur and Selukar (1991) to multiparameter case. The problem was studied by Campillo and Le Gland (1989), Konecky (1991), James and Le Gland (1993), Kutoyants and Pohlman (1994) and Kutoyants (1994) when the intensity of noise becomes small.

The problem of optimal filtering of a signal when the noise is a fractional Brownian motion was studied by Kleptsyna, Kloeden and Anh (1998a, b), Le Breton (1998a, b), Coutin and Decreusefond (1999b), Kleptsyna, Le Breton and Roubaud (1998, 1999).

We study the properties of the MLE of a parameter appearing linearly in drift coefficient of a nonlinear stochastic differential equation driven by fBm when the signal process is a nonlinear ordinary diffusion process.

On the probability space $(\Omega, \mathcal{F}, \mathbb{P})$ consider the stochastic differential equations

$$dY_t = \theta f(t, X_t)\, dt + g(t) dW_t^H,$$
$$dX_t = a(t, X_t)\, dt + b(t, X_t)\, dV_t, \ t \in [0, T],$$
$$Y_0 = \xi, X_0 = \eta,$$

where $W^H, H \in (\frac{1}{2}, 1)$ is the fBm defined above and V is a standard Brownian motion independent of W^H such that the pair (η, ξ) is independent of (V, W^H). We assume that the conditional distribution of η given ξ is some fixed π_0.

Here $\theta \in \Theta$ open in $\mathbb{R}$ is the unknown parameter to be estimated on the basis of observation $\{Y_t, 0 \le t \le T\}$.

We assume that the functions $g : [0, T] \to \mathbb{R}^+$ and $f : [0, T] \times \mathbb{R} \to \mathbb{R}$ is known and it satisfies

(B1) $|f(t, x)| \le K(1 + |x|)$ for all $t \in [0, T]$,
(B2)$|f(t, x) - f(t, y)| \le K(|x - y|)$ for all $t \in [0, T]$,
for some constant $K > 0$.

We assume that the functions $a : [0, T] \times \mathbb{R} \to \mathbb{R}$ and $b : [0, T] \times \mathbb{R} \to \mathbb{R}$ are known and satisfy

(B3) (i) $|a(t, x)| \le K_1(1 + |x|)$ for all $t \in [0, T]$,
(ii)$|b(t, x)| \le K_1(1 + |x|)$ for all $t \in [0, T]$
for some constant $K_1 > 0$.

(B4) (i)$|a(t, x) - a(t, y)| \le K_1(|x - y|)$ for all $t \in [0, T]$,
(ii) $|b(t, x) - b(t, y)| \le K_1(|x - y|)$ for all $t \in [0, T]$
for some constant $K_1 > 0$.

Under the conditions (B1) - (B4), it is known that there exists a unique solution of the SDE (1) (see Lin (1996) and Kallianpur (1980)).

In order to obtain the MLE, we proceed as follows. Consider the canonical space of the process (X, Y). Let $\Omega = \mathcal{C}([0, T]; \mathbb{R}^2)$ be the space of continuous functions from $[0, T]$ into $\mathbb{R}^2$. Consider also the canonical process $(X, W^*) = ((X_t, W_t^*), t \in [0, T])$ on Ω where $(X_t, W_t^*)(x, y) = (x_t, y_t)$ for any $(x, y) \in \Omega$.

The probability $\tilde{\mathbb{P}}$ denotes the unique probability measure on Ω such that defining the variable ξ by $\xi = W_0^*$ and $\tilde{W} = (\tilde{W}_t), t \in [0, T]$ by $\tilde{W}_t = W_t^* - W_0^*$, $t \in [0, T]$, the pair (X, ξ) is independent of $\tilde{W}$ and the process $\tilde{W}$ is a fBm with Hurst parameter H.

The canonical filtration on Ω is $(\mathcal{F}_t, t \in [0, T])$ where $\mathcal{F}_t = \sigma(\{(X_s, W_s^*), 0 \le s \le t\} \bigvee \mathcal{N}, \mathcal{N}$ denoting the set of null sets of $(\Omega, \tilde{\mathbb{P}})$.

Define for all continuous functions $x = (x_t, t \in [0, T])$ the function $h(\theta, x)$ on $[0, T]$ by

$$h(\theta, x)(t) := \frac{\theta f(t, x_t)}{g(t)}, t \in [0, T].$$

Consider, for any $t \in [0, T]$, the function $k_{h(\theta,x)}^t = (k_{h(\theta,x)}^t(s), 0 < s < t)$ defined from Theorem 2.4.1 with $h(\theta, x)$ in place of h.

Define the processes $N = (N_t, t \in [0, T])$ and $\langle N \rangle = (\langle N \rangle_t, t \in [0, T])$ from Theorem 2.1 by plugging in the process $h(\theta, x)$ in place of h, i.e.,

$$N_t := N_t^{h(\theta,X)}, \quad \langle N \rangle_t := \langle N^{h(\theta,X)} \rangle_t.$$

Notice that N_t and $\langle N \rangle_t$ depend only on the values of $X^{(t)} = (X_s, 0 \leq s \leq t)$.

Define the $(\mathcal{F}_t)$-adapted processes $\langle N, N^* \rangle = (\langle N, N^* \rangle_t, t \in [0, T])$ and $q(\theta, X) = (q_t(\theta, X), t \in [0, T])$ by

$$\langle N, N^* \rangle_t := \langle N^{h(\theta,X)}, N^* \rangle_t = \int_0^t k_*^t(s) h(\theta, X)(s) ds, t \in [0, T])$$

and

$$q_t(\theta, X) := q_t^{h(\theta,X)} = \frac{d\langle N, N^* \rangle_t}{d\langle N^* \rangle_t}, t \in [0, T].$$

Let $\tilde{q}_t(X) := \frac{q_t(\theta, X)}{\theta}$.

Define for $0 \leq t \leq T$, the processes

$$\tilde{N}_t(\theta, x) := \int_0^t k_{h(\theta,x)}^t(s) d\tilde{W}_s^H, \quad \langle \tilde{N} \rangle_t(\theta, x) := \int_0^t h(\theta, x)(s) k_h^t(s) ds.$$

$\tilde{N}_t(\theta, x)$ is a Gaussian martingale under $\tilde{\mathbb{P}}$. Define

$$\Lambda_t(\theta, x) := \exp\{\tilde{N}_t(\theta, x) - \frac{1}{2} \langle \tilde{N} \rangle_t(\theta, x)\}, t \in [0, T].$$

Define for any $t \in [0, T]$,
$$\Lambda_t(\theta) := \Lambda_t(\theta, X).$$

Let $\mathbb{P} := \Lambda_T(\theta) \tilde{\mathbb{P}}$.

Let $\mathcal{Y}_t := \sigma(\{Y_s, 0 \leq s \leq t\}), t \in [0, T]$. Define the optimal filter $\pi_t(\phi) := \mathbb{E}[\phi(X_t)|\mathcal{Y}_t], t \in [0, T]$ and the unnormalized filter $\sigma_t(\phi) := \tilde{\mathbb{E}}[\phi(X_t)\Lambda_t|\mathcal{Y}_t], t \in [0, T]$.

Then the *Kallianpur-Stribel formula* holds: for all $t \in [0, T]$, $\tilde{\mathbb{P}}$ and $\mathbb{P}$ almost surely

$$\pi_t(\phi) = \frac{\sigma_t(\phi)}{\sigma_t(1)}.$$

Introduce the processes Z and Z^* defined by

$$Z_t := \int_0^t k_{h(\theta,X)}^t(s) g^{-1}(s) dY_s, t \in [0, T]$$

and

$$Z_t^* := \int_0^t k_*^t(s)g^{-1}(s)dY_s, t \in [0,T].$$

The processes Z and Z^* are semimartingales with the following decomposition:

$$Z_t^* = \langle N, N^* \rangle_t + N_t^*, t \in [0,T],$$

$$Z_t = \langle N \rangle_t + N_t, t \in [0,T].$$

From the above, we can write

$$Z_t^* = \int_0^t q_s(\theta, X)d\langle N^* \rangle_s + N_t^*, t \in [0,T]),$$

and

$$Z_t = \int_0^t q_s^2(\theta, X)d\langle N^* \rangle_s + \int_0^t q_s(\theta, X)dN_s^*, t \in [0,T]),$$

Hence we get the integral representation of Z in terms of Z^*:

$$Z_t = \int_0^t q_s(\theta, X)dZ_t^*, t \in [0,T].$$

Introduce the process $\nu = (\nu_t, t \in [0,T])$ defined by

$$\nu_t = Z_t^* - \int_0^t \pi_s(q)d\langle N^* \rangle_s, t \in [0,T].$$

which plays the role of *innovation process* in the usual situation where the noises are Brownian motions.

Recall the notation $\pi_s(q) = \mathbb{E}[q_s(\theta, X)|\mathcal{Y}_s], s \in [0,t]$.

The particular case of unnormalized filter is

$$\tilde{\Lambda}_t(\theta) := \sigma_t(1) = \tilde{\mathbb{E}}[\Lambda_t|\mathcal{Y}_t], t \in [0,T].$$

By Proposition 5 in Kleptsyna, Le Breton and Roubaud (1999), we have

$$\tilde{\Lambda}_T(\theta, \mathcal{Y}_t) = \exp\left\{ \theta \int_0^T \pi_s(\tilde{q})dZ_s^* - \frac{\theta^2}{2} \int_0^T \pi_s^2(\tilde{q})d\langle N^* \rangle_s \right\}.$$

Thus the maximum likelihood estimator (MLE) of θ is

$$\tilde{\theta}_T := \arg\max_{\theta \in \Theta} \tilde{\Lambda}_T(\theta)$$

$$= \frac{\int_0^T \pi_s(\tilde{q})dZ_s^*}{\int_0^T \pi_s^2(\tilde{q})d\langle N^* \rangle_s}.$$

We shall prove the strong consistency and asymptotic normality of the MLE with random normalization.

Theorem 4.1 *Under the conditions (B1)-(B4), $\tilde{\theta}_T \longrightarrow \theta$ a.s. as $T \to \infty$, i.e., $\tilde{\theta}_T$ is a strongly consistent estimator of θ.*

Moreover,

$$\limsup_T \frac{A_T^{1/2}|\tilde{\theta}_T - \theta|}{(2\log\log A_T)^{1/2}} = 1 \ a.s.$$

where $A_T = \int_0^T \pi_s^2(\tilde{q})d\langle N^*\rangle_s$.

Proof. Note that

$$dZ_t^* = \pi_t(q)d\langle N^*\rangle_t + d\nu_t.$$

From Lemma 3 in Kleptsyna, Le Breton and Roubaud (1999), we have ν is a continuous Gaussian $(\mathcal{Y}_t, \mathbb{P})$ martingale such that $\langle\nu\rangle = \langle N^*\rangle$.

Hence

$$\tilde{\theta}_T = \frac{\int_0^T \pi_s(\tilde{q})dZ_s^*}{\int_0^T \pi_s^2(\tilde{q})d\langle N^*\rangle_s}$$

$$= \theta + \frac{\int_0^T \pi_s(\tilde{q})d\nu_s}{\int_0^T \pi_s^2(\tilde{q})d\langle N^*\rangle_s}.$$

Now by the strong law of large numbers for continuous martingales (see Revuz and Yor (1991) or Liptser and Shiryayev (1989, Theorem 2.6.10)), the second term in r.h.s. converges to zero a.s. as $T \to \infty$. Hence strong consistency follows. Since the deviation $\tilde{\theta}_T - \theta$ is obtained from a stochastic integral with respect to a continuous martingale, the law of the iterated logarithm follows from Corollary 1.1.12 of Revuz and Yor (1991).   □

Theorem 4.2 *Under the conditions (B1)-(B4),*

$$\sqrt{\int_0^T \pi_s^2(\tilde{q})d\langle N^*\rangle_s}(\tilde{\theta}_T - \theta) \xrightarrow{\mathcal{D}} \mathcal{N}(0,1) \ \ as \ T \to \infty.$$

Proof. Note that

$$\sqrt{\int_0^T \pi_s^2(\tilde{q})d\langle N^*\rangle_s}(\tilde{\theta}_T - \theta) = \frac{\int_0^T \pi_s(\tilde{q})d\nu_s}{\sqrt{\int_0^T \pi_s^2(\tilde{q})d\langle N^*\rangle_s}}.$$

By the central limit theorem for stochastic integrals with respect to Gaussian martingales (see Liptser and Shiryayev (1989)), the right hand side above converges in distribution to $\mathcal{N}(0,1)$ as $T \to \infty$.   □

6.5 Examples

(a) Directly Observed Fractional Ornstein-Uhlenbeck Model

The fractional Ornstein-Uhlenbeck process, is an extension of Ornstein-Uhlenbeck process with fractional Brownian motion (fBm) driving term. In finance it is known as fractional Vasicek model, and is being extensively used these days as one-factor short-term interest rate model which takes into account the long memory effect of the interest rate. The model parameter is usually unknown and must be estimated from data.

Define

$$\kappa_H := 2H\Gamma(3/2 - H)\Gamma(H + 1/2),$$
$$k_H(t, s) := \kappa_H^{-1}(s(t - s))^{\frac{1}{2} - H}$$
$$\lambda_H := \frac{2H\Gamma(3 - 2H)\Gamma(H + \frac{1}{2})}{\Gamma(3/2 - H)}$$
$$v_t \equiv v_t^H := \lambda_H^{-1} t^{2 - 2H}$$
$$M_t^H := \int_0^t k_H(t, s)dW_s^H$$

From Norros et al. (199) it is well known that M_t^H is a Gaussian martingale, called the fundamental martingale whose variance function $\langle M^H \rangle_t$ is v_t^H.

Moreover, the natural filtration of the martingale M^H coincides with the natural filtration of the fBm W^H since

$$W_t^H := \int_0^t K(t, s)dM_s^H$$

holds for $H \in (1/2, 1)$ where

$$K_H(t, s) := H(2H - 1)\int_s^t r^{h - \frac{1}{2}}(r - s)^{H - \frac{3}{2}}, \quad 0 \leq s \leq t$$

and for $H = 1/2$, the convention $K_{1/2} \equiv 1$ is used.

Define

$$Q_t := \frac{d}{dv_t}\int_0^t k_H(t, s)X_s ds.$$

It is easy to see that

$$Q_t = \frac{\lambda_H}{2(2 - 2H)}\left\{ t^{2H-1}Z_t + \int_0^t r^{2H-1}dZ_s \right\}.$$

Define the process $Z = (Z_t, t \in [0, T])$ by

$$Z_t := \int_0^t k_H(t, s) dX_s.$$

The following facts are known from Kleptsyna and Le Breton (2002):

(i) Z is the fundamental semimartingale associated with the process X.

(ii) Z is a $(\mathcal{F}_t)$ -semimartingale with the decomposition

$$Z_t = \theta \int_0^t Q_s dv_s + M_t^H.$$

(iii) X admits the representation

$$X_t = \int_0^t K_H(t, s) dZ_s.$$

(iv) The natural filtration $(\mathcal{Z}_t)$ of Z and $(\mathcal{X}_t)$ of X coincide.

Let $(\Omega, \mathcal{F}, \{\mathcal{F}_t\}_{t \geq 0}, P)$ be a stochastic basis on which is defined the Ornstein-Uhlenbeck process X_t satisfying the Itô stochastic differential equation

$$dX_t = \theta X_t dt + dW_t^H, \quad t \geq 0, \quad X_0 = 0$$

where $\{W_t^H\}$ is a fractional Brownian motion with $H > 1/2$ with the filtration $\{\mathcal{F}_t\}_{t \geq 0}$ and $\theta > 0$ is the unknown parameter to be estimated on the basis of continuous observation of the process $\{X_t\}$ on the time interval $[0, T]$.

Let the realization $\{X_t, 0 \leq t \leq T\}$ be denoted by X_0^T. Let P_θ^T be the measure generated on the space (C_T, B_T) of continuous functions on $[0, T]$ with the associated Borel σ-algebra B_T generated under the supremum norm by the process X_0^T and P_0^T be the standard Wiener measure. Applying Girsanov type formula for fBm, when θ is the true value of the parameter, P_θ^T is absolutely continuous with respect to P_0^T and the Radon-Nikodym derivative (likelihood) of P_θ^T with respect to P_0^T based on X_0^T is given by

$$L_T(\theta) := \frac{dP_\theta^T}{dP_0^T}(X_0^T) = \exp\left\{\theta \int_0^T Q_t dZ_t - \frac{\theta^2}{2} \int_0^T Q_t^2 dv_t\right\}.$$

Consider the score function, the derivative of the log-likelihood function, which is given by

$$R_T(\theta) := \int_0^T Q_t dZ_t - \theta \int_0^T Q_t^2 dv_t.$$

A solution of $R_T(\theta) = 0$ provides the maximum likelihood estimate (MLE)

$$\hat{\theta}_T := \frac{\int_0^T Q_t dZ_t}{\int_0^T Q_t^2 dv_t}.$$

Strictly speaking, $\hat{\theta}_T$ is not the maximum likelihood estimate of θ since $\hat{\theta}_T$ may take negative values where as the parameter θ is assumed to be strictly positive. For an exact definition of MLE, see Kutoyants (1994). Nevertheless, we use this terminology.

Using the fractional Itô formula, the score function $R_T(\theta)$ can be written as

$$R_T(\theta) = \frac{1}{2}\left[\frac{\lambda_H}{(2-2H)} Z_T \int_0^T t^{2H-1} dZ_t - T \right] - \theta \int_0^T Q_t^2 dv_t.$$

Consider the estimating function

$$c_T(\theta) = -\frac{T}{2} - \theta \int_0^T Q_t^2 dv_t$$

and the minimum contrast estimate (MCE)

$$\tilde{\theta}_T := \frac{-T}{2\int_0^T Q_t^2 dv_t}.$$

M-estimator is reduced to the minimum contrast estimator. Note that the MCE does not involve stochastic integral unlike the MLE.

Observe that

$$\left(\frac{T}{-2\theta} \right)^{1-H} (\tilde{\theta}_T - \theta) = \frac{\left(\frac{-2\theta}{T}\right)^{1-H} J_T}{\left(\frac{2\theta}{T}\right)^{2-2H} I_T}$$

where

$$J_T := \theta I_T - \frac{T}{2} \quad \text{and} \quad I_T := \int_0^T Q_t^2 dv_t.$$

Kleptsyna and Le Breton (2002) obtained the following Cameron-Martin type formula.

Let $\phi_T(z_1) := E \exp(z_1 I_T)$, $z_1 \in \mathbb{R}$. Then $\phi_T(z_1)$ exists for $|z_1| \leq \delta$, for some $\delta > 0$ and is given by

$$\phi_T(z_1) = \left\{ \frac{4 \sin \pi H \sqrt{\theta^2 - 2z_1} e^{-\theta T}}{\pi T D_T^H(\theta; \sqrt{\theta^2 - 2z_1})} \right\}^{1/2}$$

and we choose the principal branch of the square root, where

$$D_T^H(\theta; \alpha) := [\alpha \cosh(\frac{\alpha}{2}T) - \theta \sinh(\frac{\alpha}{2}T)]^2 J_{-H}(\frac{\alpha}{2}T) J_{H-1}(\frac{\alpha}{2}T)$$
$$- [\alpha \sinh(\frac{\alpha}{2}T) - \theta \cosh(\frac{\alpha}{2}T)]^2 J_{1-H}(\frac{\alpha}{2}T) J_H(\frac{\alpha}{2}T)$$

for $\alpha > 0$ and J_ν is the modified Bessel function of first kind of order ν.

Now using this formula and the arguments in Chapter 17 of Liptser and Shiryayev (1978), Kleptsyna and Le Breton (2002) showed that $\hat{\theta}_T$ is strongly consistent estimator of θ.

(b) Partially Observed Fractional Ornstein-Uhlenbeck Process

Consider the linear stochastic system with fBm observation noise

$$dY_t = \theta X_t dt + dW_t^H,$$

$$dX_t = -X_t dt + dV_t,\ t \in [0, T],$$

$$Y_0 = 0, X_0 = 0,$$

where $\theta < 0$. Note that here the signal process is the classical Ornstein-Uhlenbeck process.

Let $\hat{X}_t := E(X_t | \mathcal{Y}_t)$ and $\gamma_t := E([X_t - \hat{X}_t]^2 | \mathcal{Y}_t)$.
Then $\hat{X}_t$ satisfies the SDE

$$d\hat{X}_t = -\hat{X}_t dt + \theta \gamma_t d\nu_t,\ t \in [0, T], \hat{X}_0 = 0$$

and γ_t satisfies the deterministic differential equation, known as the Ricatti equation

$$d\gamma_t = dt - 2\gamma_t dt - \theta^2 \gamma_t^2 d\langle N^* \rangle_t, \gamma_0 = 0.$$

As $t \to \infty, \gamma_t \to \gamma_\theta$ where $\gamma_\theta := \frac{-1+\sqrt{1+\theta^2}}{\theta^2}$. Now we make a simplifying assumption commonly made in the literature (see Kutoyants (1984), p. 103). We assume that the system has reached the steady state, i.e., we assume that $\hat{X}^t$ satisfies the SDE

$$d\hat{X}_t = -\hat{X}_t dt + \theta \gamma_\theta d\nu_t,\ t \in [0, T], \hat{X}_0 = 0.$$

Now it is easy to verify that

$$\hat{X}_t = \theta \gamma_\theta \int_0^t \exp\{\sqrt{1+\theta^2}(t-s)\} dY_s.$$

Thus the MLE can be now explicitly calculated and it satisfies the asymptotic properties mentioned in Theorems 4.1 and 4.2.

(c) Hermite Ornstein-Uhlenbeck Process

Let $Z_t^{H,k}$ be Hermite process of order k. Consider the Hermite Ornstein-Uhlenbeck process

$$dY_t^{(H,k)} = \theta Y_t^{(H,k)} dt + \sigma dZ_t^{(H,k)},\ \ Y_0^{(H,k)} = \xi$$

has unique solution

$$Y_t^{(H,k)} = e^{\theta t}\left(\xi + \sigma \int_0^t e^{-\theta u} dZ_u^{(H,k)}\right).$$

In particular if

$$Y_0^{(H,k)} = \sigma \int_{-\infty}^0 e^{-\theta u} Z_u^{(H,k)},$$

then

$$Y_t^{(H,k)} = e^{\theta t}\sigma \int_{-\infty}^t e^{-\theta u} dZ_u^{(H,k)}.$$

The process $Y_t^{(H,k)}$ is called the Hermite Ornstein-Uhlenbeck process of order k. Because the covariance structure of the Hermite process is same as the fractional Brownian motion's one, the above integral is a Wiener integral. The covariance of the process is

$$E[Y_t^{(H,k)}Y_s^{(H,k)}] = \sigma^2 \int_{-\infty}^t \int_{-\infty}^s e^{\theta(t-u)}e^{\theta(t-v)}|u - v|^{2H-2}dvdu < \infty.$$

Remarks

(1) Rate of normal approximation, i.e, Berry-Esseen type results for the MLE remains to be investigated.

(2) Asymptotic behavior of posterior distributions and Bayes estimators remain to be investigated.

(3) Approximate maximum likelihood estimation based on discrete observations (both small Δ and fixed Δ cases) remain to be investigated. For the fractional O-U model, an approximation of the corresponding *solution* based on Euler scheme with discrete observations and its convergence was shown by Comte (1996). It remains to investigate the approximation of MLE and their asymptotics. Also it remains to investigate approximate Bayes estimation. Asymptotics of approximate ML and Bayes estimators using approximations of the continuous likelihood was studied by Bishwal (2005b) for ordinary diffusions.

(4) Extension to fractional diffusions where the parameter occurs nonlinearly in the drift coefficient should not pose much difficulties.

(5) It remains to investigate nonparametric estimators of drift coefficient. Nonparametric estimator of diffusion coefficient was studied by Leon and Ludena (2000) using local time.

(6) Fractional Cox-Ingersoll-Ross model remains to be investigated.

Part II

Discrete Sampling

7

Approximate Maximum Likelihood Estimation in Nonhomogeneous Diffusions

7.1 Introduction

In Chapters 1-6, we concentrated on the situation where the diffusion process is observed continuously throughout a time interval $[0, T]$. However, one encounters practical difficulties to obtain a complete continuous observation of the sample path. Discrete observations are much more likely. Hence drift estimation when only discrete observations at times $t_1, t_2, \ldots, t_n$ with $0 \leq t_1 < t_2 < \ldots t_n = T$ are available, has been the recent trend of investigation.

Consider the stochastic differential equation

$$dX_t = \mu(\theta, t, X_t)dt + \sigma(t, X_t)dW_t, \ t \geq 0$$

where $\{W_t, t \geq 0\}$ is a standard Wiener process, $\theta \in \Theta \subseteq \mathbb{R}$ is the unknown parameter which is to be estimated on the basis of observations of the process X at times $0 = t_0 < t_1 < \ldots < t_n$ and $\mu : \Theta \times [0, T] \times \mathbb{R} \to \mathbb{R}$ and $\sigma : [0, T] \times \mathbb{R} \to \mathbb{R}$ are known smooth functions.

Ideally, parameter estimation for diffusion processes based on discrete-time observations should be based on the likelihood function. The literature has mainly concentrated on two approaches based on likelihood function.

Since the process $\{X_t\}$ is Markov, hence if the transition densities $p(s, x, t, y; \theta)$ of X are known, one can use the log likelihood function

$$l_n(\theta) = \sum_{i=1}^{n} \log p(t_{i-1}, X_{t_{i-1}}, t_i, X_{t_i}; \theta)$$

for estimation of θ. The corresponding maximum likelihood estimate $\hat{\theta}_n$ is known to have usual good properties (see Billingsley (1961), Dacunha-Castelle and Florens-Zmirou (1986)). In the case of time-equidistant observations ($t_i = i\Delta, i = 0, 1, 2, \ldots, n$ for some fixed $\Delta > 0$) Dacunha-Castelle and Florens-Zmirou (1986) proved the consistency and asymptotic normality of $\hat{\theta}_n$ as $n \to \infty$, irrespective of the value of Δ.

Unfortunately, the transition densities of X are usually unknown except in some very special cases like the linear ones (see Le Breton (1976), Robinson (1977)).

Pedersen (1995a,b) and Ait-Sahalia (2002) used approximations of the transition densities and the methods prove quite computer intensive. Pedersen used numerical approximations based on iterations of the Gaussian transition densities emanating from the Euler scheme, while Ait-Sahalia (2002), using a specific transformation of the diffusion, was able to obtain accurate theoretical approximations based on Hermite function expansions and studied the asymptotic behaviour of the approximate MLE. Brandt and Santa-Clara (2002) studied approximation of transition densities using the Euler discretization scheme. Elerian (1998) obtained a closed form approximation to the transition density using the Milstein discretization scheme which improves an order of accuracy in estimation as compared to the Euler scheme. Poulsen (1999) and Jensen and Poulsen (1999) used numerical solution of partial differential equations for transition densities and the methods prove to be quite computer intensive.

When the transition densities of X are unknown, one alternative to approximating $l_n(\theta)$ is to approximate the log-likelihood function for θ based on continuous observations of X. Recall that, under certain conditions, the log-likelihood function for θ based on continuous observation of X in the time interval $[0, T]$ is given by

$$l_T(\theta) = \int_0^T \frac{\mu(\theta, t, X_t)}{\sigma^2(t, X_t)} dX_t - \frac{1}{2} \int_0^T \frac{\mu^2(\theta, t, X_t)}{\sigma^2(t, X_t)} dt.$$

(see Prakasa Rao (1999)). Thus to obtain an approximation to $l_T(\theta)$ one has to approximate the integrals in $l_T(\theta)$.

In $l_T(\theta)$ one can use Itô type approximation for the stochastic integral and rectangular approximation for the ordinary integral and obtain an approximate log-likelihood function

$$\widetilde{l}_n(\theta) = \sum_{i=1}^n \frac{\mu(\theta, t_{i-1}, X_{t_{i-1}})}{\sigma^2(t_{i-1}, X_{t_{i-1}})}(X_{t_i} - X_{t_{i-1}}) - \frac{1}{2} \sum_{i=1}^n \frac{\mu^2(\theta, t_{i-1}, X_{t_{i-1}})}{\sigma^2(t_{i-1}, X_{t_{i-1}})}(t_i - t_{i-1}).$$

The maximizer of $\widetilde{l}_n$ provides an approximate maximum likelihood estimator (AMLE). For the homogeneous case, i.e., $\mu(\theta, t, X_t) = f(\theta, X_t)$, Yoshida (1992) showed that the AMLE based on $\widetilde{l}_n(\theta)$ is weakly consistent as $T \to \infty$ and $\frac{T}{n} \to 0$ and asymptotically normally distributed as $T \to \infty$ and $\frac{T}{n^{2/3}} \to 0$ when the diffusion is homogeneous and ergodic.

In this chapter, we consider the nonhomogeneous SDE with drift function $\mu(\theta, t, X_t) = \theta f(t, X_t)$ for some smooth function $f(t, x)$. A relevant question is how close are the *discrete* estimators to the true continuous MLE θ_T. We fix T and obtain several approximate maximum likelihood estimators (AMLEs) using several approximations to $l_T(\theta)$. We study their rates of convergence to

θ_T, the maximum likelihood estimator based on $l_T(\theta)$. In a sense we measure the loss of information due to discretization.

We take the approach that better rates could be obtained for the AMLE if we had better approximations to $l_T(\theta)$. For this purpose we first obtain several approximations of the Fisk-Stratonovich (FS) integral and study their rates of convergence. Then we convert the Itô integral to the FS integral and apply the new approximations to the FS stochastic integral and rectangular approximation for the ordinary integrals. We thus obtain several approximate log-likelihood functions and hence several AMLEs. In Section 7.2 we give the model and assumptions, and introduce several new approximations of the FS integral. In Section 7.3 we study the L_2 rates of convergence of the approximants of Itô and FS integrals. In Section 7.4 we study the in probability rates of convergence of the approximate log-likelihood functions. In Section 7.5 we obtain the error bounds in probability of the different AMLEs and the continuous MLE. In Section 7.6 we study some examples.

7.2 Model, Assumptions and Definitions

Let $(\Omega, \mathcal{F}, \{\mathcal{F}_t\}_{t\geq 0}, P)$ be a stochastic basis satisfying the usual hypotheses on which we define the real valued diffusion process $\{X_t, t \geq 0\}$ satisfying the stochastic differential equation

$$dX_t = \theta f(t, X_t)dt + dW_t, \ t \geq 0 \qquad (2.1)$$

where $\{W_t, t \geq 0\}$ is a standard Wiener process adapted to $\{\mathcal{F}_t, t \geq 0\}$ such that for $0 \leq s < t, W_t - W_s$ is independent of $\mathcal{F}_s$. Here $\theta \in \Theta$ open in $\mathbb{R}$ is the unknown parameter to be estimated on the basis of observations of the process $\{X_t, t \geq 0\}$. Let θ_0 be the (unknown) true value of the parameter θ.

Throughout the Chapter C is a generic constant. We use the following notations:

$$f_x := \frac{\partial f}{\partial x}, f_t := \frac{\partial f}{\partial t}, f_{xx} := \frac{\partial^2 f}{\partial x^2}, f_{tt} := \frac{\partial^2 f}{\partial t^2}, f_{tx} := \frac{\partial^2 f}{\partial t \partial x}.$$

We assume the following conditions:

(A1) $f(\cdot, \cdot)$ satisfies the Lipschitz and growth condition:

$|f(t, x) - f(t, y)| \leq K|x - y|$, and $|f(t, x)| \leq K(1 + |x|)$ for all $t \in [0, T]$.

for some constant $K > 0$.

(A2) $f_x(\cdot, \cdot)$ satisfies the Lipschitz and growth condition:

$|f_x(t, x) - f_x(t, y)| \leq K_1|x - y|$ and $|f_x(t, x)| \leq K_1(1 + |x|)$ for all $t \in [0, T]$.

for some constant $K_1 > 0$.

(A3) $f(\cdot,\cdot)$ is a real valued function satisfying

$$E_{\theta_0}\left\{\int_0^T f^2(t, X_t)dt\right\} < \infty \quad \text{for all } T > 0.$$

(A4)$_j$ $f(\cdot,\cdot)$ is j times continuously differentiable in x for $j \geq 1$ and

$$\sup_{0 \leq t \leq T} E|f_x(t, X_t)|^8 < \infty, \quad \sup_{0 \leq t \leq T} E|f_{xx}(t, X_t)|^8 < \infty.$$

(A5)$_k$ $f(\cdot,\cdot)$ is k times continuously differentiable in t for $k \geq 1$ and

$$\sup_{0 \leq t \leq T} E|f_t(t, X_t)|^4 < \infty, \quad \sup_{0 \leq t \leq T} E|f_{tt}(t, X_t)|^4 < \infty.$$

(A6) $\sup_{0 \leq t \leq T} E|f_{tx}(t, X_t)|^8 < \infty.$

(A7) $E[X_0]^8 < \infty.$

It is well known that equation (2.1) has a unique solution $\{X_t\}$ under the condition (A1) (see e.g., Mao (1997)).

We first review the well known definitions of the stochastic integrals and introduce some new definitions. Let π_n be the partition $\pi_n = \{0 = t_0 < t_1 < \ldots < t_n = T\}, t_k = kh, k = 0, 1, \ldots, n$ such that $h \to 0$ as $n \to \infty$. A stochastic integral is usually defined as the L^2 limit of a partition sum and the limit depends on the choice of time points of the integrand. From (2.2)-(2.13) lim denotes limit in the mean square sense. For any non-anticipative square integrable f, (forward) Itô integral is defined as

$$\int_0^T f(t, X_t)dX_t = \lim_{h \to 0} \sum_{i=1}^n f(t_{i-1}, X_{t_{i-1}})(X_{t_i} - X_{t_{i-1}}). \tag{2.2}$$

The backward Itô integral is defined as

$$\int_0^T f(t, X_t)dX_t = \lim_{h \to 0} \sum_{i=1}^n f(t_i, X_{t_i})(X_{t_i} - X_{t_{i-1}}) \tag{2.3}$$

(see Mckean (1969), p. 34.)

Fisk (1963) and Stratonovich (1964) introduced a *symmetrized* stochastic integral known as the Fisk-Stratonovich integral (FS integral, in short) which obeys the rules of ordinary calculus (see Ikeda-Watanabe (1989) and Protter (1990).) On the other hand with Brownian integrator Itô integral is a martingale and it has many interesting properties emanating from martingale which FS integral does not have. Note that with Brownian integrand, backward Itô integral is a backward martingale. In order to distinguish between the two integrals we use $\int$ for Itô integral and $\oint$ for FS integral.

For two continuous semimartingales $\{M_t\}$ and $\{N_t\}$, Fisk (1963) defined the FS integral as

$$\oint_0^T N_t dM_t := \lim_{h \to 0} \sum_{i=1}^n \frac{N_{t_{i-1}} + N_{t_i}}{2}(M_{t_i} - M_{t_{i-1}}). \tag{2.4}$$

(see also Karatzas and Shreve (1988)).

For the nonhomogeneous diffusion process $\{X_t\}$, his definition may be adopted to define the FS integral as

$$\oint_0^T f(t, X_t)dX_t = \lim_{h \to 0} S_{1,n}$$

where

$$S_{1,n} := \sum_{i=1}^n \frac{f(t_{i-1}, X_{t_{i-1}}) + f(t_i, X_{t_i})}{2}(X_{t_i} - X_{t_{i-1}}). \tag{2.5}$$

It is easy to show that the limit exists by following Stratonovich (1964). Note that with Brownian integrator, FS integral is the arithmetic mean of a forward and a backward martingale.

In general, the Itô, the McKean and the Fisk approximations can be defined as particular cases of the following scheme: for any $\beta \in [0, 1]$, define

$$S_{A,n} := \sum_{i=1}^n [\beta f(t_{i-1}, X_{t_{i-1}}) + (1 - \beta)f(t_i, X_{t_i})](X_{t_i} - X_{t_{i-1}}). \tag{2.6}$$

The particular cases $\beta = 1$ gives Itô approximation (2.2), $\beta = 0$ gives McKean's approximation (2.3) and $\beta = \frac{1}{2}$ gives Fisk's approximation (2.5). Stratonovich (1964) defined the FS integral as

$$\oint_0^T f(t, X_t)dX_t := \lim_{h \to \infty} \sum_{i=1}^n f\left(t_{i-1}, \frac{X_{t_{i-1}} + X_{t_i}}{2}\right)(X_{t_i} - X_{t_{i-1}}). \tag{2.7}$$

We adopt the following definition of Stratonovich (1964) which does not essentially alter anything in the limit and which fits into our general definition (2.11).

$$\oint_0^T f(t, X_t)dX_t = \lim_{h \to 0} S_{2,n}$$

where

$$S_{2,n} := \sum_{i=1}^n f\left(\frac{t_{i-1} + t_i}{2}, \frac{X_{t_{i-1}} + X_{t_i}}{2}\right)(X_{t_i} - X_{t_{i-1}}). \tag{2.8}$$

The idea underlying the definitions of Fisk (1963) and Stratonovich (1964, 1966) is almost same. But Stratonovich restricts his definition to a narrow class of integrands, e.g., f satisfies $(A4)_1$.

In analogy with ordinary numerical integration, Itô's definition (2.2) resembles the *rectangular rule*, Fisk's definition (2.5) resembles the *trapezoidal rule* and Stratonovich's definition (2.8) resembles the *midpoint rule*.

We introduce several approximations of the FS integral in order to improve the accuracy of approximation. One approximation of the FS integral may also be defined as the convex combination of $S_{1,n}$ and $S_{2,n}$, more precisely $\oint_0^T f(t, X_t) dX_t = \lim_{h \to 0} S_{B,n}$ where

$$S_{B,n} :=$$
$$\sum_{i=1}^{n} \left[\alpha \left(\frac{f(t_{i-1}, X_{t_{i-1}}) + f(t_i, X_{t_i})}{2} \right) \right.$$
$$\left. + (1 - \alpha) f \left(\frac{t_{i-1} + t_i}{2}, \frac{X_{t_{i-1}} + X_{t_i}}{2} \right) \right] (X_{t_i} - X_{t_{i-1}}) \tag{2.9}$$

and $0 \leq \alpha \leq 1$.

The cases $\alpha = 1$ gives $S_{1,n}$ and $\alpha = 0$ gives $S_{2,n}$. The case $\alpha = \frac{1}{3}$ gives

$$S_{5,n}$$
$$:= \frac{1}{6} \sum_{i=1}^{n} \left[f(t_{i-1}, X_{t_{i-1}}) + 4f(\frac{t_{i-1} + t_i}{2}, \frac{X_{t_{i-1}} + X_{t_i}}{2}) + f(t_i, X_{t_i}) \right]$$
$$(X_{t_i} - X_{t_{i-1}}). \tag{2.10}$$

In analogy with ordinary numerical integration (2.10) resembles the *Simpson's rule*.

For a smooth function f, we define a new stochastic integral as

$$B_T$$
$$:= \int_0^T f(t, X_t) dW_t$$
$$= l.i.m._{n \to \infty} \sum_{i=1}^{n} \sum_{j=1}^{m} p_j \, f\left((1 - s_j) t_{i-1} + s_j t_i, \right.$$
$$(1 - s_j) X_{t_{i-1}} + s_j X_{t_i}) \, (W_{t_i} - W_{t_{i-1}}) \tag{2.11}$$

where and $p_j, \quad j \in \{1, 2, \cdots, m\}$ is a probability mass function of a discrete random variable S on $0 \leq s_1 < s_2 < \cdots < s_m \leq 1$ with $P(S = s_j) = p_j$, $j \in \{1, 2, \cdots, m\}$.

Denote the k-th moment of the random variable S as $\mu_k := \sum_{j=1}^{m} s_j^k p_j$, $k = 1, 2, \cdots$.

The new integral and the Itô integral are connected as follows:

$$B_T = I_T + \mu_1 \int_0^T f_x(t, X_t) dt$$

where $I_T = \int_0^T f(t, X_t)dW_t$ is the Itô integral.

When $\mu_1 = 0$, the new integral is the Itô integral. When $\mu_1 = \frac{1}{2}$, the new integral is the Fisk-Stratonovich integral.

This is a generalization of a proposal in Yor (1976) who a general definition of stochastic integral as the L_2-limit of the following approximant.

Let μ be a probability measure on $([0,1], \mathcal{B}([0,1]))$ with $\mu_k = \int_0^1 x^k d\mu(x), k = 1, 2, \ldots$ and put

$$S_{Y,n} := \sum_{i=1}^{n} \int_0^1 f(X_{t_{i-1}} + s(X_{t_i} - X_{t_{i-1}}))d\mu(s)(W_{t_i} - W_{t_{i-1}}).$$

Our main contribution is to construct probability distributions in order to obtain higher order rate of convergence of the approximations to the stochastic integral. This problem is connected to the truncated Hausdorff moment problem.

Let conv (A) be the convex hull of a set A, i.e. the smallest set containing all convex combinations of elements in A. Define $\mathcal{P} = \text{conv}\left(\{\delta(s)\}_{s \in [0,1]}\right)$ and

$$\alpha_n^- = \min_{P \in \mathcal{P}} \left\{ \int_0^1 s^n dP(s) : \int_0^1 s^j dP(s) = \frac{1}{1+j}, j = 1, 2, \cdots, n-1 \right\}, \quad (1)$$

$$\alpha_n^+ = \max_{P \in \mathcal{P}} \left\{ \int_0^1 s^n dP(s) : \int_0^1 s^j dP(s) = \frac{1}{1+j}, j = 1, 2, \cdots, n-1 \right\}. \quad (2)$$

The problems (1) and (2) are special cases of the truncated Hausdorff moment problem. It is well known that Hausdorff moment problem is ill-posed.

The above approximants (2.5) - (2.11) converge in L_2 to the FS integral as $n \to \infty$. It will turn out that the order of approximation error in the new integral is determined by the first k for which $\mu_k \neq \frac{1}{1+k}$.

The order of mean square approximation error (rate of convergence) in the new integral is $n^{-\nu}$ where

$$\nu := \inf\left\{ k : \mu_k \neq \frac{1}{1+k}, \ \mu_j = \frac{1}{1+j}, j = 0, 1, \cdots, k-1 \right\}.$$

We construct approximation schemes with rate of convergence up to $\nu = 6$.

If one chooses the probability distribution as uniform distribution for which the moments are a harmonic sequence $(\mu_1, \mu_2, \mu_3, \mu_4, \mu_5, \mu_6, \cdots) = (\frac{1}{2}, \frac{1}{3}, \frac{1}{4}, \frac{1}{5}, \frac{1}{6}, \frac{1}{7}, \cdots)$ then there is no change in rate of convergence than second order. If one can construct a probability distribution for which the harmonic sequence is truncated at a point, then there is a rate of convergence improvement at the point of truncation.

Putting the mass 1 at the point $s = 0$ we obtain the Itô approximant in (2.2) for which $\mu_1 = 0$. Thus the rate is $\nu = 1$ since $\mu_1 \neq \frac{1}{2}$.

Putting the mass 1 at the point $s = 1$ we obtain the backward Itô approximant in (2.3) for which $\mu_1 = 1$. Thus the rate is $\nu = 1$ since $\mu_1 \neq \frac{1}{2}$.

Putting the masses $\frac{1}{2}, \frac{1}{2}$ respectively at the points $0, 1$ we obtain the approximant $S_{1,n}$ for which $\mu_1 = \frac{1}{2}$ and $\mu_2 = \frac{1}{4}$. Thus the rate is $\nu = 2$.

Putting the mass 1 at the point $\frac{1}{2}$ we obtain the approximant $S_{2,n}$ for which $\mu_1 = \frac{1}{2}$ and $\mu_2 = \frac{1}{2}$. Thus the rate is $\nu = 2$.

Putting the masses $\frac{1}{6}, \frac{2}{3}, \frac{1}{6}$ respectively at the points $0, \frac{1}{2}, 1$ we obtain the approximant $S_{5,n}$ for which $\mu_1 = \frac{1}{2}, \mu_2 = \frac{1}{3}, \mu_3 = \frac{1}{4}$ and $\mu_4 = \frac{5}{25}$. Thus the rate is $\nu = 4$.

Now we obtain several special cases of (2.11) and determine the first k for which $\mu_k \neq \frac{1}{1+k}$ in order to have faster rate of convergence to the FS integral.

Putting the masses $\frac{1}{4}, \frac{3}{4}$ respectively at the points $0, \frac{2}{3}$ we obtain the asymmetric approximant

$$S_{3,n} := \frac{1}{4} \sum_{i=1}^{n} \left[f(t_{i-1}, X_{t_{i-1}}) + 3f(\frac{t_{i-1} + 2t_i}{3}, \frac{X_{t_{i-1}} + 2X_{t_i}}{3}) \right] (X_{t_i} - X_{t_{i-1}})$$

$$(2.12)$$

for which $\mu_1 = \frac{1}{2}, \mu_2 = \frac{1}{3}, \mu_3 = \frac{2}{9}$. Thus the rate is $\nu = 3$.

Putting the masses $\frac{3}{4}, \frac{1}{4}$ respectively at the points $\frac{1}{3}, 1$ we obtain another approximant

$$S_{4,n} := \frac{1}{4} \sum_{i=1}^{n} \left[3f(\frac{2t_{i-1} + t_i}{3}, \frac{2X_{t_{i-1}} + X_{t_i}}{3}) + f(t_i, X_{t_i}) \right] (X_{t_i} - X_{t_{i-1}})$$

$$(2.13)$$

for which $\mu_1 = \frac{1}{2}, \mu_2 = \frac{1}{3}$ and $\mu_3 = \frac{10}{36}$. Thus the rate is $\nu = 3$.

Putting the masses $\frac{1}{8}, \frac{3}{8}, \frac{3}{8}, \frac{1}{8}$ respectively at the points $0, \frac{1}{3}, \frac{2}{3}, 1$ we obtain the symmetric approximant

$$S_{6,n} := \frac{1}{8} \sum_{i=1}^{n} \left[f(t_{i-1}, X_{t_{i-1}}) + 3f(\frac{2t_{i-1} + t_i}{3}, \frac{2X_{t_{i-1}} + X_{t_i}}{3}) \right.$$
$$\left. + 3f(\frac{t_{i-1} + 2t_i}{3}, \frac{X_{t_{i-1}} + 2X_{t_i}}{3}) + f(t_i, X_{t_i}) \right] (X_{t_i} - X_{t_{i-1}}) \qquad (2.14)$$

for which $\mu_1 = \frac{1}{2}, \mu_2 = \frac{1}{3}, \mu_3 = \frac{1}{4}$ and $\mu_4 = \frac{11}{54}$. Thus the rate is $\nu = 4$.

Putting the masses $(\frac{1471}{24192}, \frac{6925}{24192}, \frac{1475}{12096}, \frac{2725}{12096}, \frac{5675}{24192}, \frac{1721}{24192})$ respectively at the points
$(0, \frac{1}{5}, \frac{2}{5}, \frac{3}{5}, \frac{4}{5}, 1)$ we obtain the scheme

$$S_{7,n}$$
$$:= \frac{1}{24192} \sum_{i=1}^{n} \left[1471 f(t_{i-1}, X_{t_{i-1}}) + 6925 f(\frac{t_{i-1} + t_i}{5}, \frac{X_{t_{i-1}} + X_{t_i}}{5}) \right.$$
$$+ 2950 f(\frac{2t_{i-1} + 2t_i}{5}, \frac{2X_{t_{i-1}} + 2X_{t_i}}{5}) + 5450 f(\frac{3t_{i-1} + 3t_i}{5}, \frac{3X_{t_{i-1}} + 3X_{t_i}}{5})$$
$$\left. + 5675 f(\frac{4t_{i-1} + 4t_i}{5}, \frac{4X_{t_{i-1}} + 4X_{t_i}}{5}) + 1721 f(t_i, X_{t_i}) \right] (X_{t_i} - X_{t_{i-1}}) \quad (2.14a)$$

for which $\mu_1 = \frac{1}{2}, \mu_2 = \frac{1}{3}, \mu_3 = \frac{1}{4}, \mu_4 = \frac{1}{5}$ and $\mu_5 = \frac{841}{5040}$. Thus the rate is $\nu = 5$.

Putting the masses $\frac{7}{90}, \frac{16}{45}, \frac{2}{15}, \frac{16}{45}, \frac{7}{90}$ respectively at the points $0, \frac{1}{4}, \frac{1}{2}, \frac{3}{4}, 1$ we obtain the symmetric approximant

$$
S_{8,n} := \frac{1}{90} \sum_{i=1}^{n} \left[7f(t_{i-1}, X_{t_{i-1}}) + 32f\left(\frac{3t_{i-1} + t_i}{4}, \frac{3X_{t_{i-1}} + X_{t_i}}{4}\right) \right.
$$
$$
+12f\left(\frac{t_{i-1} + t_i}{2}, \frac{X_{t_{i-1}} + X_{t_i}}{2}\right) + 32f\left(\frac{t_{i-1} + 3t_i}{4}, \frac{X_{t_{i-1}} + 3X_{t_i}}{4}\right)
$$
$$
\left. +7f(t_i, X_{t_i})\right] (X_{t_i} - X_{t_{i-1}}) \tag{2.15}
$$

for which $\mu_1 = \frac{1}{2}, \mu_2 = \frac{1}{3}, \mu_3 = \frac{1}{4}, \mu_4 = \frac{1}{5}, \mu_5 = \frac{1}{6}$ and $\mu_6 = \frac{110}{768}$. Thus the rate is $\nu = 6$.

Putting the masses $\frac{19}{288}, \frac{75}{288}, \frac{50}{288}, \frac{50}{288}, \frac{75}{288}, \frac{19}{288}$ respectively at the points $0, \frac{1}{5}, \frac{2}{5}, \frac{3}{5}, \frac{4}{5}, 1$ we obtain another symmetric approximant

$$
S_{9,n} := \frac{1}{288} \sum_{i=1}^{n} \left[19f(t_{i-1}, X_{t_{i-1}}) + 75f\left(\frac{4t_{i-1} + t_i}{5}, \frac{4X_{t_{i-1}} + X_{t_i}}{5}\right) \right.
$$
$$
+50f\left(\frac{3t_{i-1} + 2t_i}{5}, \frac{3X_{t_{i-1}} + 2X_{t_i}}{5}\right)
$$
$$
+50f\left(\frac{2t_{i-1} + 3t_i}{5}, \frac{2X_{t_{i-1}} + 3X_{t_i}}{5}\right) + 75f\left(\frac{t_{i-1} + 4t_i}{5}, \frac{X_{t_{i-1}} + 4X_{t_i}}{5}\right)
$$
$$
\left. +19f(t_i, X_{t_i})\right] (X_{t_i} - X_{t_{i-1}}) \tag{2.16}
$$

for which $\mu_1 = \frac{1}{2}, \mu_2 = \frac{1}{3}, \mu_3 = \frac{1}{4}, \mu_4 = \frac{1}{5}, \mu_5 = \frac{1}{6}$ and $\mu_6 = \frac{3219}{22500}$. Thus the rate is $\nu = 6$.

Now we are ready to define several approximate maximum likelihood estimators (AMLEs) using the above approximation schemes of stochastic integrals. Let P_θ be the measure generated by the process $X_0^T \equiv \{X_t, 0 \le t \le T\}$ on the measurable space (C_T, B_T) of continuous functions on $[0, T]$ with the associated Borel σ-algebra B_T under the supremum norm. Let E_θ be the expectation with respect to the measure P_θ and P_W be the standard Wiener measure on (C_T, B_T). Under assumption (A3) the measures P_θ and P_W are equivalent and the Radon-Nikodym derivative of P_θ w.r.t. P_W is given by

$$
\frac{dP_\theta}{dP_W}(X_0^T) = \exp\left\{ \theta \int_0^T f(t, X_t) dX_t - \frac{\theta^2}{2} \int_0^T f^2(t, X_t) dt \right\} \tag{2.17}
$$

along the sample path X_0^T (see Liptser and Shiryayev (1977)). Let

$$
L_T(\theta) = \log \frac{dP_\theta}{dP_W} \tag{2.18}
$$

be the log-likelihood function. The maximum likelihood estimate (MLE) of θ is defined as

$$\theta_T := \arg\max_{\theta \in \Theta} L_T(\theta)$$

$$= \left\{ \int_0^T f(t, X_t) dX_t \right\} \left\{ \int_0^T f^2(t, X_t) dt \right\}^{-1} \qquad (2.19)$$

MLE θ_T possesses good asymptotic properties (see Chapter 2). Since we have discrete observations of the diffusion, we need to find good approximations of the MLE. For that purpose, we convert the Itô integral to the FS integral appearing in the numerator of θ_T. The Itô and the FS integrals are connected by (see Ikeda and Watanabe (1989))

$$\oint_0^T f(t, X_t) dX_t = \int_0^T f(t, X_t) dX_t + \frac{1}{2} \int_0^T f_x(t, X_t) dt \quad a.s. \qquad (2.20)$$

Now replacing the Itô integral with the FS integral (by using (2.20) in (2.19))

$$\theta_T = \left\{ \oint_0^T f(t, X_t) dX_t - \frac{1}{2} \int_0^T f_x(t, X_t) dt \right\} \left\{ \int_0^T f^2(t, X_t) dt \right\}^{-1}. \qquad (2.21)$$

We study the approximation of the MLE θ_T when the process X is observed at the points $0 = t_0 < t_1 < \ldots < t_n = T$ with $t_i = ih, i = 0, 1, 2, \ldots, n$ such that $h \to 0$ as $n \to \infty$. Itô approximation of the stochastic integral and rectangular approximation of the ordinary integral in the log-likelihood (2.18) yields the approximate log-likelihood function

$$L_{n,T}(\theta) = \theta \sum_{i=1}^n f(t_{i-1}, X_{t_{i-1}})(X_{t_i} - X_{t_{i-1}}) - \frac{\theta^2}{2} \sum_{i=1}^n f^2(t_{i-1}, X_{t_{i-1}})(t_i - t_{i-1})$$

$$(2.22)$$

with the corresponding approximate maximum likelihood estimator (AMLE)

$$\theta_{n,T} = \left\{ \sum_{i=1}^n f(t_{i-1}, X_{t_{i-1}})(X_{t_i} - X_{t_{i-1}}) \right\} \left\{ \sum_{i=1}^n f^2(t_{i-1}, X_{t_{i-1}})(t_i - t_{i-1}) \right\}^{-1}.$$

$$(2.23)$$

Let

$$J_{n,T} := \sum_{i=1}^n f^2(t_{i-1}, X_{t_{i-1}})(t_i - t_{i-1}) \text{ and } D_{n,T} := \sum_{i=1}^n f_x(t_{i-1}, X_{t_{i-1}})(t_i - t_{i-1}).$$

$$(2.24)$$

Using the transformation (2.20) in the log-likelihood (2.17) and using approximations $S_{j,n}, j = 1, 2, \cdots, 9$ for the FS stochastic integral and rectangular approximation for the ordinary integral we obtain the approximate log-likelihood functions

$$L_{n,T,j}(\theta) := \theta[S_{j,n} - \frac{1}{2}D_{n,T}] - \frac{\theta^2}{2}J_{n,T}, j = 1, 2, \cdots, 9. \tag{2.25}$$

and the corresponding AMLEs as

$$\theta_{n,T,j} := [S_{j,n} - \frac{1}{2}D_{n,T}][J_{n,T}]^{-1}, j = 1, 2, \cdots, 9. \tag{2.26}$$

Clearly

$$\theta_{n,T,1} := \left[\frac{1}{2}\sum_{i=1}^{n}[f(t_{i-1}, X_{t_{i-1}}) + f(t_i, X_{t_i})](X_{t_i} - X_{t_{i-1}})\right.$$

$$\left. -\frac{1}{2}\sum_{i=1}^{n}f_x(t_{i-1}, X_{t_{i-1}})(t_i - t_{i-1})\right]$$

$$\left[\sum_{i=1}^{n}f^2(t_{i-1}, X_{t_{i-1}})(t_i - t_{i-1})\right]^{-1}.$$

We shall study the properties of $\theta_{n,T}$ and $\theta_{n,T,1}$ only.

7.3 Accuracy of Approximations of the Itô and FS Integrals

We establish the following L^2 rates of convergence of the approximants of the Itô and the FS integral to the corresponding integrals when the integrator is the standard Wiener process. This is an extension of Theorem 2.1 in Prakasa Rao and Rubin (2000) to nonhomogeneous case who obtained it for stationary homogeneous diffusions with some stronger regularity conditions.

Theorem 3.1 *Under the assumptions (A1) - (A3), (A4)₁, (A5)₁, (A6) and (A7), we have*

$$(a)\ E\left|\sum_{i=1}^{n}f(t_{i-1}, X_{t_{i-1}})(W_{t_i} - W_{t_{i-1}}) - \int_0^T f(t, X_t)dW_t\right|^2 \le C\frac{T^2}{n}.$$

Under the assumptions (A1) - (A3), (A4)₂, (A5)₂, (A6) and (A7), we have (b)

$$E\left|\sum_{i=1}^{n}\frac{f(t_{i-1}, X_{t_{i-1}}) + f(t_i, X_{t_i})}{2}(W_{t_i} - W_{t_{i-1}}) - \oint_0^T f(t, X_t)dW_t\right|^2 \le C\frac{T^3}{n^2}.$$

Proof. We use the arguments in Prakasa Rao and Rubin (1998) to prove the theorem. First we shall prove (b). Let π_n be the partition $\pi_n := \{0 = t_0 < t_1 < \cdots < t_n = T\}, t_i = ih, i = 0, 1, \cdots, n$ such that $h \to 0$. Define F and F_{π_n} as:

$$F := \oint_0^T f(t, X_t) dW_t \qquad (3.1)$$

and

$$F_{\pi_n} := \sum_{i=1}^n \frac{f(t_{i-1}, X_{t_{i-1}}) + f(t_i, X_{t_i})}{2} (W_{t_i} - W_{t_{i-1}}). \qquad (3.2)$$

Let π'_n be a partition, finer than π_n, obtained by choosing the mid point $\tilde{t}_{i-1}$ from each of the intervals $t_{i-1} < \tilde{t}_{i-1} < t_i$, $i = 1, 2, \ldots, n$. Let $0 = t'_0 < t'_1 < \ldots < t'_{2n} = T$ be the points of subdivision of the refined partition π'_n. Define the approximating sum $F_{\pi'_n}$ as before. We shall first obtain bounds on $E|F_{\pi_n} - F_{\pi'_n}|^2$ in order to get bounds on $E|F_{\pi_n} - F|^2$.

Let $0 \leq t^*_0 < t^*_1 < t^*_2 \leq T$ be three equally spaced points on $[0, T]$ and let us denote $X_{t^*_i}$ by X_i and $W_{t^*_i}$ by $W_i, i = 0, 1, 2$. Define

$$Z := \frac{f(t^*_2, X_2) + f(t^*_0, X_0)}{2} (W_2 - W_0)$$

$$- \left\{ \frac{f(t^*_2, X_2) + f(t^*_1, X_1)}{2} (W_2 - W_1) + \frac{f(t^*_1, X_1) + f(t^*_0, X_0)}{2} (W_1 - W_0) \right\}$$

$$= \left(\frac{W_1 - W_0}{2} \right) (f(t^*_2, X_2) - f(t^*_1, X_1))$$

$$+ \left(\frac{W_2 - W_1}{2} \right) (f(t^*_0, X_0) - f(t^*_1, X_1)). \qquad (3.3)$$

Let

$$I_1 := \int_{t^*_0}^{t^*_1} f(t, X_t) dt, \quad I_2 := \int_{t^*_1}^{t^*_2} f(t, X_t) dt. \qquad (3.4)$$

Clearly by Taylor expansion

$$f(t^*_2, X_2) - f(t^*_1, X_1)$$
$$= (X_2 - X_1) f_x(t^*_1, X_1) + (t^*_2 - t^*_1) f_t(t^*_1, X_1) + \frac{1}{2} (X_2 - X_1)^2 f_{xx}(\tau_1, \nu_1)$$
$$+ \frac{1}{2} (t^*_2 - t^*_1)^2 f_{tt}(\tau_1, \nu_1) + (t^*_2 - t^*_1)(X_2 - X_1) f_{tx}(\tau_1, \nu_1)$$
$$= (W_2 - W_1 + I_2) f_x(t^*_1, X_1) + (t^*_2 - t^*_1) f_t(t^*_1, X_1) + \frac{1}{2} (X_2 - X_1)^2 f_{xx}(\tau_1, \nu_1)$$
$$+ \frac{1}{2} (t^*_2 - t^*_1)^2 f_{tt}(\tau_1, \nu_1) + (t^*_2 - t^*_1)(X_2 - X_1) f_{tx}(\tau_1, \nu_1)$$

$$(3.5)$$

where $|X_1 - \nu_1| < |X_2 - X_1|, |t^*_1 - \tau_1| \leq |t^*_2 - t^*_1|,$

and

$$f(t_0^*, X_0) - f(t_1^*, X_1)$$
$$= (X_0 - X_1)f_x(t_1^*, X_1) + (t_0^* - t_1^*)f_t(t_1^*, X_1) + \frac{1}{2}(X_0 - X_1)^2 f_{xx}(\tau_2, \nu_2)$$
$$+ \frac{1}{2}(t_0^* - t_1^*)^2 f_{tt}(\tau_2, \nu_2) + (t_0^* - t_1^*)(X_0 - X_1)f_{tx}(\tau_2, \nu_2)$$
$$= -(W_1 - W_0 + I_1)f_x(t_1^*, X_1) + (t_0^* - t_1^*)f_t(t_1^*, X_1) + \frac{1}{2}(X_0 - X_1)^2 f_{xx}(\tau_2, \nu_2)$$
$$+ \frac{1}{2}(t_0^* - t_1^*)^2 f_{tt}(\tau_2, \nu_2) + (t_0^* - t_1^*)(X_0 - X_1)f_{tx}(\tau_2, \nu_2)$$

$$(3.6)$$

where $|X_1 - \nu_2| < |X_0 - X_1|, |t_1^* - \tau_2| < |t_0^* - t_1^*|$.

Relations (3.3) to (3.6) show that

$$Z = \left(\frac{W_1 - W_0}{2}\right) I_2 f_x(t_1^*, X_1) + \left(\frac{W_1 - W_0}{2}\right)(t_2^* - t_1^*)f_t(t_1^*, X_1)$$
$$+ \left(\frac{W_1 - W_0}{4}\right)(X_2 - X_1)^2 f_{xx}(\tau_1, \nu_1) + \left(\frac{W_1 - W_0}{4}\right)(t_2^* - t_1^*)^2 f_{tt}(\tau_1, \nu_1)$$
$$+ \left(\frac{W_1 - W_0}{2}\right)(X_2 - X_1)(t_2^* - t_1^*)f_{tx}(\tau_1, \nu_1) - \left(\frac{W_2 - W_1}{2}\right) I_1 f_x(t_1^*, X_1)$$
$$+ \left(\frac{W_2 - W_1}{2}\right)(t_0^* - t_1^*)f_t(t_1^*, X_1) + \left(\frac{W_2 - W_1}{4}\right)(X_0 - X_1)^2 f_{xx}(\tau_2, \nu_2)$$
$$+ \left(\frac{W_2 - W_1}{4}\right)(t_0^* - t_1^*)^2 f_{tt}(\tau_2, \nu_2)$$
$$+ \left(\frac{W_2 - W_1}{2}\right)(X_0 - X_1)(t_0^* - t_1^*)f_{tx}(\tau_2, \nu_2)$$
$$=: M_1 + M_2$$

$$(3.7)$$

where

$$M_1 :=$$
$$(W_1 - W_0)\left\{\frac{I_2}{2}f_x(t_1^*, X_1) + \frac{(t_2^* - t_1^*)}{2}f_t(t_1^*, X_1) + \frac{(X_2 - X_1)^2}{4}f_{xx}(\tau_1, \nu_1)\right.$$
$$\left. + \frac{(t_2^* - t_1^*)^2}{4}f_{tt}(\tau_1, X_1) + \frac{1}{2}(X_2 - X_1)(t_2^* - t_1^*)f_{tx}(\tau_1, \nu_1)\right\},$$

$$(3.8)$$

$$M_2 :=$$
$$(W_2 - W_1)\left\{-\frac{I_1}{2}f_x(t_1^*, X_1) + \frac{(t_0^* - t_1^*)}{2}f_t(t_1^*, X_1) + \frac{(X_1 - X_0)^2}{4}f_{xx}(\tau_2, \nu_2)\right.$$
$$\left. + \frac{(t_0^* - t_1^*)^2}{4}f_{tt}(\tau_2, X_2) + \frac{1}{2}(X_0 - X_1)(t_0^* - t_1^*)f_{tx}(\tau_2, \nu_2)\right\}.$$

$$(3.9)$$

Clearly,
$$E(Z^2) \leq 2\left[E(M_1^2) + E(M_2^2)\right].$$
Notice that M_2's corresponding to different subintervals of $[0, T]$-generated by π_n form a martingale difference sequence and M_1's corresponding to different subintervals of $[0, T]$ generated by π_n form a reverse martingale difference sequence.

Observe that

$$E|M_2|^2$$

$$= E(W_2 - W_1)^2 E\left\{-\frac{I_1}{2}f_x(t_1^*, X_1) + \frac{(t_0^* - t_1^*)}{2}f_t(t_1^*, X_1) + \frac{(X_1 - X_0)^2}{4}\right.$$

$$\times f_{xx}(\tau_2, \nu_2)$$

$$\left.+ \frac{(t_0^* - t_1^*)^2}{4}f_{tt}(\tau_2, \nu_2) + \frac{1}{2}(X_0 - X_1)(t_0^* - t_1^*)f_{tx}(\tau_2, \nu_2)\right\}^2$$

$$\leq 4(t_2^* - t_1^*)\left\{E(-\frac{I_1}{2}f_x(t_1^*, X_1))^2 + \frac{(t_0^* - t_1^*)^2}{4}E(f_t(t_1^*, X_1))^2\right.$$

$$+ E\left\{\frac{(X_1 - X_0)^4}{16}(f_{xx}(\tau_2, \nu_2))^2\right\} + \frac{(t_0^* - t_1^*)^2}{16}E(f_{tt}(\tau_2, \nu_2))^2$$

$$\left.+ \frac{1}{4}(t_0^* - t_1^*)^2 E\{(X_0 - X_1)f_{tx}(\tau_2, \nu_2)\}^2\right\}$$

$$\leq 4(t_2^* - t_1^*)\left\{\frac{E(I_1)^4}{16}E(f_x(t_1^*, X_1))^4\right\}^{1/2} + \frac{(t_0^* - t_1^*)^2}{4}E(f_t(t_1^*, X_1))^2$$

$$+ \left\{\frac{E(X_1 - X_0)^8}{256}E(f_{xx}(\tau_2, \nu_2))^2\right\}^{1/2} + \frac{(t_0^* - t_1^*)^2}{16}E(f_{tt}(\tau_2, \nu_2))^2$$

$$+ \frac{1}{4}(t_0^* - t_1^*)^2 E(X_0 - X_1)^4 E(f_{tx}(\tau_2, \nu_2))^4\right\}^{1/2}$$

$$\leq C\ (t_2^* - t_1^*)\left[\{E(I_1)^4\}^{1/2} + \{E(X_1 - X_0)^8\}^{1/2}\right] \tag{3.10}$$

by $(A4)_2$.

By Theorem 4 of Gikhman and Skorohod (1972, p. 48) there exists $C > 0$ such that for $0 \leq s < t \leq T$,

$$E(X_t - X_s)^{2m} \leq C(E(X_0^{2m}) + 1)(t - s)^m, m \geq 1 \tag{3.11}$$

Hence
$$E(X_1 - X_0)^8 \leq C(E(X_0^8) + 1)(t_1^* - t_0^*)^4.$$

Further by $(A4)_2$

$$E(I_1^4) = E\left(\int_{t_0^*}^{t_1^*} f(t, X_t)dt\right)^4$$

$$\leq K^4 E \left(\int_{t_0^*}^{t_1^*} (1 + |X_t|) dt \right)^4$$

$$\leq 4K^4 (t_1^* - t_0^*)^4 \sup_{0 \leq t \leq T} E(1 + |X_t|)^4$$

$$\leq C(t_1^* - t_0^*)^4. \tag{3.12}$$

Relations (3.10) - (3.12) prove that

$$E(M_2^2) \leq C(t_2^* - t_1^*)(t_1^* - t_0^*)^2 \tag{3.13}$$

for some constant $C > 0$ independent of t_0^*, t_1^* and t_2^*. Let us estimate $E(M_1^2)$. Observe that

$$E(M_1^2)$$

$$= E \left[(W_1 - W_0) \left\{ \frac{I_2}{2} f_x(t_1^*, X_1) + \frac{(t_2^* - t_1^*)}{2} f_t(t_1^*, X_1) \right. \right.$$

$$+ \frac{(X_2 - X_1)^2}{4} f_{xx}(\tau_1, \nu_1) + \frac{(t_2^* - t_1^*)^2}{4} f_{tt}(\tau_1, X_1)$$

$$\left. \left. + \frac{1}{2}(X_2 - X_1)(t_2^* - t_1^*)f_{tx}(\tau_1, \nu_1) \right\} \right]^2$$

$$\leq E \left[(W_1 - W_0)^4 E \left\{ \frac{I_2}{2} f_x(t_1^*, X_1) + \frac{(t_2^* - t_1^*)}{2} f_t(t_1^*, X_1) \right. \right.$$

$$+ \frac{(X_2 - X_1)^2}{4} f_{xx}(\tau_1, \nu_1) + \frac{(t_2^* - t_1^*)^2}{4} f_{tt}(\tau_1, X_1)$$

$$\left. \left. + \frac{1}{2}(X_2 - X_1)(t_2^* - t_1^*)f_{tx}(\tau_1, \nu_1) \right\}^4 \right]^{1/2}$$

(by the Cauchy-Schwarz inequality)

$$\leq C(t_1^* - t_0^*)^2 \left\{ \frac{E(I_2 f_x(t_1^*, X_1)^4}{16} + \frac{(t_2^* - t_1^*)^4}{16} E(f_t(t_1^*, X_1))^4 \right.$$

$$+ \frac{E((X_2 - X_1)f_{xx}(\tau_1, \nu_1))^4}{256} + \frac{(t_2^* - t_1^*)^8}{256} E(f_{tt}(\tau_1, X_1))^4$$

$$\left. + \frac{1}{16}(t_2^* - t_1^*)E((X_2 - X_1)f_{tx}(\tau_1, \nu_1))^4 \right\}^{1/2}$$

(by C_r - inequality and the fact that $E(W_1 - W_0)^4 = 3(t_1^* - t_0^*)^2$)

$$\leq C(t_1^* - t_0^*)^2 \left\{ \frac{E(I_2 f_x(t_1^*, X_1))^4}{16} + \frac{(t_2^* - t_1^*)^4}{16} E(f_t(t_1^*, X_1))^4 \right.$$

$$+ \frac{E((X_2 - X_1)f_{xx}(\tau_1, \nu_1))^4}{256} + \frac{(t_2^* - t_1^*)^8}{256} E(f_{tt}(\tau_1, X_1))^4$$

$$\left. + \frac{1}{16}(t_2^* - t_1^*)^4 E((X_2 - X_1)f_{tx}(\tau_1, \nu_1))^4 \right\}^{1/2} \tag{3.14}$$

$$\leq C(t_1^* - t_0^*)^2 \left\{ \frac{\{E(I_2)^8 E(f_x(t_1^*, X_1))^8\}^{1/2}}{16} + \frac{(t_2^* - t_1^*)^4}{16} E(f_t(t_1^*, X_1))^4 \right.$$

$$+ \frac{\{E(X_2 - X_1))^8 E(f_{xx}(\tau_1, \nu_1))^8\}^{1/2}}{256} + \frac{(t_2^* - t_1^*)^8}{256} E(f_{tt}(\tau_1, X_1))^4$$

$$\left. + \frac{1}{16}(t_2^* - t_1^*)^4 \{E((X_2 - X_1)^8 E(f_{tx}(\tau_1, \nu_1))^8\}^{1/2} \right\}^{1/2}$$

(by the Cauchy-Schwartz inequality)

From (3.11) it follows that there exists a constant $C > 0$ such that

$$E(X_2 - X_1)^8 \leq C(t_2^* - t_1^*)^4 \tag{3.15}$$

Furthermore, by $(A5)_1$

$$E(I_2)^8 = E \left[\int_{t_1^*}^{t_2^*} f(t, X_t) dt \right]^8$$

$$\leq CE \left[\int_{t_1^*}^{t_2^*} (1 + |X_t|) dt \right]^8 \text{ by (A1)}$$

$$\leq CE \left[\left\{ \int_{t_1^*}^{t_2^*} (1 + |X_t|) dt \right\}^2 \right]^4$$

$$\leq CE \left[(t_2^* - t_1^*) \left\{ \int_{t_1^*}^{t_2^*} (1 + |X_t|)^2 dt \right\} \right]^4 \tag{3.16}$$

$$\leq C(t_2^* - t_1^*)^4 E \left[\int_{t_1^*}^{t_2^*} (1 + |X_t|^2) dt \right]^4$$

$$\leq C(t_2^* - t_1^*)^7 \int_{t_1^*}^{t_2^*} E(1 + |X_t|^8) dt$$

$$= C(t_2^* - t_1^*)^8 \text{ (by } (A4)_2)$$

Relations (3.14) - (3.16) prove that

$$E(M_1^2) \leq C(t_1^* - t_0^*)(t_2^* - t_1^*)^2 \tag{3.17}$$

for some constant $C > 0$ independent of t_0^*, t_1^* and t_2^*. Inequalities (3.13) and (3.17) prove that there exists a constant $C > 0$, independent of t_0^*, t_1^* and t_2^* such that

$$E(M_i^2) \leq C(t_2^* - t_1^*)^3, \quad i = 1, 2. \tag{3.18}$$

Using the property that M_2 corresponding to different subintervals form a martingale difference sequence and M_1 form a reverse martingale difference sequence, it follows that

$$E|F_{\pi_n} - F_{\pi_n'}|^2 \leq C\frac{T^3}{n^2} \tag{3.19}$$

for some constant $C > 0$.

Let $\{\pi_n^{(p)}, p \geq 0\}$ be the sequence of partitions such that $\pi_n^{(i+1)}$ is a refinement of $\pi_n^{(i)}$ by choosing the midpoints of the subintervals generated by $\pi_n^{(i)}$. Note that $\pi_n^{(0)} = \pi_n$ and $\pi_n^{(1)} = \pi_n'$. The analysis given above proves that

$$E|F_{\pi_n}(p) - F_{\pi_n}(p+1)|^2 \leq C\frac{T^3}{2^p n^2}, p \geq 0 \tag{3.20}$$

where $F_{\pi_n}(p)$ is the approximant corresponding to $\pi_n^{(p)}$ and $F_{\pi_n}(0) = F_{\pi_n}$. Therefore,

$$E|F_{\pi_n} - F_{\pi_n}(p+1)|^2 = E\left\{\sum_{k=0}^p [F_{\pi_n}(k) - F_{\pi_n}(k+1)]\right\}^2$$

$$\leq \left\{\sum_{k=0}^p \left(E|F_{\pi_n}(k) - F_{\pi_n}(k+1)|^2\right)^{\frac{1}{2}}\right\}^2$$

$$\leq \left\{\sum_{k=0}^p \left(\frac{CT^3}{2^k n^2}\right)^{1/2}\right\}^2 \leq C\frac{T^3}{n^2}$$

for all $p \geq 0$. Let $p \to \infty$. Since the integral F exists, $F_{\pi_n}(p+1)$ converges in L_2 to F as $p \to \infty$. Note that $\{\pi_n(p+1), p \geq 0\}$ is a sequence of partitions such that the mesh of the partition tends to zero as $p \to \infty$ for any fixed n. Therefore,

$$E|F_{\pi_n} - F|^2 \leq C\frac{T^3}{n^2} \tag{3.21}$$

where

$$F = \lim_{n\to\infty} F_{\pi_n} = \oint_0^T f(t, X_t)dW_t.$$

This proves (b). To prove (a), let π_n be the partition as defined previously, and let I_{π_n} and I be defined by

$$I_{\pi_n} := \sum_{i=1}^n f(t_{i-1}, X_{t_{i-1}})(W_{t_i} - W_{t_{i-1}}), \quad I := \int_0^T f(t, X_t)dW_t.$$

By arguments used to establish (3.21) and by noting that $\{I_{\pi_n}, n \geq 1\}$ is a martingale, it can be easily shown that (in this case we need the existence and finite moments conditions for first derivative of f only)

$$E|I_{\pi_n} - I|^2 \leq C\frac{T^2}{n}. \tag{3.22}$$

Note that while computing Z after Taylor expansion, in the Itô case the first order terms remain which contribute lower order rate but in the FS case, due to symmetry, the first order term cancels and the second order terms contribute higher order rate This completes the proof of the theorem. $\qquad\square$

7.4 Accuracy of Approximations of the Log-likelihood Function

In this Section we obtain rate of convergence of the approximate log-likelihood functions $L_{n,T}(\theta)$ and $L_{n,T,1}(\theta)$ to the continuous log-likelihood function $L_T(\theta)$ as $n \to \infty$. Let $L_{n,T}^{(i)}, L_{n,T,1}^{(i)}$ and $L_T^{(i)}$ denote the derivative of order i w.r.t. θ of $L_{n,T}, L_{n,T,1}$ and L_T respectively. First we obtain the following L_2-bound on the approximation of ordinary integral.

Theorem 4.1 *Under (A1)-(A3), (A4)$_2$ and (A5)$_2$, for $g = f, f^2$ or f_x, we have*

$$E \left| \sum_{i=1}^n g(t_{i-1}, X_{t_{i-1}})(t_i - t_{i-1}) - \int_0^T g(t, X_t)dt \right|^2 \leq C \frac{T^4}{n^2}.$$

Proof: By Itô formula, we have

$$g(t, X_t) - g(t_{i-1}, X_{t_{i-1}})$$

$$= \int_{t_{i-1}}^t g_x(u, X_u)dW_u$$

$$+ \int_{t_{i-1}}^t \left[g_u(u, X_u) \, \theta f(u, X_u) g_x(u, X_u) + \frac{1}{2} g_{xx}(u, X_u) \right] du$$

$$=: \int_{t_{i-1}}^t g_x(u, X_u)dW_u + \int_{t_{i-1}}^t G(u, X_u)du.$$

On substitution

$$E \left| \sum_{i=1}^n g(t_{i-1}, X_{t_{i-1}})(t_i - t_{i-1}) - \int_0^T g(t, X_t)dt \right|^2$$

$$= E \left| \sum_{i=1}^n \int_{t_{i-1}}^{t_i} \left[g(t_{i-1}, X_{t_{i-1}}) - g(t, X_t) \right] dt \right|^2$$

$$= E \left| \sum_{i=1}^n \int_{t_{i-1}}^{t_i} \left[\int_{t_{i-1}}^t g_x(u, X_u)dW_u + \int_{t_{i-1}}^t G(u, X_u)du \right] dt \right|^2$$

$$\leq 2E \left| \sum_{i=1}^n \int_{t_{i-1}}^{t_i} \int_{t_{i-1}}^t g_x(u, X_u)dW_u dt \right|^2 + 2E \left| \sum_{i=1}^n \int_{t_{i-1}}^{t_i} \int_{t_{i-1}}^t G(u, X_u)du dt \right|^2$$

$$=: 2H_1 + 2H_2.$$

With

$$B_{i,t} := \int_{t_{i-1}}^t g_x(u, X_u)dW_u, \quad t_{i-1} \leq t < t_i$$

and

$$B_{j,s} := \int_{t_{j-1}}^{s} g_x(u, X_u)dW_u, t_{j-1} \le s < t_j, \quad j \neq i, \ 1 \le i < j \le n,$$

we have

$$H_1 = E \left| \sum_{i=1}^{n} \int_{t_{i-1}}^{t_i} B_{i,t}dt \right|^2$$

$$\le E \sum_{i=1}^{n} \left(\int_{t_{i-1}}^{t_i} B_{i,t}dt \right)^2 + \sum_{j \neq i=1}^{n} E \left(\int_{t_{i-1}}^{t_i} B_{i,t}dt \right) \left(\int_{t_{j-1}}^{t_j} B_{j,s}ds \right)$$

$$\le \sum_{i=1}^{n} (t_i - t_{i-1}) \int_{t_{i-1}}^{t_i} E(B_{i,t})^2 dt$$

(the last term being zero due to orthogonality of $B_{i,t}$ and $B_{j,s}$)

$$\le \sum_{i=1}^{n} (t_i - t_{i-1}) \int_{t_{i-1}}^{t_i} \int_{t_{i-1}}^{t} E[g_x(u, X_u)]^2 dudt$$

$$\le C \sum_{i=1}^{n} (t_i - t_{i-1}) \int_{t_{i-1}}^{t_i} (t - t_{i-1}) dt \text{ by } (A4)_2$$

$$\le C \frac{T}{n} \sum_{i=1}^{n} (t_i - t_{i-1})^2$$

$$\le C \frac{T^3}{n^2}.$$

With

$$\psi_{i,t} := \int_{t_{i-1}}^{t} G(u, X_u)du, \quad t_{i-1} \le t < t_i$$

and

$$\psi_{j,s} := \int_{t_{j-1}}^{s} G(u, X_u)du, \quad t_{j-1} \le s < t_j, \ j \neq i, \ 1 \le i < j \le n,$$

we have

$$H_2 = E \left| \sum_{i=1}^{n} \int_{t_{i-1}}^{t_i} \psi_{i,t}dt \right|^2$$

$$\le E \sum_{i=1}^{n} \left(\int_{t_{i-1}}^{t_i} \psi_{i,t}dt \right)^2 + \sum_{j \neq i=1}^{n} E \left(\int_{t_{i-1}}^{t_i} \psi_{i,t}dt \right) \left(\int_{t_{j-1}}^{t_j} \psi_{j,s}ds \right)$$

$$\le \sum_{i=1}^{n} (t_i - t_{i-1})^{3/2} \int_{t_{i-1}}^{t_i} E(\psi_{i,t})^2 dt$$

$$+ \sum_{j\neq i=1}^{n} \left\{ E\left(\int_{t_{i-1}}^{t_i} \psi_{i,t} dt\right)^2 E\left(\int_{t_{j-1}}^{t_j} \psi_{j,s} ds\right)^2 \right\}^{1/2}$$

$$\leq \sum_{i=1}^{n} (t_i - t_{i-1})^{3/2} \int_{t_{i-1}}^{t_i} E(\psi_{i,t})^2 dt$$

$$+ \sum_{j\neq i=1}^{n} \left\{ \left((t_i - t_{i-1}) \int_{t_{i-1}}^{t_i} E[\psi_{i,t}]^2 dt \right) \left((t_j - t_{j-1}) \int_{t_{j-1}}^{t_j} E[\psi_{j,s}]^2 ds \right) \right\}^{1/2}$$

$$\leq C \sum_{i=1}^{n} (t_i - t_{i-1})^3 + C \sum_{j\neq i=1}^{n} (t_i - t_{i-1})^2 (t_j - t_{j-1})^2$$

(since $E[\psi_{i,t}]^2 \leq C(t_i - t_{i-1})^2$ by $(A4)_2$ and $(A5)_2$)

$$\leq Cn\frac{T^3}{n^3} + Cn(n-1)\frac{T^4}{n^2}$$

$$\leq C\frac{T^4}{n^2}.$$

Combining bounds for H_1 and H_2, completes the proof of the theorem. $\square$

Theorem 4.2 *For some constant $K > 0$ and for $i = 0, 1, 2$ there exist two sequence $\{H_n^i(K), n \geq 1\}$ and $\{G_n^i(K), n \geq 1\}$ of positive random variables which are bounded in P_{θ_0}-probability for all $\theta \in \Theta$ such that under the assumptions (A1) - (A3), $(A4)_1$, $(A5)_1$, (A6) and (A7), we have*

(a) $\quad \sup_{\{\theta\in\Theta:|\theta|\leq K\}} |L_{n,T}^{(i)}(\theta) - L_T^{(i)}(\theta)| \leq \frac{T}{n^{1/2}} H_n^i(K),$

and under the assumptions (A1) - (A3), $(A4)_2$, $(A5)_1$, (A6) and (A7), we have

(b) $\quad \sup_{\{\theta\in\Theta:|\theta|\leq K\}} |L_{n,T,1}^{(i)}(\theta) - L_T^{(i)}(\theta)| \leq \frac{T^2}{n} G_n^i(K).$

Proof: By using the arguments to prove Theorem 4.1, it can be shown that

$$E\left| \sum_{i=1}^{n} \int_{t_{i-1}}^{t_i} f(t, X_t)[f(t_{i-1}, X_{t_{i-1}}) - f(t, X_t)]dt \right|^2 \leq C\frac{T^4}{n^2}. \qquad (4.1)$$

Note that

$$|L_{n,T}(\theta) - L_T(\theta)|$$

$$= \left| \theta \left[\sum_{i=1}^{n} f(t_{i-1}, X_{t_{i-1}})(X_{t_i} - X_{t_{i-1}}) - \int_0^T f(t, X_t)dX_t \right] \right.$$

$$\left. - \frac{\theta^2}{2} \left[\sum_{i=1}^{n} f^2(t_{i-1}, X_{t_{i-1}})(t_i - t_{i-1}) - \int_0^T f^2(t, X_t)dt \right] \right|$$

$$\leq |\theta| \left| \sum_{i=1}^{n} f(t_{i-1}, X_{t_{i-1}})(W_{t_i} - W_{t_{i-1}}) - \int_0^T f(t, X_t)dW_t \right|$$

$$+ |\theta| |\sum_{i=1}^{n} \int_{t_{i-1}}^{t_i} f(t, X_t)[f(t_{i-1}, X_{t_{i-1}}) - f(t, X_t)]dt|$$

$$+ \frac{\theta^2}{2} |\sum_{i=1}^{n} f^2(t_{i-1}, X_{t_{i-1}})(t_i - t_{i-1}) - \int_0^T f^2(t, X_t)dt|$$

$$\leq K \frac{T}{n^{1/2}} V_n + K \frac{T^2}{n} A_n + \frac{K^2}{2} \frac{T^2}{n} D_n$$

(by Theorem 4.3.1 (a) and Theorem 4.4.1 and (4.1)).

$$= \frac{T}{n^{1/2}} H_n^0(K)$$

where $H_n^0(K) := KV_n + K\frac{T}{n^{1/2}} A_n + \frac{K^2}{2} \frac{T}{n^{1/2}} D_n$.

Next,

$$|L_{n,T}^{(1)}(\theta) - L_T^{(1)}(\theta)|$$

$$= \left| \sum_{i=1}^{n} f(t_{i-1}, X_{t_{i-1}})(W_{t_i} - W_{t_{i-1}}) - \int_0^T f(t, X_t)dW_t \right.$$

$$+ \theta \sum_{i=1}^{n} \int_{t_{i-1}}^{t_i} f(t, X_t)[f(t_{i-1}, X_{t_{i-1}}) - f(t, X_t)]dt$$

$$\left. - \theta \left[\sum_{i=1}^{n} f^2(t, X_{t_{i-1}})(t_i - t_{i-1}) - \int_0^T f^2(t, X_t)dt \right] \right|$$

$$\leq \frac{T}{n^{1/2}} V_n + \frac{T^2}{n} A_n + K \frac{T^2}{n} D_n$$

(by Theorem 3.1(a) and Theorem 4.4.1 and (4.1))

$$= \frac{T}{n^{1/2}} H_n^1(K)$$

where $H_n^1(K) := V_n + A_n + K\frac{T}{n^{1/2}} D_n$.

Further

$$|L_{n,T}^{(2)}(\theta) - L_T^{(2)}(\theta)|$$

$$\leq |\theta| |\sum_{i=1}^{n} f^2(t_{i-1}, X_{t_{i-1}})(t_i - t_{i-1}) - \int_0^T f^2(t, X_t)dt|$$

$$\leq K \frac{T^2}{n} D_n$$

$$= \frac{T^2}{n} H_n^{(2)}(K) \quad \text{(by Theorem 4.1)}$$

where $H_n^{(2)}(K) := KD_n$. Similarly,

$$|L_{n,T,1}(\theta) - L_T(\theta)|$$

$$\leq K\frac{T^{3/2}}{n}U_n + \frac{K}{2}\frac{T^2}{n}R_n + \frac{K^2}{2}\frac{T^2}{n}D_n \quad \text{(by Theorems 3.1 (b) and 4.1)}$$

$$= \frac{T^2}{n}G_n^0(K)$$

where $G_n^0(K) = KU_n + \frac{K}{2}T^{1/2}R_n + \frac{K^2}{2}T^{1/2}D_n$. Next

$$|L_{n,T,1}^{(1)}(\theta) - L_T^{(1)}(\theta)|$$

$$\leq \frac{T^2}{n}U_n + \frac{1}{2}\frac{T^2}{n}R_n + K\frac{T^2}{n}D_n \quad \text{(by Theorem 3.1 (b) and 4.4.1)}$$

$$= \frac{T^2}{n}G_n^1(K)$$

where $G_n^1(K) := U_n + \frac{1}{2}R_n + KD_n$.
Finally

$$|L_{n,T,1}^{(2)}(\theta) - L_T^{(2)}(\theta)|$$

$$\leq \frac{T^2}{n}D_n \quad \text{(by Theorem 4.1)}$$

$$= \frac{T^2}{n}G_n^2(K)$$

where $G_n^2(K) := D_n$. This completes the proof of the theorem. $\qquad\square$

7.5 Accuracy of Approximations of the Maximum Likelihood Estimate

In this Section we obtain the rate of convergence of the AMLEs $\theta_{n,T}$ and $\theta_{n,T,1}$ to the continuous MLE θ_T as $h \to 0$ and $n \to \infty$. To obtain the rate of convergence of the AMLEs we need the following general theorem on approximate maximum likelihood estimation.

Theorem 5.1 *(Le Breton (1976, page 138)). Let $(\Omega, \mathcal{A}, \{P_\theta; \theta \in \mathbb{R}\})$ be a statistical structure dominated by P, with a log-likelihood function $L(\theta, \cdot)$. Let $\{\mathcal{A}_n, n \geq 1\}$ be a sequence of sub-σ-algebras of $\mathcal{A}$ and, for all $n \geq 1$, $L_n(\theta, \cdot)$ be the log-likelihood function on the statistical structure $(\Omega, \mathcal{A}_n, \{P_{\theta|\mathcal{A}_n}; \theta \in \mathbb{R}\})$ or any $\mathcal{A}_n$ - measurable function. Let us suppose that the following assumptions are satisfied.*

(C1) L and L_n are twice continuously differentiable with derivatives $L^{(i)}$ and $L_n^{(i)}$ respectively, $i = 0, 1, 2$.

(C2) $L^{(2)}$ does not depend on θ and P-almost surely strictly negative.

(C3) $L^{(1)}(\theta, \cdot) = 0$ admits P almost surely a unique solution $\hat{\theta}$.

(C4) There exists a sequence $\{\gamma_n, n \geq 1\}$ of positive numbers converging to zero and for $i = 0, 1, 2$ and all $K > 0$ there exists a sequence $\{\nabla_n^i(K), n \geq 1\}$ of positive random variables such that $\theta \in \mathbb{R}$,

    (a) $\{\nabla_n^i(K), n \geq 1\}$ is bounded in P_θ probability,

    (b) $\sup_{|\theta| \leq K} |L_n^{(i)}(\theta, \cdot) - L^{(i)}(\theta, \cdot)| \leq \gamma_n \nabla_n^i(K) P_\theta$ almost surely.

Then there exists a sequence $\{\theta_n, n \geq 1\}$ of random variables satisfying

    (i) θ_n is $\mathcal{A}_n$-measurable,

and for all $\theta \in \mathbb{R}$

    (ii) $\lim_{n \to \infty} P_\theta[L_n^{(1)}(\theta_n) = 0] = 1$

    (iii) $P_\theta - \lim_{n \to \infty} \theta_n = \hat{\theta}$, where $\hat{\theta}$ is the MLE based on L.

Furthermore if $\{\theta_n', n \geq 1\}$ is another sequence of random variables satisfying (i), (ii) and (iii), then for all $\theta \in \mathbb{R}$

$$\lim_{n \to \infty} P_\theta[\theta_n = \theta_n'] = 1.$$

Lastly if $\{\theta_n, n \geq 1\}$ is a sequence satisfying (i), (ii) and (iii) then for all $\theta \in \mathbb{R}$, the sequence $\{\gamma_n^{-1}(\theta_n - \hat{\theta}), n \geq 1\}$ is bounded in P_θ probability.   □

We establish the following probability bounds on the accuracy of different approximations of the MLE.

Theorem 5.2 *Under the assumptions (A1)-(A3), (A4)$_1$, (A5)$_1$, (A6) and (A7)*

    (i) $P_{\theta_0} - \lim_{n \to \infty} \theta_{n,T} = \theta_T$.

    (ii) $|\theta_{n,T} - \theta_T| = O_{P_{\theta_0}}(\frac{T}{n^{1/2}})$.

Under the assumptions (A1) - (A3), (A4)$_2$, (A5)$_2$, (A6) and (A7)

    (iii) $P_{\theta_0} - \lim_{n \to \infty} \theta_{n,T,1} = \theta_T$.

    (iv) $|\theta_{n,T,1} - \theta_T| = O_{P_{\theta_0}}(\frac{T^2}{n})$.

Proof: We use Theorem 5.1 to prove these results. It is easily seen that the assumptions (C1) - (C3) are satisfied for $\{L_T, L_{n,T}; n \geq 1\}$ and $\{L_T, L_{n,T,1}; n \geq 1\}$. By Theorem 5.1 assumption (C4) is satisfied for $\{L_T, L_{n,T}; n \geq 1\}$ with $\gamma_n = (\frac{T}{n^{1/2}}), \nabla_n^i(K) = H_n^i(K)$ and for $\{L_T, L_{n,T,1}; n \geq 1\}$ with $\gamma_n = \frac{T^2}{n}, \nabla_n^i(K) = G_n^i(K)$. Hence all the results of Theorem 5.2 follow immediately from Theorem 5.1.   □

Remarks

(1) Theorem 5.2 (i) and (iii) were shown by Le Breton (1976) for the particular case $f(t, x) = x$, i.e., the Ornstein-Uhlenbeck process. However his proofs have some technical errors.

(2) Extension of results of this Chapter to general nonlinear SDE, i.e., with drift $b(\theta, t, X_t)$ remains open. It seems impossible by the present method because to apply Theorem 4.1, assumption (C2), that the second derivative of the log likelihood is independent θ, may not be satisfied in general. Some other methods are needed.

(3) So far as the rate of convergence to the MLE is concerned, the AMLE $\theta_{n,T,1}$ is the best because eventhough there is scope for improving the order of approximation of the FS integral, there is no scope of improving the order of approximation of the ordinary integrals. We conjecture that one can not find an AMLE having faster rate of convergence $\theta_{n,T,1}$.

(4) Since the FS integral obeys the rules of ordinary calculus hence we conjecture that the approximant in (2.7) will converge to the FS integral at the same rate as in (2.7) i.e., of the order $O(h^2)$ in L_2. We also conjecture that the approximant in (2.9) will converge to the FS integral faster than those in (2.6) and (2.7). The rate of convergence of the approximants in (2.8)-(2.16) to the FS integral remains open.

(5) It would be interesting to obtain limit distributions of the approximant of the Itô integral and the different approximants of the FS integral centered at the corresponding integral with suitable normalizations. Also it remains to investigate the limit distributions of $\theta_{n,T}$ and $\theta_{n,T,j}, j = 1, 2, 3, \ldots, 8$ centered at the MLE θ_T with suitable normalizations. To obtain rates of convergence of the estimators $\theta_{n,T}$ and $\theta_{n,T,j}, j = 1, 2, 3, \ldots, 8$ to θ remains open.

(6) Extension of the results of this Chapter to multidimensional process and multidimensional parameter remains to be investigated.

7.6 Example: Chan-Karloyi-Longstaff-Sanders Model

The Chan-Karloyi-Longstaff-Sanders (CKLS) model is used as an short rate model in term structure of interest rates. The general one factor interest rate model comprises of a linear drift with constant elasticity of variance.

CKLS process $\{X_t\}$ satisfies the Itô stochastic differential equation

$$dX_t = \theta(\kappa - X_t)\, dt + \sigma\, X_t^\gamma\, dW_t, \ t \geq 0, \ X_0 = x_0.$$

The unknown parameters are θ which is the speed of mean reversion, θ is central tendency parameter or the level of mean reversion, σ is the standard deviation of volatility and γ is the elasticity of variance.

Sometimes γ is called the *elasticity of volatility*. The case $\gamma = 0$ gives Vasicek, $\gamma = 1$ gives Black-Scholes and $\gamma = 1/2$ gives Cox-Ingersoll-Ross model. CKLS demonstrated that γ should be in fact 1.5. We assume γ to know known and obtain estimators of θ and κ.

Let the continuous realization $\{X_t, 0 \leq t \leq T\}$ be denoted by X_0^T. Let P_θ^T be the measure generated on the space (C_T, B_T) of continuous functions on

$[0, T]$ with the associated Borel σ-algebra B_T generated under the supremum norm by the process X_0^T and let P_0^T be the standard Wiener measure. It is well known that when κ is the true value of the parameter P_θ^T is absolutely continuous with respect to P_0^T and the Radon-Nikodym derivative (likelihood) of $P_{\theta,\kappa}^T$ with respect to P_0^T based on X_0^T is given by

$$L_T(\theta, \kappa) := \frac{dP_{\theta,\kappa}^T}{dP_0^T}(X_0^T)$$

$$= \exp\left\{\int_0^T \theta(\kappa - X_t)\sigma^{-2}X_t^{1-2\gamma}dX_t - \frac{1}{2}\int_0^T \theta(\kappa - X_t)^2\sigma^{-2}X_t^{-2\gamma}dt\right\}.$$

Consider the score function, the derivative of the log-likelihood function, which is given by

$$l_T(\theta, \kappa) := \sigma^{-2}\left\{\int_0^T \theta(\kappa - X_t)X_t^{1-2\gamma}dX_t - \frac{1}{2}\int_0^T \theta(\kappa - X_t)^2X_t^{-2\gamma}dt\right\}.$$

A solution of the estimating equation $l_T(\theta, \kappa) = 0$ provides the conditional maximum likelihood estimate (MLEs)

$$\hat{\theta}_T := \frac{\int_0^T X_t^{-2\gamma}dX_t \int_0^T X_t^{1-2\gamma}dt - \int_0^T X_t^{1-2\gamma}dX_t \int_0^T X_t^{-2\gamma}dt}{\int_0^T X_t^{-2\gamma}dt \int_0^T X_t^{2-2\gamma}dt - \int_0^T X_t^{1-2\gamma}dt \int_0^T X_t^{1-2\gamma}dt}.$$

and

$$\hat{\kappa}_T := \frac{\int_0^T X_t^{-2\gamma}dX_t \int_0^T X_t^{2-2\gamma}dt - \int_0^T X_t^{1-2\gamma}dX_t \int_0^T X_t^{1-2\gamma}dt}{\int_0^T X_t^{-2\gamma}dX_t \int_0^T X_t^{1-2\gamma}dt - \int_0^T X_t^{1-2\gamma}dX_t \int_0^T X_t^{-2\gamma}dt}.$$

We transform the Itô integrals $\int$ to the Stratonovich integrals $\oint$ as follows: For a smooth function $f(\cdot)$ we have

$$\int_0^T f(X_t)dX_t = \oint_0^T f(X_t)dX_t - \frac{\sigma^2}{2}\int_0^T f'(X_t)X_t^{2\gamma}dt.$$

For simplicity we assume $\sigma = 1$.
Thus

$$\int_0^T X_t^{-2\gamma}dX_t = \oint_0^T X_t^{-2\gamma}dX_t + \gamma\int_0^T X_t^{-1}dt.$$

$$\int_0^T X_t^{1-2\gamma}dX_t = \oint_0^T X_t^{1-2\gamma}dX_t - \frac{1-2\gamma}{2}T.$$

$$\int_0^T X_t^{2-2\gamma}dX_t = \oint_0^T X_t^{2-2\gamma}dX_t - (1-\gamma)\int_0^T X_t dt.$$

Thus Stratonovich integral based MLEs are

$$
\hat{\theta}_T := \left[\left(\oint_0^T X_t^{-2\gamma} dX_t + \gamma \int_0^T X_t^{-1} dt \right) \int_0^T X_t^{1-2\gamma} dt \right.
$$
$$
\left. - \left(\oint_0^T X_t^{1-2\gamma} dX_t - \frac{1-2\gamma}{2} T \right) \int_0^T X_t^{-2\gamma} dt \right]
$$
$$
\left[\int_0^T X_t^{-2\gamma} dt \int_0^T X_t^{2-2\gamma} dt - \int_0^T X_t^{1-2\gamma} dt \int_0^T X_t^{1-2\gamma} dt \right]^{-1}
$$

and

$$
\hat{\kappa}_T := \left[\left(\oint_0^T X_t^{-2\gamma} dX_t + \gamma \int_0^T X_t^{-1} dt \right) \int_0^T X_t^{2-2\gamma} dt \right.
$$
$$
\left. - \left(\oint_0^T X_t^{1-2\gamma} dX_t - \frac{1-2\gamma}{2} T \right) \int_0^T X_t^{1-2\gamma} dt \right]
$$
$$
\left[\left(\oint_0^T X_t^{-2\gamma} dX_t + \gamma \int_0^T X_t^{-1} dt \right) \int_0^T X_t^{1-2\gamma} dt \right.
$$
$$
\left. - \left(\oint_0^T X_t^{1-2\gamma} dX_t - \frac{1-2\gamma}{2} T \right) \int_0^T X_t^{-2\gamma} dt \right]^{-1}.
$$

For a weight function $w_{t_i} \geq 0$, define weighted AMLEs

$$
\tilde{\theta}_{n,T} := \left(\left[\left\{ \sum_{i=1}^n w_{t_i} X_{t_{i-1}}^{-2\gamma} + \sum_{i=2}^{n+1} w_{t_i} X_{t_{i-1}}^{-2\gamma} \right\} (X_{t_i} - X_{t_{i-1}}) \right. \right.
$$
$$
\left. + \gamma \left\{ \sum_{i=1}^n w_{t_i} X_{t_{i-1}}^{-1} + \sum_{i=2}^{n+1} w_{t_i} X_{t_{i-1}}^{-1} \right\} (t_i - t_{i-1}) \right]
$$
$$
\left[\left\{ \sum_{i=1}^n w_{t_i} X_{t_{i-1}}^{1-2\gamma} + \sum_{i=2}^{n+1} w_{t_i} X_{t_{i-1}}^{1-2\gamma} \right\} (t_i - t_{i-1}) \right.
$$
$$
\left. - \left\{ \sum_{i=1}^n w_{t_i} X_{t_{i-1}}^{1-2\gamma} + \sum_{i=2}^{n+1} w_{t_i} X_{t_{i-1}}^{1-2\gamma} \right\} (X_{t_i} - X_{t_{i-1}}) - \frac{1-2\gamma}{2} T \right]
$$
$$
\left. \left[\left\{ \sum_{i=1}^n w_{t_i} X_{t_{i-1}}^{2-2\gamma} + \sum_{i=2}^{n+1} w_{t_i} X_{t_{i-1}}^{2-2\gamma} \right\} (t_i - t_{i-1}) \right] \right)
$$
$$
\left(\left[\left\{ \sum_{i=1}^n w_{t_i} X_{t_{i-1}}^{-2\gamma} + \sum_{i=2}^{n+1} w_{t_i} X_{t_{i-1}}^{-2\gamma} \right\} (t_i - t_{i-1}) \right] \right.
$$
$$
\left[\left\{ \sum_{i=1}^n w_{t_i} X_{t_{i-1}}^{2-2\gamma} + \sum_{i=2}^{n+1} w_{t_i} X_{t_{i-1}}^{2-2\gamma} \right\} (t_i - t_{i-1})
$$

$$- \left\{ \sum_{i=1}^{n} w_{t_i} X_{t_{i-1}}^{1-2\gamma} + \sum_{i=2}^{n+1} w_{t_i} X_{t_{i-1}}^{1-2\gamma} \right\} (t_i - t_{i-1}) \right]$$

$$\left[\left\{ \sum_{i=1}^{n} w_{t_i} X_{t_{i-1}}^{1-2\gamma} + \sum_{i=2}^{n+1} w_{t_i} X_{t_{i-1}}^{1-2\gamma} \right\} (t_i - t_{i-1}) \right] \right)^{-1}.$$

$$\tilde{\kappa}_{n,T} := \left(\left[\left\{ \sum_{i=1}^{n} w_{t_i} X_{t_{i-1}}^{-2\gamma} + \sum_{i=2}^{n+1} w_{t_i} X_{t_{i-1}}^{-2\gamma} \right\} (X_{t_i} - X_{t_{i-1}}) \right.\right.$$

$$+ \gamma \left\{ \sum_{i=1}^{n} w_{t_i} X_{t_{i-1}}^{-1} + \sum_{i=2}^{n+1} w_{t_i} X_{t_{i-1}}^{-1} \right\} (t_i - t_{i-1}) \right]$$

$$\left[\left\{ \sum_{i=1}^{n} w_{t_i} X_{t_{i-1}}^{2-2\gamma} + \sum_{i=2}^{n+1} w_{t_i} X_{t_{i-1}}^{2-2\gamma} \right\} (t_i - t_{i-1}) \right.$$

$$- \left\{ \sum_{i=1}^{n} w_{t_i} X_{t_{i-1}}^{1-2\gamma} + \sum_{i=2}^{n+1} w_{t_i} X_{t_{i-1}}^{1-2\gamma} \right\} (X_{t_i} - X_{t_{i-1}}) - \frac{1-2\gamma}{2} T \right]$$

$$\left[\left\{ \sum_{i=1}^{n} w_{t_i} X_{t_{i-1}}^{1-2\gamma} + \sum_{i=2}^{n+1} w_{t_i} X_{t_{i-1}}^{1-2\gamma} \right\} (t_i - t_{i-1}) \right] \right)$$

$$\left(\left[\left\{ \sum_{i=1}^{n} w_{t_i} X_{t_{i-1}}^{-2\gamma} + \sum_{i=2}^{n+1} w_{t_i} X_{t_{i-1}}^{-2\gamma} \right\} (X_{t_i} - X_{t_{i-1}}) \right] \right.$$

$$\left[\left\{ \sum_{i=1}^{n} w_{t_i} X_{t_{i-1}}^{1-2\gamma} + \sum_{i=2}^{n+1} w_{t_i} X_{t_{i-1}}^{1-2\gamma} \right\} (t_i - t_{i-1}) \right.$$

$$- \left\{ \sum_{i=1}^{n} w_{t_i} X_{t_{i-1}}^{1-2\gamma} + \sum_{i=2}^{n+1} w_{t_i} X_{t_{i-1}}^{1-2\gamma} \right\} (X_{t_i} - X_{t_{i-1}}) \right]$$

$$\left[\left\{ \sum_{i=1}^{n} w_{t_i} X_{t_{i-1}}^{-2\gamma} + \sum_{i=2}^{n+1} w_{t_i} X_{t_{i-1}}^{-2\gamma} \right\} (t_i - t_{i-1}) \right] \right)^{-1}.$$

With $w_{t_i} = 1$, we obtain the forward AMLE as

$$\tilde{\theta}_{n,T,F} := \left[\sum_{i=1}^{n} X_{t_{i-1}}^{1-2\gamma} (X_{t_i} - X_{t_{i-1}}) - \frac{1-\gamma}{2} \sum_{i=1}^{n} X_{t_{i-1}}^{\gamma} (t_i - t_{i-1}) \right]$$

$$\left[\sum_{i=1}^{n} X_{t_{i-1}}^{2-2\gamma} (t_i - t_{i-1}) \right]^{-1}.$$

$$\tilde{\kappa}_{n,T,F} := \left[\sum_{i=1}^{n} X_{t_{i-1}}^{1-2\gamma}(X_{t_i} - X_{t_{i-1}}) - \frac{1-\gamma}{2} \sum_{i=1}^{n} X_{t_{i-1}}^{\gamma}(t_i - t_{i-1}) \right]$$
$$\left[\sum_{i=1}^{n} X_{t_{i-1}}^{2-2\gamma}(t_i - t_{i-1}) \right]^{-1}.$$

With $w_{t_i} = 0$, we obtain the backward AMLE as

$$\tilde{\theta}_{n,T,B} := \left[\sum_{i=1}^{n} X_{t_i}^{1-2\gamma}(X_{t_i} - X_{t_{i-1}}) - \frac{1-\gamma}{2} \sum_{i=1}^{n} X_{t_i}^{\gamma}(t_i - t_{i-1}) \right]$$
$$\left[\sum_{i=1}^{n} X_{t_i}^{2-2\gamma}(t_i - t_{i-1}) \right]^{-1}.$$

$$\tilde{\kappa}_{n,T,B} := \left[\sum_{i=1}^{n} X_{t_i}^{1-2\gamma}(X_{t_i} - X_{t_{i-1}}) - \frac{1-\gamma}{2} \sum_{i=1}^{n} X_{t_i}^{\gamma}(t_i - t_{i-1}) \right]$$
$$\left[\sum_{i=1}^{n} X_{t_i}^{2-2\gamma}(t_i - t_{i-1}) \right]^{-1}.$$

Analogous to the estimators for the discrete AR (1) model, we define the simple symmetric and weighted symmetric estimators (see Fuller (1996)):

With $w_{t_i} = 1/2$, the simple symmetric AMLE is defined as

$$\tilde{\theta}_{n,T,z} := \left[\left\{ \sum_{i=2}^{n} X_{t_{i-1}}^{1-2\gamma} + \frac{1}{2}(X_{t_0}^{1-2\gamma} + X_{t_n}^{1-2\gamma}) \right\} (X_{t_i} - X_{t_{i-1}}) \right.$$
$$\left. - \frac{1-\gamma}{2} \left\{ \sum_{i=2}^{n} X_{t_{i-1}}^{\gamma} + 0.5(X_{t_0}^{\gamma} + X_{t_n}^{\gamma}) \right\} (t_i - t_{i-1}) \right]$$
$$\left[\left\{ \sum_{i=2}^{n} X_{t_{i-1}}^{2-2\gamma} + 0.5(X_{t_0}^{2-2\gamma} + X_{t_n}^{2-2\gamma}) \right\} (t_i - t_{i-1}) \right]^{-1}$$

$$\tilde{\kappa}_{n,T,z} := \left[\left\{ \sum_{i=2}^{n} X_{t_{i-1}}^{1-2\gamma} + \frac{1}{2}(X_{t_0}^{1-2\gamma} + X_{t_n}^{1-2\gamma}) \right\} (X_{t_i} - X_{t_{i-1}}) \right.$$
$$\left. - \frac{1-\gamma}{2} \left\{ \sum_{i=2}^{n} X_{t_{i-1}}^{\gamma} + 0.5(X_{t_0}^{\gamma} + X_{t_n}^{\gamma}) \right\} (t_i - t_{i-1}) \right]$$
$$\left[\left\{ \sum_{i=2}^{n} X_{t_{i-1}}^{2-2\gamma} + 0.5(X_{t_0}^{2-2\gamma} + X_{t_n}^{2-2\gamma}) \right\} (t_i - t_{i-1}) \right]^{-1}$$

With the weight function

$$
w_{t_i} = \begin{cases} 0 & : \quad i = 1 \\ \frac{i-1}{n} & : \quad i = 2, 3, \cdots, n \\ 1 & : \quad i = n+1 \end{cases}
$$

the weighted symmetric AMLE is defined as

$$
\tilde{\theta}_{n,T,w} := \left[\left\{ \sum_{i=2}^{n} X_{t_{i-1}}^{1-2\gamma} + \sum_{i=1}^{n} X_{t_{i-1}}^{1-2\gamma} \right\} (X_{t_i} - X_{t_{i-1}}) \right.
$$
$$
\left. - \frac{1-\gamma}{2} \left\{ \sum_{i=2}^{n} X_{t_{i-1}}^{\gamma} + \sum_{i=1}^{n} X_{t_{i-1}}^{\gamma} \right\} (t_i - t_{i-1}) \right]
$$
$$
\left[\left\{ \sum_{i=2}^{n} X_{t_{i-1}}^{2-2\gamma} + \sum_{i=1}^{n} X_{t_{i-1}}^{2-2\gamma} \right\} (t_i - t_{i-1}) \right]^{-1}.
$$

$$
\tilde{\kappa}_{n,T,w} := \left[\left\{ \sum_{i=2}^{n} X_{t_{i-1}}^{1-2\gamma} + \sum_{i=1}^{n} X_{t_{i-1}}^{1-2\gamma} \right\} (X_{t_i} - X_{t_{i-1}}) \right.
$$
$$
\left. - \frac{1-\gamma}{2} \left\{ \sum_{i=2}^{n} X_{t_{i-1}}^{\gamma} + \sum_{i=1}^{n} X_{t_{i-1}}^{\gamma} \right\} (t_i - t_{i-1}) \right]
$$
$$
\left[\left\{ \sum_{i=2}^{n} X_{t_{i-1}}^{2-2\gamma} + \sum_{i=1}^{n} X_{t_{i-1}}^{2-2\gamma} \right\} (t_i - t_{i-1}) \right]^{-1}.
$$

Note that estimator (1.13) is analogous to the trapezoidal rule in numerical analysis. One can instead use the midpoint rule to define another estimator

$$
\tilde{\theta}_{n,T,A} := \left[\sum_{i=1}^{n} \left(\frac{X_{t_{i-1}} + X_{t_i}}{2} \right)^{1-2\gamma} (X_{t_i} - X_{t_{i-1}}) \right.
$$
$$
\left. - \frac{1-\gamma}{2} \sum_{i=1}^{n} \left(\frac{X_{t_{i-1}} + X_{t_i}}{2} \right)^{\gamma} (t_i - t_{i-1}) \right]
$$
$$
\left[\sum_{i=1}^{n} \left(\frac{X_{t_{i-1}} + X_{t_i}}{2} \right)^{2-2\gamma} (t_i - t_{i-1}) \right]^{-1}.
$$

$$\tilde{\kappa}_{n,T,A} := \left[\sum_{i=1}^{n} \left(\frac{X_{t_{i-1}} + X_{t_i}}{2} \right)^{1-2\gamma} (X_{t_i} - X_{t_{i-1}}) \right.$$
$$\left. - \frac{1-\gamma}{2} \sum_{i=1}^{n} \left(\frac{X_{t_{i-1}} + X_{t_i}}{2} \right)^{\gamma} (t_i - t_{i-1}) \right]$$
$$\left[\sum_{i=1}^{n} \left(\frac{X_{t_{i-1}} + X_{t_i}}{2} \right)^{2-2\gamma} (t_i - t_{i-1}) \right]^{-1}.$$

One can use the Simpson's rule to define another estimator where the denominator is a convex combination of the denominators in (1.11) and (1.12)

$$\tilde{\theta}_{n,T,S} :=$$
$$\left[\frac{1}{3} \sum_{i=1}^{n} \left\{ X_{t_{i-1}}^{1-2\gamma} + 4 \left(\frac{X_{t_{i-1}} + X_{t_i}}{2} \right)^{1-2\gamma} + X_{t_i}^{1-2\gamma} \right\} (X_{t_i} - X_{t_{i-1}}) \right.$$
$$\left. - \frac{1-\gamma}{6} \sum_{i=1}^{n} \left\{ X_{t_{i-1}}^{\gamma} + 4 \left(\frac{X_{t_{i-1}} + X_{t_i}}{2} \right)^{\gamma} + X_{t_i}^{-1} \right\} (t_i - t_{i-1}) \right]$$
$$\left[\frac{1}{3} \sum_{i=1}^{n} \left\{ X_{t_{i-1}}^{2-2\gamma} + 4 \left(\frac{X_{t_{i-1}} + X_{t_i}}{2} \right)^{2-2\gamma} + X_{t_i}^{2-2\gamma} \right\} (t_i - t_{i-1}) \right]^{-1}.$$

$$\tilde{\kappa}_{n,T,S} :=$$
$$\left[\frac{1}{3} \sum_{i=1}^{n} \left\{ X_{t_{i-1}}^{1-2\gamma} + 4 \left(\frac{X_{t_{i-1}} + X_{t_i}}{2} \right)^{1-2\gamma} + X_{t_i}^{1-2\gamma} \right\} (X_{t_i} - X_{t_{i-1}}) \right.$$
$$\left. - \frac{1-\gamma}{6} \sum_{i=1}^{n} \left\{ X_{t_{i-1}}^{\gamma} + 4 \left(\frac{X_{t_{i-1}} + X_{t_i}}{2} \right)^{\gamma} + X_{t_i}^{-1} \right\} (t_i - t_{i-1}) \right]$$
$$\left[\frac{1}{3} \sum_{i=1}^{n} \left\{ X_{t_{i-1}}^{2-2\gamma} + 4 \left(\frac{X_{t_{i-1}} + X_{t_i}}{2} \right)^{2-2\gamma} + X_{t_i}^{2-2\gamma} \right\} (t_i - t_{i-1}) \right]^{-1}.$$

In general, one can generalize Simpson's rule for any $0 \leq \alpha \leq 1$ as

$$\tilde{\theta}_{n,T,GS} :=$$
$$\left[\sum_{i=1}^{n} \left\{ \alpha \frac{X_{t_{i-1}}^{1-2\gamma} + X_{t_i}^{1-2\gamma}}{2} + (1-\alpha) \left(\frac{X_{t_{i-1}} + X_{t_i}}{2} \right)^{1-2\gamma} \right\} (X_{t_i} - X_{t_{i-1}}) \right.$$
$$\left. - \frac{1-\gamma}{2} \sum_{i=1}^{n} \left\{ \alpha \frac{X_{t_{i-1}}^{-1} + X_{t_i}^{\gamma}}{2} + (1-\alpha) \left(\frac{X_{t_{i-1}} + X_{t_i}}{2} \right)^{\gamma} \right\} (t_i - t_{i-1}) \right]$$
$$\left[\sum_{i=1}^{n} \left\{ \alpha \frac{X_{t_{i-1}}^{2-2\gamma} + X_{t_i}^{2-2\gamma}}{2} + (1-\alpha) \left(\frac{X_{t_{i-1}} + X_{t_i}}{2} \right)^{2-2\gamma} \right\} (t_i - t_{i-1}) \right]^{-1}$$

$\tilde{\kappa}_{n,T,GS} :=$

$$\left[\sum_{i=1}^{n} \left\{ \alpha \frac{X_{t_{i-1}}^{1-2\gamma} + X_{t_i}^{1-2\gamma}}{2} + (1-\alpha) \left(\frac{X_{t_{i-1}} + X_{t_i}}{2} \right)^{1-2\gamma} \right\} (X_{t_i} - X_{t_{i-1}}) \right.$$

$$\left. - \frac{1-\gamma}{2} \sum_{i=1}^{n} \left\{ \alpha \frac{X_{t_{i-1}}^{\gamma} + X_{t_i}^{\gamma}}{2} + (1-\alpha) \left(\frac{X_{t_{i-1}} + X_{t_i}}{2} \right)^{\gamma} \right\} (t_i - t_{i-1}) \right]$$

$$\left[\sum_{i=1}^{n} \left\{ \alpha \frac{X_{t_{i-1}}^{2-2\gamma} + X_{t_i}^{2-2\gamma}}{2} + (1-\alpha) \left(\frac{X_{t_{i-1}} + X_{t_i}}{2} \right)^{2-2\gamma} \right\} (t_i - t_{i-1}) \right]^{-1}$$

The case $\alpha = 0$ produces the estimator (1.18). The case $\alpha = 1$ produces the estimator (1.17). The case $\alpha = \frac{1}{3}$ produces the estimator (1.19).

We propose a very general form of the quadrature based estimator as

$$\tilde{\kappa}_{n,T,Q} := \left[\sum_{i=1}^{n} \sum_{j=1}^{m} \left[(1-s_j)X_{t_{i-1}} + s_j X_{t_i} \right]^{1-2\gamma} p_j \, (X_{t_i} - X_{t_{i-1}}) \right.$$

$$\left. - \frac{1-\gamma}{2} \left\{ \sum_{i=1}^{n} \sum_{j=1}^{m} \left[(1-s_j)X_{t_{i-1}} + s_j X_{t_i} \right]^{\gamma} p_j \right\} (t_i - t_{i-1}) \right]$$

$$\left[\left\{ \sum_{i=1}^{n} \sum_{j=1}^{m} \left[(1-s_j)X_{t_{i-1}} + s_j X_{t_i} \right]^{1-2\gamma} p_j \right\} (t_i - t_{i-1}) \right]^{-1}$$

where p_j, $j \in \{1, 2, \cdots, m\}$ is a probability mass function of a discrete random variable S on $0 \le s_1 < s_2 < \cdots < s_m \le 1$ with $P(S = s_j) := p_j$, $j \in \{1, 2, \cdots, m\}$. Denote the k-th moment of the random variable S as $\mu_k := \sum_{j=1}^{m} s_j^k p_j$, $k = 1, 2, \cdots$.

Now using the distributions in section 7.2, one can obtain several higher order estimators.

7.7 Summary of Truncated Distributions

Itô's Distribution
Support: $s_1 = 0$
Probability: $p_1 = 1$
Moment: $\mu_1 = 0$
Rate: $\nu = 1$

McKean's Distribution
Support: $s_1 = 1$
Probability: $p_1 = 1$
Moment: $\mu_1 = 1$
Rate: $\nu = 1$

Distribution 1
Support: $(s_1, s_2) = (0, 1)$
Probability: $(p_1, p_2) = (\frac{1}{2}, \frac{1}{2})$
Moment: $(\mu_1, \mu_2) = (\frac{1}{2}, \frac{1}{4})$
Rate: $\nu = 2$

Distribution 2
Support: $s_1 = \frac{1}{2}$
Probability: $p_1 = 1$
Moment: $(\mu_1, \mu_2) = (\frac{1}{2}, \frac{1}{2})$
Rate: $\nu = 2$

Distribution 3
Support: $(s_1, s_2) = (0, \frac{2}{3})$
Probability: $(p_1, p_2) = (\frac{1}{4}, \frac{3}{4})$
Moment: $(\mu_1, \mu_2, \mu_3) = (\frac{1}{2}, \frac{1}{3}, \frac{2}{9})$
Rate: $\nu = 3$

Distribution 4
Support: $(s_1, s_2) = (\frac{1}{3}, 1)$
Probability: $(p_1, p_2) = \frac{3}{4}, \frac{1}{4}$
Moment: $(\mu_1, \mu_2, \mu_3) = (\frac{1}{2}, \frac{1}{3}, \frac{10}{36})$
Rate: $\nu = 3$

Distribution 5
Support: $(s_1, s_2, s_3) = (0, \frac{1}{2}, 1)$
Probability: $(p_1, p_2, p_3) = (\frac{1}{6}, \frac{2}{3}, \frac{1}{6})$
Moment: $(\mu_1, \mu_2, \mu_3, \mu_4) = (\frac{1}{2}, \frac{1}{3}, \frac{1}{4}, \frac{5}{24})$
Rate: $\nu = 4$

Distribution 6
Support: $(s_1, s_2, s_3, s_4) = (0, \frac{1}{3}, \frac{2}{3}, 1)$
Probability: $(p_1, p_2, p_3, p_4) = (\frac{1}{8}, \frac{3}{8}, \frac{3}{8}, \frac{1}{8})$
Moment: $(\mu_1, \mu_2, \mu_3, \mu_4) = (\frac{1}{2}, \frac{1}{3}, \frac{1}{4}, \frac{11}{54})$
Rate: $\nu = 4$

Distribution 7
Support: $(s_1, s_2, s_3, s_4, s_5) = (0, \frac{1}{5}, \frac{2}{5}, \frac{3}{5}, \frac{4}{5}, 1)$
Probability: $(p_1, p_2, p_3, p_4, p_5) = (\frac{1471}{24192}, \frac{6925}{24192}, \frac{1475}{12096}, \frac{2725}{12096}, \frac{5675}{24192}, \frac{1721}{24192})$
Moment: $(\mu_1, \mu_2, \mu_3, \mu_4, \mu_5, \mu_6) = (\frac{1}{2}, \frac{1}{3}, \frac{1}{4}, \frac{1}{5}, \frac{841}{5040})$.
Rate: $\nu = 5$

Distribution 8

Support: $(s_1, s_2, s_3, s_4, s_5) = (0, \frac{1}{4}, \frac{1}{2}, \frac{3}{4}, 1)$

Probability: $(p_1, p_2, p_3, p_4, p_5) = (\frac{7}{90}, \frac{16}{45}, \frac{2}{15}, \frac{16}{45}, \frac{7}{90})$

Moment: $(\mu_1, \mu_2, \mu_3, \mu_4, \mu_5, \mu_6) = (\frac{1}{2}, \frac{1}{3}, \frac{1}{4}, \frac{1}{5}, \frac{1}{6}, \frac{110}{768})$

Rate: $\nu = 6$

Distribution 9

Support: $(s_1, s_2, s_3, s_4, s_5) = (0, \frac{1}{5}, \frac{2}{5}, \frac{3}{5}, \frac{4}{5}, 1)$

Probability: $(p_1, p_2, p_3, p_4, p_5) = (\frac{19}{288}, \frac{75}{288}, \frac{50}{288}, \frac{50}{288}, \frac{75}{288}, \frac{19}{288})$

Moment: $(\mu_1, \mu_2, \mu_3, \mu_4, \mu_5, \mu_6) = (\frac{1}{2}, \frac{1}{3}, \frac{1}{4}, \frac{1}{5}, \frac{1}{6}, \frac{3219}{22500})$

Rate: $\nu = 6$

8

Rates of Weak Convergence of Estimators in the Ornstein-Uhlenbeck Process

8.1 Introduction

In Chapter 7 we studied the rates of convergence of two approximate maximum likelihood estimators (AMLEs) to the continuous MLE of the parameter appearing linearly in the drift coefficient of Itô SDE when T remains fixed, T being the length of the observation time interval $[0, T]$. As far as we know no results are known about the rate of convergence of AMLEs to the true value of the parameter as $T \to \infty$. In this Chapter for the Ornstein-Uhlenbeck process, we obtain Berry-Esseen type bounds using both random and nonrandom normings of the two AMLEs $\theta_{n,T}$ and $\theta_{n,T,1}$ defined in (2.23) and (2.26) of Chapter 7. First we identify one AMLE $\theta_{n,T}$ as the conditional least squares estimator (CLSE). We show that the other AMLE $\theta_{n,T,1}$ has sharper Berry-Esseen bound than than that for $\theta_{n,T}$. The Berry-Esseen bounds of the AMLEs $\theta_{n,T}$ and $\theta_{n,T,1}$ show that the *rapidly increasing experimental design* (RIED) condition i.e. $T \to \infty$ and $\frac{T}{n^{1/2}} \to 0$ is not essential for its asymptotic normality AMLEs. Then we study the Berry-Esseen bounds for approximate minimum contrast estimators (AMCEs) and approximate Bayes estimators (ABEs) under the same sampling desing conditions.

The Chapter is organised as follows: In Section 8.2 we prepare notations and preliminaries. Section 8.3 contains Berry-Esseen type bounds for one AMLE. Section 8.4 contains Berry-Esseen type bounds for another AMLE. Section 8.5 contains Berry-Esseen type bounds for another AMCEs. Section 8.6 contains Berry-Esseen type bounds for another ABEs. We use purely non-random and various random normings (both sample dependent and parameter dependent)in order to obtain Berry-Esseen bounds. We find probabilistic bound on the difference between the discrete and the corresponding continuous estimators.

This Chapter is adapted from Bishwal and Bose (2001), Bishwal (2006a) and Bishwal (2006).

8.2 Notations and Preliminaries

Let $\{X_t, t \geq 0\}$ be the Ornstein-Uhlenbeck process satisfying the stochastic differential equation

$$dX_t = \theta X_t dt + dW_t, \; t \geq 0, X_0 = 0 \tag{2.1}$$

where $\{W_t, \; t \geq 0\}$ is a standard Wiener process and let $\theta \in (-\infty, 0)$ be the unknown parameter to be estimated from the observations of the process $\{X_t, t \geq 0\}$.

Recall that based on continuous observation of $\{X_t\}$ on $[0, T]$ the log-likelihood function and the maximum likelihood estimator (MLE) are given respectively by

$$L_T = \theta \int_0^T X_t dX_t - \frac{\theta^2}{2} \int_0^T X_t^2 dt \tag{2.2}$$

and

$$\theta_T = \frac{\int_0^T X_t dX_t}{\int_0^T X_t^2 dt.} \tag{2.3}$$

Large deviation probability bound and Berry-Esseen bounds for θ_T using different normings are obtained in Chapter 1.

We assume that the process $\{X_t\}$ is observed at the points $0 \leq t_0 < t_1 < \ldots < t_n = T$ with $\Delta t_i = t_i - t_{i-1} = \frac{T}{n}, i = 1, 2, \ldots, n$. For simplicity only we assume equidistant time points.

The conditional least squares estimator (CLSE) based on $X_{t_0}, \ldots, X_{t_n}$ is defined as

$$\theta_{n,T} := \arg\inf_{\theta} \sum_{i=1}^n \frac{[X_{t_i} - X_{t_{i-1}} - \theta X_{t_{i-1}} \Delta t_i]^2}{\Delta t_i} \tag{2.4}$$

and is given by

$$\theta_{n,T} = \frac{\displaystyle\sum_{i=1}^n X_{t_{i-1}}(X_{t_i} - X_{t_{i-1}})}{\displaystyle\sum_{i=1}^n X_{t_{i-1}}^2 \Delta t_i}. \tag{2.5}$$

Note that the estimator $\theta_{n,T}$ may be viewed as an approximate maximum likelihood estimator (AMLE1) which maximizes the approximate log-likelihood given by

$$L_{n,T}(\theta) = \theta \sum_{i=1}^n X_{t_{i-1}}(X_{t_i} - X_{t_{i-1}}) - \frac{\theta^2}{2} \sum_{i=1}^n X_{t_{i-1}}^2 \Delta t_i. \tag{2.6}$$

$L_{n,T}$ is obtained by an Itô approximation of the stochastic integral and rectangular rule approximation of the ordinary integral in L_T.

Le Breton (1976) studied the convergence of the estimator $\theta_{n,T}$ to θ_T as $n \to \infty$ and T remains fixed. In particular, he showed that $|\theta_{n,T} - \theta_T| = 0_P(\frac{T^2}{n})^{1/2}$. Dorogovcev (1976) and Kasonga (1988) respectively proved the weak and strong consistency of the estimator $\theta_{n,T}$ as $T \to \infty$ and $\frac{T}{n} \to 0$. Under the more restrictive conditions $T \to \infty$ and $\frac{T}{n^{1/2}} \to 0$, called the *rapidly increasing experimental design* (RIED) condition, Prakasa Rao (1983) proved the asymptotic normality and asymptotic efficiency of the estimator $\theta_{n,T}$.

Using Itô formula for the stochastic integral and rectangular rule approximation for the ordinary integral in (2.2) we obtain the approximate likelihood

$$L_{n,T,1}(\theta) = \frac{\theta}{2}(X_T^2 - T) - \frac{\theta^2}{2}\sum_{i=1}^{n} X_{t_{i-1}}^2 \Delta t_i \tag{2.7}$$

Note that if we transform the Itô integral to the FS integral in L_T using (2.20) of Chapter 4 and then apply FS type approximation for the FS integral and rectangale rule approximation for the ordinary in L_T, then also we obtain the approximate likelihood $L_{n,T,1}$.

Maximizing $L_{n,T,1}$ provides another approximate maximum likelihood estimate (AMLE2) $\theta_{n,T,1}$ given by

$$\theta_{n,T,1} = \frac{\frac{1}{2}(X_T^2 - T)}{\sum_{i=1}^{n} X_{t_{i-1}}^2 \Delta t_i}. \tag{2.8}$$

In this Chapter, we obtain the Berry-Esseen bounds, i.e., rates of convergence to normality of the two estimators given in (2.5) and (2.8).

In Section 8.3 we obtain the Berry-Esseen bounds the estimator $\theta_{n,T}$. Using purely non-random and various random normings (both sample dependent and parameter dependent) we obtain different Berry-Esseen bounds for the estimator $\theta_{n,T}$. Then we obtain probabilistic bounds on $|\theta_{n,T} - \theta_T|$.

In Section 8.4 we derive Berry-Esseen bounds for the estimators $\theta_{n,T,1}$ using purely random and various random normings (both sample dependent and parameter dependent). Then we find probabilistic bounds on $|\theta_{n,T,1} - \theta_T|$.

We use the following notations: $\Phi(\cdot)$ denotes the standard normal distribution function. C is a generic constant (which may depend on the parameter) throughout the Chapter. P and E denote probability and expectation under the true value of the parameter.

Introduce the following notations:

$$Y_{n,T} := \sum_{i=1}^{n} X(t_{i-1})\left[W_{t_i} - W(t_{i-1})\right], \ Y_T := \int_0^T X_t dW_t,$$

$$Z_{n,T} := \sum_{i=1}^{n} X(t_{i-1})\left[X_{t_i} - X(t_{i-1})\right], \ Z_T := \int_0^T X_t dX_t,$$

$$I_{n,T} := \sum_{i=1}^{n} X^2(t_{i-1})(t_i - t_{i-1}), \ I_T := \int_0^T X_t^2 dt,$$

$$V_{n,T} := \sum_{i=1}^{n} \int_{t_{i-1}}^{t_i} X_{t_{i-1}} \left[X_t - X_{t_{i-1}} \right] dt.$$

We shall use the following lemma in the sequel whose proof is elementary.

Lemma 2.1 *Let* Q_n, R_n, Q *and* R *be random variables on the same prob. space* $(\Omega, \mathcal{F}, P)$ *with* $P(R > 0) > 0$. *Suppose* $|Q_n - Q| = O_P(\delta_{1n})$ *and* $|R_n - R| = O_P(\delta_{1n})$ *where* $\delta_{1n}, \delta_{2n} \to 0$ *as* $n \to \infty$. *Then*

$$\left| \frac{Q_n}{Q} - \frac{R_n}{R} \right| = O_P(\delta_{1n} \bigvee \delta_{2n}).$$

8.3 Berry-Esseen Type Bounds for AMLE1

We will use the following preliminary lemma in the sequel.
Lemma 3.1 *(Wick's Lemma)*
Let $(\xi_1, \xi_2, \xi_3, \xi_4)$ *be a Gaussian random vector with zero mean. Then*

$$E(\xi_1\xi_2\xi_3\xi_4) = E(\xi_1\xi_2)E(\xi_3\xi_4) + E(\xi_1\xi_3)E(\xi_2\xi_4) + E(\xi_1\xi_4)E(\xi_2\xi_3).$$

Lemma 3.2

(a) $E\,|Y_{n,T} - Y_T|^2 = O(\dfrac{T^2}{n})$,

(b) $E\,|Z_{n,T} - Z_T|^2 = O(\dfrac{T^2}{n})$,

(c) $E\,|I_{n,T} - I_T|^2 = O(\dfrac{T^4}{n^2})$.

Proof. Let $g_i(t) = X_{t_{i-1}} - X_t$ for $t_{i-1} \le t < t_i$, $i = 1, 2, \ldots, n$. Since

$$E|X_{t_{i-1}} - X_t|^{2k} \le C(t_{i-1} - t)^k, k = 1, 2, \ldots \qquad (3.1)$$

(by (3.11) of Chapter 4), hence

$$E|Y_{n,T} - Y_T|^2$$

$$= E|\sum_{i=1}^{n} X_{t_{i-1}}[W_{t_i} - W_{t_{i-1}}] - \int_0^T X_t dW_t|^2$$

$$= E|\int_0^T g_i(t)dW_t|^2$$

$$= \int_0^T E(g_i^2(t))dt$$

$$\le C\sum_{i=1}^{n} \int_{t_{i-1}}^{t_i} |t_{i-1} - t|dt$$

$$= Cn\frac{(t_i - t_{i-1})^2}{2} = C\frac{T^2}{n}.$$

This completes the proof of (a).

Note that (b) and (c) are given in Le Breton (1976, Lemma 6). However, we give a complete proof since Le Breton's proofs have some technical errors.

Using (2.1) and the fact that

$$X_{t_i} - X_{t_{i-1}} = \int_{t_{i-1}}^{t_i} \theta X_t dt + W_{t_i} - W_{t_{i-1}}$$

we obtain

$$E|Z_{n,T} - Z_T|^2$$

$$= E|\sum_{i=1}^{n} X_{t_{i-1}}[X_{t_i} - X_{t_{i-1}}] - \int_0^T X_t dW_t|^2$$

$$= E|\sum_{i=1}^{n} \int_{t_{i-1}}^{t_i} \theta X_t X_{t_{i-1}} dt + \sum_{i=1}^{n} X_{t_{i-1}}[W_{t_i} - W_{t_{i-1}}]$$

$$- \int_0^T \theta X_t^2 dt - \int_0^T X_t dW_t|^2$$

$$\leq 2E|\sum_{i=1}^{n} X_{t_{i-1}}[W_{t_i} - W_{t_{i-1}}] - \int_0^T X_t dW_t|^2$$

$$+ 2\theta^2 E|\sum_{i=1}^{n} \int_{t_{i-1}}^{t_i} X_t[X_{t_{i-1}} - X_t] dt|^2.$$

$$=: N_1 + N_2.$$

N_1 is $O(\frac{T^2}{n})$ by Lemma 3.2(a). To estimate N_2 let $\psi_i(t) := X_t[X_{t_{i-1}} - X_t]$ for $t_{i-1} \leq t < t_i$, $i = 1, 2, \ldots, n$. Then

$$E|\sum_{i=1}^{n} \int_{t_{i-1}}^{t_i} \psi_i(t) dt|^2$$

$$= \sum_{i=1}^{n} E|\int_{t_{i-1}}^{t_i} \psi_i(t) dt|^2 + 2 \sum_{i,j=1,i<j}^{n} E\left[\int_{t_{i-1}}^{t_i} \psi_i(t) dt \int_{t_{i-1}}^{t_i} \psi_j(s) ds\right]$$

$$=: M_1 + M_2.$$

By the boundedness of $E(X_t^4)$ and (2.1) we have

$$E(\psi_i^2(t))$$

$$= E\{X_t^2[X_{t_{i-1}} - X_t]^2\}$$

$$\leq \{E(X_t^4)\}^{1/2}\{E[X_{t_{i-1}} - X_t]^4\}^{1/2}$$

$$\leq C(t_{i-1} - t). \tag{3.2}$$

Note that

$$M_1 = \sum_{i=1}^{n} E | \int_{t_{i-1}}^{t_i} \psi_i(t) dt |^2$$

$$\leq \sum_{i=1}^{n} (t_i - t_{i-1}) \int_{t_{i-1}}^{t_i} E(\psi^2(t)) dt$$

$$\leq C \frac{T}{n} \sum_{i=1}^{n} \int_{t_{i-1}}^{t_i} (t - t_{i-1}) dt$$

$$\leq C \frac{T}{n} \sum_{i=1}^{n} (t_i - t_{i-1})^2 = C \frac{T^3}{n^2}$$

and

$$M_2 = 2 \sum_{i,j=1,i<j}^{n} E \int_{t_{i-1}}^{t_i} \int_{t_{j-1}}^{t_i} [\psi_i(t) \psi_j(s)] dt ds$$

$$= 2 \sum_{i,j=1,i<j}^{n} \int_{t_{i-1}}^{t_i} \int_{t_{j-1}}^{t_j} E[\psi_i(t) \psi_j(s)] dt ds.$$

By Lemma 3.1, we have

$$E[\psi_i(t) \psi_j(s)]$$
$$= E[X_t(X_{t_{i-1}} - X_t) X_s(X_{t_{j-1}} - X_s)]$$
$$= E[X_t(X_{t_{i-1}} - X_t)] E[X_s(X_{t_{j-1}} - X_s)]$$
$$+ E[X_t X_s] E[(X_{t_{i-1}} - X_t)(X_{t_{j-1}} - X_s)]$$
$$+ E[X_t(X(t_{j-1}) - X_s)] E[X_s(X_{t_{i-1}} - X_t)]$$
$$=: A_1 + A_2 + A_3.$$

Note that
$$X_t = \int_0^t e^{\theta(t-u)} dW_u, t \geq 0.$$

Let $a := e^\theta$. For $s \geq t$, we have

$$E(X_t X_s)$$
$$= E(\int_0^t e^{\theta(t-u)} dW_u)(\int_0^s e^{\theta(s-u)} dW_u)$$
$$= \int_0^t e^{\theta(t+s-2u)} du$$
$$= \frac{1}{2\theta} [a^{s+t} - a^{s-t}]$$

Observe that

$$
E(X_t - X_{t_{i-1}})(X_s - X_{t_{j-1}})
$$
$$
= E(X_t X_s) - E(X_t X_{t_{j-1}}) - E(X_{t_{i-1}} X_s) + E(X_{t_{i-1}} X_{t_{j-1}})
$$
$$
= \frac{1}{2\theta}(a^s - a^{t_{j-1}})[(a^t - a^{t_{i-1}}) + (a^{-t_{i-1}} - a^{-t})]
$$
$$
= \frac{1}{2\theta}(s - t_{j-1})a^{t^*}[(t - t_{i-1}a^{t^{**}} + (t - t_{i-1})a^{-t^{***}}]
$$
$$
(\text{where } t_{j-1} < t^* < s, t_{i-1} < t^{**}, t^{***} < t)
$$
$$
\leq \frac{1}{2\theta}(s - t_{j-1})a^t(t - t_{i-1})a^{t_{i-1}} + (s - t_{j-1})a^t(t - t_{i-1})a^{-t}]
$$
$$
\leq C(s - t_{j-1})(t - t_{i-1}).
$$

Thus $A_2 \leq C(s - t_{j-1})(t - t_{i-1})$ since $|E(X_t X_s)|$ is bounded. Next

$$
|E[X_t(X_{t_{i-1}} - X_t)]|
$$
$$
= \frac{1}{2|\theta|}[a^{t+t_{i-1}} - a^{t-t_{i-1}} - a^{2t} + 1]
$$
$$
= \frac{1}{2|\theta|}a^t[a^{t_{i-1}} - a^{-t_{i-1}} - a^t + a^{-t}]
$$
$$
\leq \frac{1}{2|\theta|}a^t(t - t_{i-1})[a^{t_{i-1}} + a^{-t}]
$$
$$
\leq C(t - t_{i-1})
$$

and

$$
|E[X_s(X_s - X_{t_{j-1}})]|
$$
$$
= \frac{1}{2|\theta|}[a^{2s} - 1 - a^{s+t_{j-1}} + a^{s+t_{j-1}}]
$$
$$
= \frac{1}{2|\theta|}a^s[a^s - a^{-s} - a^{t_{j-1}} + a^{-t_{j-1}}]
$$
$$
\leq \frac{1}{2|\theta|}a^s(s - t_{j-1})[a^{t_{j-1}} + a^{-s}]
$$
$$
\leq C(s - t_{j-1}).
$$

Thus $A_1 \leq C(s - t_{j-1})(t - t_{i-1})$.
    Next

$$
|E[X_t(X_s - X_{t_{j-1}})]|
$$
$$
= \frac{1}{2|\theta|}[a^{s+t} - a^{s-t} - a^{t+t_{j-1}} + a^{t_{j-1}-t}]
$$
$$
= \frac{1}{2|\theta|}a^t(a^s - a^{t_{j-1}})
$$

$$\leq \frac{1}{2|\theta|}a^t(1 - a^{-2t})(s - t_{j-1})a^t$$

$$\leq (a^{2t} - 1)(s - t_{j-1})$$

$$\leq C(s - t_{j-1})$$

and

$$|E[X_s(X_t - X_{t_{i-1}})]|$$

$$= \frac{1}{2|\theta|}[a^{t+s} - a^{s-t} - a^{s+t_{i-1}} + a^{s-t_{i-1}}]$$

$$= \frac{1}{2|\theta|}a^s[a^t - a^{-t} - a^{t_{i-1}} + a^{-t_{i-1}}]$$

$$\leq \frac{1}{2|\theta|}a^s(t - t_{i-1})[a^{t_{i-1}} + a^{-t}]$$

$$\leq C(t - t_{i-1}).$$

Thus $A_3 \leq C(s - t_{j-1})(t - t_{i-1})$.
Hence $E[f_i(t)f_j(s)] \leq C(s - t_{j-1})(t - t_{i-1})$.
Thus

$$M_2 = 2 \sum_{i,j=1,i<j}^{n} \int_{t_{i-1}}^{t_i} \int_{t_{j-1}}^{t_j} E[f_i(t)f_j(s)]dtds$$

$$\leq C \sum_{i,j=1,i<j}^{n} \int_{t_{i-1}}^{t_i} \int_{t_{j-1}}^{t_j} (t - t_{i-1})(s - t_{j-1})dtds$$

$$= C \sum_{i,j=1,i<j}^{n} (t_{i-1} - t_i)^2(t_{j-1} - t_i)^2$$

$$= Cn^2(\frac{T}{n})^4 = C\frac{T^4}{n^2}.$$

Hence, N_2 is $O(\frac{T^3}{n^2})$. Combining N_1 and N_2 completes the proof of (b). We next prove (c). Let $\chi_i(t) := X_{t_{i-1}}^2 - X_t^2, t_{i-1} \leq t < t_i, i = 1, 2, \ldots, n$. Then

$$E|I_{n,T} - I_t|^2$$

$$= E|\sum_{i=1}^{n} X_{t_{i-1}}^2(t_i - t_{i-1}) - \int_0^T X_t^2 dt|^2$$

$$= E|\sum_{i=1}^{n} \int_{t_{i-1}}^{t_i} [X_{t_{i-1}}^2 - X_t^2]dt|^2$$

$$= E|\sum_{i=1}^{n} \int_{t_{i-1}}^{t_i} \chi_i(t)dt|^2$$

$$= \sum_{i=1}^{n} E|\int_{t_{i-1}}^{t_i} \chi_i(t)dt|^2 + 2 \sum_{i,j=1,i<j}^{n} E\int_{t_{i-1}}^{t_i} \int_{t_{j-1}}^{t_j} \chi_i(t)\chi_j(s)dtds$$

$$=: B_1 + B_2.$$

$$
\begin{aligned}
E\chi_i^2(t) &= E[X_{t_{i-1}}^2 - X_t^2]^2 \\
&= E[X_{t_{i-1}} - X_t]^2[X_{t_{i-1}} + X_t]^2 \\
&\leq \{E[X_{t_{i-1}} - X_t]^4\}^{1/2}\{\{E[X_{t_{i-1}} + X_t]^4\}^{1/2} \\
&\leq C(t - t_{i-1})
\end{aligned}
$$

(by (3.1) and the boundedness of the second term)

$$
\begin{aligned}
B_1 &= \sum_{i=1}^n E|\int_{t_{i-1}}^{t_i} \chi_i(t)dt|^2 \\
&\leq \sum_{i=1}^n (t_i - t_{i-1}) \int_{t_{i-1}}^{t_i} E(\chi_i^2(t))dt \\
&\leq C\frac{T}{n}\sum_{i=1}^n \int_{t_{i-1}}^{t_i} (t - t_{i-1})dt \\
&= C\frac{T^3}{n^2}.
\end{aligned}
$$

Note that

$$
\begin{aligned}
&E[\chi_i(t)\chi_j(s)] \\
&= E(X_{t_{i-1}}^2 - X_t^2)(X_{t_{j-1}}^2 - X_s^2) \\
&= E(X_{t_{i-1}} - X_t)(X_{t_{i-1}} + X_t)(X_{t_{j-1}} - X_s)(X_{t_{j-1}} + X_s).
\end{aligned}
$$

Now using Lemma 3.1 and proceeding similar to the estimation of M_2 it is easy to see that

$$B_2 \leq C\frac{T^4}{n^2}.$$

Combining B_1 and B_2, (c) follows.   □

Theorem 3.3 *Let* $\alpha_{n,T} := \max(T^{-1/2}(\log T)^{1/2}, \frac{T^2}{n}(\log T)^{-1}, \frac{T^4}{n^2}(\log T)^{-1})$. *We have,*

(a) $\sup_{x\in\mathbb{R}} \left| P\left\{ \left(\frac{T}{-2\theta}\right)^{1/2} (\theta_{n,T} - \theta) \leq x \right\} - \Phi(x) \right| = O(\alpha_{n,T})$.

(b) $\sup_{x\in\mathbb{R}} \left| P\left\{ I_{n,T}^{1/2}(\theta_{n,T} - \theta) \leq x \right\} - \Phi(x) \right| = O(\alpha_{n,T})$.

(c) $\sup_{x\in\mathbb{R}} \left| P\left\{ \left(\frac{T}{|2\theta_{n,T}|}\right)^{1/2} (\theta_{n,T} - \theta) \leq x \right\} - \Phi(x) \right| = O(\alpha_{n,T})$.

Proof: (a)   It is easy to see that

$$\theta_{n,T} - \theta = \frac{Y_{n,T}}{I_{n,T}} + \theta \frac{V_{n,T}}{I_{n,T}} \tag{3.3}$$

Hence

$$\sup_{x \in \mathbb{R}} \left| P \left\{ \left(\frac{T}{-2\theta} \right)^{1/2} (\theta_{n,T} - \theta) \le x \right\} - \Phi(x) \right|$$

$$= \sup_{x \in \mathbb{R}} \left| P \left\{ \left(\frac{T}{-2\theta} \right)^{1/2} \frac{Y_{n,T}}{I_{n,T}} + \left(\frac{T}{-2\theta} \right)^{1/2} \theta \frac{V_{n,T}}{I_{n,T}} \le x \right\} - \Phi(x) \right|$$

$$\le \sup_{x \in \mathbb{R}} \left| P \left\{ \left(\frac{T}{-2\theta} \right)^{1/2} \frac{Y_{n,T}}{I_{n,T}} \le x \right\} - \Phi(x) \right|$$

$$+ P \left\{ \left| \left(\frac{T}{-2\theta} \right)^{1/2} \frac{V_{n,T}}{I_{n,T}} \right| > \epsilon \right\} + \epsilon.$$

$$=: K_1 + K_2 + \epsilon. \tag{3.4}$$

Note that by Lemma 2.1.1 (b)

$$K_1 = \sup_{x \in \mathbb{R}} \left| P \left\{ \left(\frac{T}{-2\theta} \right)^{1/2} \frac{Y_{n,T}}{I_{n,T}} \le x \right\} - \Phi(x) \right|$$

$$= \sup_{x \in \mathbb{R}} \left| P \left\{ \frac{\left(-\frac{2\theta}{T} \right)^{1/2} Y_{n,T}}{\left(\frac{-2\theta}{T} \right) I_{n,T}} \le x \right\} - \Phi(x) \right| \tag{3.5}$$

$$\le \sup_{x \in \mathbb{R}} \left| P \left\{ \left(\frac{-2\theta}{T} \right)^{1/2} Y_{n,T} \le x \right\} - \Phi(x) \right|$$

$$+ P \left\{ \left(\frac{-2\theta}{T} \right) I_{n,T} - 1 > \epsilon \right\} + \epsilon$$

$$=: J_1 + J_2 + \epsilon.$$

$$J_1 = \sup_{x \in \mathbb{R}} \left| P \left\{ \left(\frac{-2\theta}{T} \right)^{1/2} (Y_{n,T} - Y_T + Y_T) \le x \right\} - \Phi(x) \right|$$

$$\le \sup_{x \in \mathbb{R}} \left| P \left\{ \left(\frac{-2\theta}{T} \right)^{1/2} Y_T \le x \right\} - \Phi(x) \right|$$

$$+ P \left\{ \left(\frac{-2\theta}{T} \right)^{1/2} |Y_{n,T} - Y_T| > \epsilon \right\} + \epsilon$$

$$\le CT^{-1/2} + \left(\frac{-2\theta}{T} \right) \frac{E|Y_{n,T} - Y_T|^2}{\epsilon^2} + \epsilon$$

$$\le CT^{-1/2} + C \frac{T/n}{\epsilon^2} + \epsilon. \text{ (by Corollary 1.2.3(a) and Lemma 3.2(a).)} \tag{3.6}$$

$$J_2 = P\left\{\left|\left(\frac{-2\theta}{T}\right)(I_{n,T} - I_T + I_T) - 1\right| > \epsilon\right\}$$

$$\leq P\left\{\left|\left(\frac{-2\theta}{T}\right)I_T - 1\right| > \frac{\epsilon}{2}\right\} + P\left\{\left(-\frac{2\theta}{2}\right)|I_{n,T} - I_T| > \frac{\epsilon}{2}\right\}$$

$$\leq C\exp\left(\frac{T\theta}{16}\epsilon^2\right) + \frac{16\theta^2}{T^2}\frac{E|I_{n,T} - I_T|^2}{\epsilon^2}$$

$$\leq C\exp\left(\frac{T\theta}{16}\epsilon^2\right) + C\frac{T^2/n^2}{\epsilon^2}. \tag{3.8}$$

Here the bound for the first term in (3.8) comes from Lemma 2.2.4(a) and that for the second term from Lemma 3.2(c). From the proof of Lemma 3.2(b) we have

$$E|V_{n,T}|^2 \leq C\frac{T^4}{n^2} \tag{3.9}$$

Next

$$K_2 = P\left\{\left|\left(\frac{T}{-2\theta}\right)^{1/2}\theta\frac{V_{n,T}}{I_{n,T}}\right| > \epsilon\right\}$$

$$= P\left\{\left|\frac{\left(\frac{-2\theta}{T}\right)^{1/2}\theta V_{n,T}}{\left(-\frac{2\theta}{T}\right)I_{n,T}}\right| > \epsilon\right\}$$

$$= P\left\{\left|\left(\frac{-2\theta}{T}\right)^{1/2}\theta V_{n,T}\right| > \delta\right\} + P\left\{\left(-\frac{2\theta}{T}\right)I_{n,T} < \frac{\delta}{\epsilon}\right\}$$

$$\text{(where we choose } \delta = \epsilon - C\epsilon^2) \tag{3.10}$$

$$\leq P\left\{\left|\left(\frac{-2\theta}{T}\right)^{1/2}\theta V_{n,T}\right| > \delta\right\} + P\left\{\left|\left(-\frac{2\theta}{T}\right)I_{n,T} - 1\right| > \delta_1\right\}$$

$$\text{(where } \delta_1 = \frac{\epsilon - \delta}{\delta} = C\epsilon)$$

$$\leq -\frac{2\theta}{T}\theta^2\frac{E|V_{n,T}|^2}{\delta^2} + C\exp\left(\frac{T\theta}{16}\delta_1^2\right) + C\frac{T^2/n^2}{\delta_1^2}$$

$$\leq C\frac{T^3/n^2}{\delta^2} + C\exp\left(\frac{T\theta}{16}\delta_1^2\right) + C\frac{T^2/n^2}{\delta_1^2} \quad \text{(by (3.9) and (3.8))}.$$

Now combining bounds from J_1, J_2, K_1 and K_2, we have since $T/n \to 0$

$$\sup_{x\in\mathbb{R}}\left|P\left\{\left(\frac{T}{-2\theta}\right)^{1/2}(\theta_{n,T} - \theta) \leq x\right\} - \Phi(x)\right|$$

$$\leq CT^{-1/2} + C\exp\left(\frac{T\theta}{16}\epsilon^2\right) + C\frac{T/n}{\epsilon^2} + C\frac{T^2/n^2}{\epsilon^2} + C\frac{T^3/n^2}{\delta^2}$$

$$+ C\exp(\frac{T\theta}{16}\delta_1^2) + C(\frac{T^2/n^2}{\delta_1^2}) + \epsilon. \tag{3.11}$$

Choosing $\epsilon = CT^{-1/2}(\log T)^{1/2}$, C large, the terms of (3.11) are of the order $O(\max(T^{-1/2}(\log T)^{1/2}, \frac{T^2}{n}(\log T)^{-1}, \frac{T^4}{n^2}(\log T)^{-1}))$. This proves (a).

(b)   Using the expression (3.3), we have

$$
\sup_{x \in \mathbb{R}} \left| P\left\{ I_{n,T}^{1/2}(\theta_{n,T} - \theta) \leq x \right\} - \Phi(x) \right|
$$

$$
= \sup_{x \in \mathbb{R}} \left| P\left\{ \frac{Y_{n,T}}{I_{n,T}^{1/2}} + \theta \frac{V_{n,T}}{I_{n,T}^{1/2}} \leq x \right\} - \Phi(x) \right| \tag{3.12}
$$

$$
\leq \sup_{x \in \mathbb{R}} \left| P\left\{ \frac{Y_{n,T}}{I_{n,T}^{1/2}} \leq x \right\} - \Phi(x) \right| + P\left\{ \left| \theta \frac{V_{n,T}}{I_{n,T}^{1/2}} \right| > \epsilon \right\} + \epsilon.
$$

$$
=: H_1 + H_2 + \epsilon.
$$

Note that

$$
H_1 = \sup_{x \in \mathbb{R}} \left| P\left\{ \frac{Y_{n,T} - Y_T + Y_T}{I_{n,T}^{1/2}} \leq x \right\} - \Phi(x) \right|
$$

$$
= \sup_{x \in \mathbb{R}} \left| P\left\{ \frac{Y_T}{I_{n,T}^{1/2}} \leq x \right\} - \Phi(x) \right| + P\left\{ \frac{|Y_{n,T} - Y_T|}{I_{n,T}^{1/2}} > \epsilon \right\} + \epsilon \tag{3.13}
$$

$$
=: F_1 + F_2 + \epsilon.
$$

Now

$$
F_1 = \sup_{x \in \mathbb{R}} \left| P\left\{ \frac{Y_T}{I_{n,T}^{1/2}} \leq x \right\} - \Phi(x) \right|
$$

$$
\leq \sup_{x \in \mathbb{R}} \left| P\left\{ \left(\frac{-2\theta}{T} \right)^{1/2} Y_T \leq x \right\} - \Phi(x) \right|
$$

$$
+ P\left\{ \left| \left(\frac{-2\theta}{T} \right)^{1/2} I_{n,T}^{1/2} - 1 \right| > \epsilon \right\} + \epsilon \text{ (by Lemma 1.2.1(b))}
$$

$$
\leq C_T^{-1/2} + P\left\{ \left| \left(\frac{-2\theta}{T} \right) I_{n,T} - 1 \right| > \epsilon \right\} + \epsilon \text{ (byCorollary 1.2.3(a))}
$$

$$
\leq CT^{-1/2} + C \exp\left(-\frac{T\theta}{16} \epsilon^2 \right) + C \frac{T^2/n^2}{\epsilon^2} + \epsilon. \text{ (by (3.8))} \tag{3.14}
$$

On the other hand,

$$
F_2 = P\left\{ \frac{|Y_{n,T} - Y_T|}{I_{n,T}^{1/2}} > \epsilon \right\}
$$

$$
\leq P\left\{ \left(\frac{-2\theta}{T} \right)^{1/2} |Y_{n,T} - Y_T| > \delta \right\} + P\left\{ \left| \left(\frac{-2\theta}{T} \right)^{1/2} I_{n,T}^{1/2} - 1 \right| > \delta_1 \right\}
$$

$$
\text{(where } \delta = \epsilon - C\epsilon^2 \text{ and } \delta_1 = (\epsilon - \delta)/\epsilon > 0)
$$

$$
\leq \frac{\left(\frac{-2\theta}{T} \right) E|Y_{n,T} - Y_T|^2}{\delta^2} + P\left\{ \left| \left(\frac{-2\theta}{T} \right) I_{n,T} - 1 \right| > \delta_1 \right\}
$$

$$
\leq C\frac{T/n}{\delta^2} + C \exp\left(\frac{T\theta}{16} \delta_1^2 \right) + C \frac{T^2/n^2}{\delta_1^2} \text{ (from Lemma 3.2(a) and (3.8).)} \tag{3.15}
$$

Using (3.15) and (3.14) in (3.13), we obtain

$$
\begin{aligned}
H_1 = \sup_{x \in \mathbb{R}} \left| P\left\{ \frac{Y_T}{I_{n,T}^{1/2}} \le x \right\} - \Phi(x) \right| \\
\le CT^{-1/2} + C\exp\left(\frac{T\theta}{16}\epsilon^2\right) + C\frac{T/n}{\delta^2} + C\frac{T^2/n^2}{\delta_1^2} \\
+ C\exp\left(\frac{T\theta}{16}\delta_1^2\right) + C\frac{T^2/n^2}{\epsilon^2} + \epsilon.
\end{aligned} \tag{3.16}
$$

$$
\begin{aligned}
H_2 &= P\left\{ \left| \theta\frac{V_{n,T}}{I_{n,T}^{1/2}} \right| > \epsilon \right\} \\
&= P\left\{ \frac{\left| \left(\frac{-2\theta}{T}\right)^{1/2}\theta V_{n,T} \right|}{\left| \left(\frac{-2\theta}{T}\right)^{1/2} I_{n,T}^{1/2} \right|} > \epsilon \right\} \\
&\le P\left\{ \left| \left(\frac{-2\theta}{T}\right)^{1/2}\theta V_{n,T} \right| > \delta \right\} + P\left\{ \left| \left(\frac{-2\theta}{T}\right)^{1/2} I_{n,T}^{1/2} \right| < \delta/\epsilon \right\} \\
&\le \left(-\frac{2\theta}{T}\right)\theta^2\frac{E|V_{n,T}|^2}{\delta^2} + P\left\{ \left| \left(-\frac{2\theta}{T}\right) I_{n,T} - 1 \right| > \delta_1 \right\} \\
&\quad \text{(where } 0 < \delta < \epsilon \text{ and } \delta_1 = (\epsilon - \delta)/\epsilon = C\epsilon > 0) \\
&\le C\frac{T^3/n^2}{\delta^2} + C\exp\left(\frac{T\theta}{16}\delta_1^2\right) + C\frac{T^2/n^2}{\delta_1^2}. \quad \text{(from (3.9) and (3.8))}
\end{aligned} \tag{3.17}
$$

Using (3.17) and (3.16) in (3.12) and choosing $\epsilon = CT^{-1/2}(\log T)^{1/2}, C$ large, the terms of (3.12) are of the order $O(\max(T^{-1/2}(\log T)^{1/2}, \frac{T^2}{n}(\log T)^{-1}, \frac{T^4}{n^2}(\log T)^{-1}))$. This proves (b).

(c) Let $D_T = \{ |\theta_{n,T} - \theta| \le d_T \}$ and $d_T = CT^{-1/2}(\log T)^{1/2}$.
On the set D_T, expanding $(2|\theta_{n,T}|)^{-1/2}$, we obtain

$$
(-2\theta_{n,T})^{-1/2} = (-2\theta)^{-1/2}\left[1 - \frac{\theta - \theta_{n,T}}{\theta}\right]^{-1/2}
$$

$$
= (-2\theta)^{-1/2}\left[1 + \frac{1}{2}\left(\frac{\theta - \theta_{n,T}}{\theta}\right) + O(d_T^2)\right].
$$

Then

$$
\begin{aligned}
&\sup_{x \in \mathbb{R}} \left| P\left\{ \left(\frac{T}{2|\theta_{n,T}|}\right)^{1/2}(\theta_{n,T} - \theta) \le x \right\} - \Phi(x) \right| \\
&\le \sup_{x \in \mathbb{R}} \left| P\left\{ \left(\frac{T}{2|\theta_{n,T}|}\right)^{1/2}(\theta_{n,T} - \theta) \le x, D_T \right\} - \Phi(x) \right| + P(D_T^c).
\end{aligned} \tag{3.18}
$$

$$P(D_T^c)$$
$$= P\left\{ |\theta_{n,T} - \theta| > CT^{-1/2}(\log T)^{1/2} \right\}$$
$$= P\left\{ \left(\frac{T}{-2\theta}\right)^{1/2} |\theta_{n,T} - \theta| > C(\log T)^{1/2}(-2\theta)^{-1/2} \right\}$$
$$\leq C(\max(T^{-1/2}(\log T)^{1/2}, \frac{T^2}{n}(\log T)^{-1}, \frac{T^4}{n^2}(\log T)^{-1})$$
$$\quad + 2(1 - \Phi\left((\log T)^{1/2}(-2\theta)^{-1/2}\right) \quad \text{(by Theorem 3.3(a))}$$
$$\leq C(\max(T^{-1/2}(\log T)^{1/2}, \frac{T^2}{n}(\log T)^{-1}, \frac{T^4}{n^2}(\log T)^{-1})).$$

On the set D_T,

$$\left| \left(\frac{\theta_{n,T}}{\theta}\right)^{1/2} - 1 \right| \leq CT^{-1/2}(\log T)^{1/2}.$$

Hence upon choosing $\epsilon = CT^{-1/2}(\log T)^{1/2}, C$ large we obtain

$$\left| P\left\{ \left(\frac{T}{-2\theta_{n,T}}\right)^{1/2} (\theta_{n,T} - \theta) \leq x, D_T \right\} - \Phi(x) \right|$$
$$\leq \left| P\left\{ \left(\frac{T}{-2\theta}\right)^{1/2} (\theta_{n,T} - \theta) \leq x, D_T \right\} - \Phi(x) \right|$$
$$\quad + P\left\{ \left| \left(\frac{\theta_{n,T}}{\theta}\right)^{1/2} - 1 \right| > \epsilon, D_T \right\} + \epsilon \quad (3.20)$$
$$\text{(by Lemma 1.2.1(b))}$$
$$\leq C(\max(T^{-1/2}(\log T)^{1/2}, \frac{T^2}{n}(\log T)^{-1}, \frac{T^4}{n^2}(\log T)^{-1}))$$
$$\text{(by Theorem 3.3(a)).}$$

(c) follows from (3.18) - (3.20). □

Theorem 3.4

$$\sup_{x \in \mathbb{R}} \left| P\left\{ I_{n,T} \left(-\frac{2\theta}{T}\right)^{1/2} (\theta_{n,T} - \theta) \leq x \right\} - \Phi(x) \right| = O\left(T^{-1/2} \bigvee \left(\frac{T}{n}\right)^{1/3} \right).$$

Proof: Let $a_{n,T} := Z_{n,T} - Z_T, \quad b_{n,T} := I_{n,T} - I_T$.

By Lemma 3.2 $E|a_{n,T}|^2 = O\left(\frac{T^2}{n}\right)$ and $E|b_{n,T}|^2 = O\left(\frac{T^4}{n^2}\right).$ (3.21)

From (3.5), we have

$$I_{n,T}\theta_{n,T} = \sum_{i=1}^n X_{t_{i-1}} \left[X_{t_i} - X_{t_{i-1}} \right]$$
$$= \int_0^T X_t dX_t + a_{n,T}$$
$$= \int_0^T X_t dW_t + \theta \int_0^T X_t^2 dt + a_{n,T}.$$

Hence $I_{n,T}(\theta_{n,T} - \theta) = -\theta b_{n,T} + a_{n,T}$.

Thus

$$
\sup_{x \in \mathbb{R}} \left| P \left\{ I_{n,T} \left(-\frac{2\theta}{T} \right)^{1/2} (\theta_{n,T} - \theta) \le x \right\} - \Phi(x) \right|
$$

$$
= \sup_{x \in \mathbb{R}} \left| P \left\{ \left(-\frac{2\theta}{T} \right)^{1/2} [Y_T - \theta b_{n,T} + a_{n,T}] \le x \right\} - \Phi(x) \right|
$$

$$
\le \sup_{x \in \mathbb{R}} \left| P \left\{ \left(-\frac{2\theta}{T} \right)^{1/2} Y_T \le x \right\} - \Phi(x) \right|
$$

$$
+ P \left\{ \left| \left(-\frac{2\theta}{T} \right)^{1/2} [-\theta b_{n,T} + a_{n,T}] \right| > \epsilon \right\} + \epsilon
$$

$$
\le CT^{-1/2} + \left(-\frac{2\theta}{T} \right) \frac{E|-\theta b_{n,T} + a_{n,T}|^2}{\epsilon^2} + \epsilon
$$

$$
\le CT^{-1/2} + C \frac{T/n}{\epsilon^2} + \epsilon \text{ (by Corollary 1.2.3(a) and (3.21)).}
$$

Choosing $\epsilon = \left(\frac{T}{n} \right)^{1/3}$, the rate is $O \left(T^{-1/2} \bigvee \left(\frac{T}{n} \right)^{1/3} \right)$. $\square$

Theorem 3.5

$$
|\theta_{n,T} - \theta_T| = O_P(\frac{T^2}{n})^{1/2}.
$$

Proof: Note that $\theta_{n,T} - \theta_T = \frac{Z_{n,T}}{I_{n,T}} - \frac{Z_T}{I_T}$.

From Lemma 3.2 it follows that $|Z_{n,T} - Z_T| = O_P(\frac{T^2}{n})^{1/2}$ and $|I_{n,T} - I_T| = O_P(\frac{T^4}{n^2})^{1/2}$. Now the theorem follows easily from the from the Lemma 2.1. $\square$

8.4 Berry-Esseen Type Bounds for AMLE2

Theorem 4.1 Let $\beta_{n,T} = O \left(T^{-1/2} (\log T)^{1/2} \bigvee \frac{T^4}{n^2} (\log T)^{-1} \right)$.

(a) $\sup_{x \in \mathbb{R}} \left| P \left\{ \left(-\frac{T}{2\theta} \right)^{1/2} (\theta_{n,T,1} - \theta) \le x \right\} - \Phi(x) \right| = O(\beta_{n,T})$,

(b) $\sup_{x \in \mathbb{R}} \left| P \left\{ I_{n,T}^{1/2} (\theta_{n,T,1} - \theta) \le x \right\} - \Phi(x) \right| = O(\beta_{n,T})$,

(c) $\sup_{x \in \mathbb{R}} \left| P \left\{ \left(\frac{T}{2|\theta_{n,T,1}|} \right)^{1/2} (\theta_{n,T,1} - \theta) \le x \right\} - \Phi(x) \right| = O(\beta_{n,T})$.

Proof. (a) From (1.8), we have

$$\begin{aligned}
I_{n,T}\theta_{n,T,1} &= \frac{1}{2}(X_T^2 - T) \\
&= \int_0^T X_t dX_t \\
&= \int_0^T X_t dW_t + \theta \int_0^T X_t^2 dt \\
&= Y_T + \theta I_T.
\end{aligned}$$

Thus

$$\begin{aligned}
& \left(-\tfrac{T}{2\theta}\right)^{1/2}(\theta_{n,T,1} - \theta) \\
&= \frac{\left(-\tfrac{T}{2\theta}\right)^{1/2}Y_T + \theta(-\tfrac{T}{2\theta})^{1/2}(I_T - I_{n,T})}{I_{n,T}} \\
&= \frac{\left(-\tfrac{2\theta}{T}\right)^{1/2}Y_T + (-\tfrac{2\theta}{T})^{1/2}(I_T - I_{n,T})}{(-\tfrac{2\theta}{T})I_{n,T}}.
\end{aligned} \tag{4.1}$$

Now

$$\begin{aligned}
& \sup_{x\in\mathbb{R}} \left| P\left\{ \left(-\frac{T}{2\theta}\right)^{1/2}(\theta_{n,T,1} - \theta) \le x \right\} - \Phi(x) \right| \\
&= \sup_{x\in\mathbb{R}} \left| P\left\{ \frac{(-\tfrac{2\theta}{T})^{1/2}Y_T + (-\tfrac{2\theta}{T})^{1/2}(I_T - I_{n,T})}{(-\tfrac{2\theta}{T})I_{n,T}} \le x \right\} - \Phi(x) \right| \\
&\le \sup_{x\in\mathbb{R}} \left| P\left\{ (-\tfrac{2\theta}{T})^{1/2}Y_T \le x \right\} - \Phi(x) \right| + P\left\{ \left| \theta(-\frac{2\theta}{T})^{1/2}(I_{n,T} - I_T) \right| > \epsilon \right\} \\
& \quad + P\left\{ \left| \left(-\frac{2\theta}{T}\right)I_{n,T} - 1 \right| > \epsilon \right\} + 2\epsilon \\
&\le CT^{-1/2} + \theta^2 \frac{(-\tfrac{2\theta}{T})E\,|I_{n,T} - I_T|^2}{\epsilon^2} + C\exp(\frac{T\theta}{4}\epsilon^2) + C\frac{T^2}{n^2\epsilon^2} + 2\epsilon \tag{4.2} \\
&\le CT^{-1/2} + C\frac{T^3}{n^2\epsilon^2} + C\exp(\frac{T\theta}{4}\epsilon^2) + C\frac{T^2}{n^2\epsilon^2} + \epsilon. \tag{4.3}
\end{aligned}$$
(by Lemma 3.2 (c))

Here the bound for the 3rd term in the r.h.s. of (4.2) is from (3.8).

Choosing $\epsilon = CT^{-1/2}(\log T)^{1/2}$, the terms in the r.h.s. of (4.3) are of the order
$O(T^{-1/2}(\log T)^{1/2} \vee \frac{T^4}{n^2}(\log T)^{-1})$.

(b) From (4.1), we have

$$I_{n,T}^{1/2}(\theta_{n,T,1} - \theta) = \frac{Y_T + \theta(I_T - I_{n,T})}{I_{n,T}^{1/2}}.$$

Now

$$
\sup_{x \in \mathbb{R}} \left| P \left\{ I_{n,T}^{1/2}(\theta_{n,T,1} - \theta) \leq x \right\} - \Phi(x) \right|
$$

$$
= \sup_{x \in \mathbb{R}} \left| P \left\{ \frac{Y_T}{I_{n,T}^{1/2}} + \theta \frac{I_T - I_{n,T}}{I_{n,T}^{1/2}} \leq x \right\} - \Phi(x) \right|
$$

$$
\leq \sup_{x \in \mathbb{R}} \left| P \left\{ \frac{Y_T}{I_{n,T}^{1/2}} \leq x \right\} - \Phi(x) \right| + P \left\{ \left| \frac{\theta(I_T - I_{n,T})}{I_{n,T}^{1/2}} \right| > \epsilon \right\} + \epsilon
$$

$$
=: U_1 + U_2 + \epsilon. \tag{4.4}
$$

We have from (3.8),

$$
U_1 \leq CT^{-1/2} + C \exp(\frac{T\theta}{16}\epsilon^2) + C \frac{T^2}{n^2 \epsilon^2} + \epsilon. \tag{4.5}
$$

Now

$$
U_2 = P \left\{ |\theta| \left| \frac{I_{n,T} - I_T}{I_{n,T}^{1/2}} \right| > \epsilon \right\}
$$

$$
= P \left\{ |\theta| \frac{|(-\frac{2\theta}{T})^{1/2}(I_{n,T} - I_T)|}{|(-\frac{2\theta}{T})^{1/2} I_{n,T}^{1/2}|} > \epsilon \right\}
$$

$$
\leq P \left\{ \left| (-\frac{2\theta}{T})^{1/2} \right| |I_{n,T} - I_T| > \delta \right\} + P \left\{ \left| (-\frac{2\theta}{T})^{1/2} I_{n,T}^{1/2} - 1 \right| > \delta_1 \right\}
$$

$$
\text{(where } \delta = \epsilon - C\epsilon^2 \text{ and } \delta_1 = (\epsilon - \delta)/\epsilon > 0)
$$

$$
\leq (-\frac{2\theta}{T}) \frac{E|I_{n,T} - I_T|^2}{\delta^2} + P \left\{ \left| (-\frac{2\theta}{T}) I_{n,T} - 1 \right| > \delta_1 \right\} \tag{4.6}
$$

$$
\leq C \frac{T^3}{n^2 \delta^2} + C \exp(\frac{T\theta}{16}\delta_1^2) + C \frac{T^2/n^2}{\delta_1^2}. \tag{4.7}
$$

Here the bound for the first term in the r.h.s. of (4.6) comes from Lemma 3.2(c) and that for the second term is from J_2 in (3.8).

Now using the bounds (4.5) and (4.7) in (4.4) with $\epsilon = CT^{-1/2}(\log T)^{1/2}$, we obtain that the terms in (4.4) is of the order $O(T^{-1/2}(\log T)^{1/2} \bigvee \frac{T^4}{n^2}(\log T)^{-1})$.

(c) Let $G_T = \{|\theta_{n,T,1} - \theta| \leq d_T\}$ and $d_T = CT^{-1/2}(\log T)^{1/2}$. On the set G_T, expanding $(2|\theta_{n,T,1}|)^{1/2}$, we obtain,

$$
(-2\theta_{n,T,1})^{-1/2} = (-2\theta)^{1/2} \left[1 - \frac{\theta - \theta_{n,T,1}}{\theta} \right]^{-1/2}
$$

$$
= (-2\theta)^{1/2} [1 + \frac{1}{2}(\frac{\theta - \theta_{n,T,1}}{\theta}) + O(d_T^2).
$$

Then

$$\sup_{x\in\mathbb{R}}\left|P\left\{(\frac{T}{2|\theta_{n,T,1}|})^{1/2}(\theta_{n,T,1}-\theta)\leq x\right\}-\Phi(x)\right|$$
$$\leq \sup_{x\in\mathbb{R}}\left\{P(\frac{T}{2|\theta_{n,T,1}|})^{1/2}(\theta_{n,T,1}-\theta)\leq x,\ G_T\right\}+P(G_T^c).$$

Now

$$P(G_T^c)$$
$$= P\left\{|\theta_{n,T,1}-\theta| > CT^{-1/2}(\log T)^{1/2}\right\}$$
$$= P\left\{\left(-\frac{T}{2\theta}\right)^{1/2}|\theta_{n,T,1}-\theta| > C(\log T)^{1/2}(-2\theta)^{-1/2}\right\}$$
$$\leq C\left(T^{-1/2}(\log T)^{1/2}\bigvee\frac{T^4}{n^2}(\log T)^{-1})+2(1-\Phi\log T^{1/2}(-2\theta)^{-1/2}\right)$$
$$\text{(by Theorem 4.1(a))}$$
$$\leq C(T^{-1/2}(\log T)^{1/2}\bigvee\frac{T^4}{n^2}(\log T)^{-1}).$$

On the set G_T,

$$\left|\left(\frac{\theta_{n,T,1}}{\theta}\right)^{1/2}-1\right|\leq CT^{-1/2}(\log T)^{1/2}$$

Hence upon choosing $\epsilon = CT^{-1/2}(\log T)^{1/2}, C$ large

$$\left|P\left\{\left(\frac{T}{-2\theta_{n,T,1}}\right)^{1/2}(\theta_{n,T,1}-\theta)\leq x, G_T\right\}-\Phi(x)\right|$$
$$\leq\left|P\left\{\left(\frac{T}{-2\theta}\right)^{1/2}(\theta_{n,T,1}-\theta)\leq x, G_T\right\}\right|$$
$$+P\left\{\left|\left(\frac{\theta_{n,T}}{\theta}\right)^{1/2}-1\right| > \epsilon, G_T\right\}+\epsilon$$
$$\text{(by Lemma 1.2.1 (b))}$$
$$\leq C(T^{-1/2}(\log T)^{1/2}\bigvee\frac{T^4}{n^2}(\log T)^{-1})\ \text{(by Theorem 4.1(a))}. \qquad \square$$

Theorem 4.2

$$\sup_{x\in\mathbb{R}}|P\left\{I_{n,T}(-\frac{2\theta}{T})^{1/2}(\theta_{n,T,1}-\theta)\leq x\right\}-\Phi(x)| = O(T^{-1/2}\bigvee(\frac{T^3}{n^2})^{1/3}).$$

Proof: From (4.1) we have

$$I_{n,T}\left(-\frac{2\theta}{T}\right)^{1/2}(\theta_{n,T,1}-\theta) = (-\frac{2\theta}{T})^{1/2}Y_T+\theta(-\frac{2\theta}{T})^{1/2}(I_T-I_{n,T}).$$

Hence by Corollary 1.2.3(a) and Lemma 3.2(c)

$$\sup_{x\in\mathbb{R}} |P\left\{ I_{n,T}(-\frac{2\theta}{T})^{1/2}(\theta_{n,T,1} - \theta) \le x \right\} - \Phi(x)|$$

$$= \sup_{x\in\mathbb{R}} |P\left\{ (-\frac{2\theta}{T})^{1/2} Y_T + \theta(-\frac{2\theta}{T})^{1/2}(I_T - I_{n,T}) \le x \right\} - \Phi(x)|$$

$$\le \sup_{x\in\mathbb{R}} |P\left\{ (-\frac{2\theta}{T})^{1/2} Y_T \le x \right\} - \Phi(x)| + P\left\{ |\theta(-\frac{2\theta}{T})^{1/2}(I_T - I_{n,T})| > \epsilon \right\} + \epsilon \cdot$$

$$\le CT^{-1/2} + C\frac{E|I_T - I_{n,T}|^2}{T\epsilon^2} + \epsilon$$

$$\le CT^{-1/2} + C\frac{T^3}{n^2\epsilon^2} + \epsilon$$

Choosing $\epsilon = (\frac{T^3}{n^2})^{1/3}$, the theorem follows.   □

Theorem 4.3 $|\theta_{n,T,1} - \theta_T| = O_P(\frac{T^2}{n})$.

Proof. We have from (2.3) $\theta_T = Z_T/I_T$. By Itô formula it is easy to see that

$$\theta_{n,T,1} = Z_T/I_{n,T}$$

Hence applying Lemma 3.7 with the aid of Lemma 3.2(c) the theorem follows.
□

Remarks

(1) The bounds in Theorem 3.3, 3.4, 4.1, 4.2 are uniform over compact subsets of the parameter space.

(2) Theorems 3.3 and 3.4 are useful for testing hypothesis about θ. They do not necessarily give confidence intervals. Theorems 3.3(b) and 3.3(c) are useful for computation of a confidence interval.

(3) It may appear from Theorem 3.3 (a) that $\frac{T}{n^{1/2}} \to 0$ is essential for the asymptotic normality of $\theta_{n,T}$. However it is not the case. Consider the bound given in (3.11). For this to converge to 0, it is necessary that $\frac{T}{n} \to 0$. Let $\gamma_n = \frac{T}{n}$. It is easy to see from this bound that if there exists an ϵ_n such that $n\gamma_n\epsilon_n^2 \to \infty$ and $\gamma^{-1}\epsilon_n^2 \to \infty$, then normal convergence holds and a rate may also be found out. Consider in particular, $T = n^\tau, \tau > 0$. Then, $\gamma_n = n^{\tau-1}$. Hence we require $n^\tau \epsilon_n^2 \to \infty$ and $n^{1-\tau}\epsilon_n^2 \to \infty$. This is possible for any $\tau > 1$. Thus if $T \approx n^\tau, \tau > 1$, then normal convergence in Theorem 3.3 (a) holds. Similar comment also holds for Theorem 3.3 (b), (c) and Theorem 4.1.

(4) If $\frac{T}{n^{1/2}} \to 0$ and $T \to \infty$, then to obtain bound of the order $T^{-1/2}(\log T)^{1/2}$ in Theorem 3.3 one needs $n > T^{5/2}(\log T)^{-3/2}$. To obtain bound of the order $T^{-1/2}(\log T)^{1/2}$ in Theorem 4.1 one needs $n > T^{9/4}(\log T)^{-3/4}$. To

obtain bound of the order $T^{-1/2}$ in Theorem 3.4 one needs $n \geq T^{10}$. To obtain bound of the order $T^{-1/2}$ in Theorem 4.2 one needs $n > T^{9/4}$. To obtain bound of the order $\frac{1}{T}$ in Theorem 3.5 one needs $n \geq T$.

(5) The norming in Theorem 3.4 and 4.2 are random which also contains the unknown parameter. It would be interesting to find the order $(T^{-1/2} \bigvee (T/n)^{1/3})$ and $(T^{-1/2} \bigvee (\frac{T^3}{n^2})^{1/3})$ respectively for the estimators $\theta_{n,T}$ and $\theta_{n,T,1}$ with completely random or completely non-random norming. The problem of obtaining rates of convergence of the CLSE in the case of non-linear drift remains open.

(6) The bounds in Theorem 3.5 and Theorem 4.3 are identical with the bounds obtained in Theorem 7.5.2 (ii) and Theorem 7.5.2 (iv) respectively.

(7) The Berry-Esseen bounds for the estimator $\theta_{n,T,1}$ in Theorem 4.1 is sharper than the corresponding bound for $\theta_{n,T}$ in Theorem 3.3. The Berry-Esseen bounds for the estimator $\theta_{n,T,1}$ in Theorem 4.2 is sharper than the corresponding bound for $\theta_{n,T}$ in Theorem 3.4. Also $\theta_{n,T,1}$ converges to θ_T at a faster rate than the rate at which $\theta_{n,T}$ converges to θ_T for fixed T when $n \to \infty$. This shows that $\theta_{n,T,1}$ is a better estimator of θ than $\theta_{n,T}$. Extension of this problem to nonlinear drift case now remains open. Berry-Esseen bounds in Theorems 3.4 and 4.2 show that *rapidly increasing experimental design* condition ($T \to \infty$ and $\frac{T}{\sqrt{n}} \to 0$) is not essential for the asymptotic normality of the estimators.

(8) In this Chapter we have only studied the rates of convergence of AMLEs in the stable case i.e., when $\theta < 0$. It would be interesting to investigate the rates of convergence of AMLEs in the explosive case i.e., when $\theta > 0$ and the unstable (critical) case, i.e., when $\theta = 0$. Note that in the explosive case with nonrandom norming the limit distribution is Cauchy but with random norming the limit distribution is normal.

(9) Here we have obtained uniform rates of convergence to normality. It would be interesting to obtain nonuniform rates of convergence to normality which are more useful.

(10) Levy driven Ornstein-Uhlenbeck process is the building block in stochastic volatility modelling, see Barndorff-Neilsen and Shephard (2001). Lot of work needs to be done in this direction.

8.5 Berry-Esseen Type Bounds for Approximate Minimum Contrast Estimators

This section introduces some new approximate minimum contrast estimators of the drift parameter in the Ornstein-Uhlenbeck process based on discretely sampled data and obtains rates of weak convergence of the distributions of the estimators to the standard normal distribution using random, nonrandom and mixed normings.

The process $\{X_t\}$ is observed at times $0 = t_0 < t_1 < \cdots t_n = T$ with $t_i - t_{i-1} = \frac{T}{n}, i = 1, 2 \cdots, n$. We assume two types of high frequency data with long observation time:

1) $T \to \infty, n \to \infty, \frac{T}{\sqrt{n}} \to 0$,   2) $T \to \infty, n \to \infty, \frac{T}{n^{2/3}} \to 0$.

Let the continuous realization $\{X_t, 0 \le t \le T\}$ be denoted by X_0^T. Let P_θ^T be the measure generated on the space (C_T, B_T) of continuous functions on $[0, T]$ with the associated Borel σ-algebra B_T generated under the supremum norm by the process X_0^T and let P_0^T be the standard Wiener measure. It is well known that when θ is the true value of the parameter P_θ^T is absolutely continuous with respect to P_0^T and the Radon-Nikodym derivative (likelihood) of P_θ^T with respect to P_0^T based on X_0^T is given by

$$L_T(\theta) := \frac{dP_\theta^T}{dP_0^T}(X_0^T) = \exp\left\{\theta \int_0^T X_t dX_t - \frac{\theta^2}{2}\int_0^T X_t^2 dt\right\}. \qquad (5.1)$$

Consider the score function, the derivative of the log-likelihood function, which is given by

$$\gamma_T(\theta) := \int_0^T X_t dX_t - \theta \int_0^T X_t^2 dt. \qquad (5.2)$$

A solution of the estimating equation

$$\gamma_T(\theta) = 0 \qquad (5.3)$$

provides the conditional maximum likelihood estimate (MLE)

$$\hat\theta_T := \frac{\int_0^T X_t dX_t}{\int_0^T X_t^2 dt}. \qquad (5.4)$$

Strictly speaking, $\hat\theta_T$ is not the maximum likelihood estimate of θ since $\hat\theta_T$ may take positive values whereas the parameter θ is assumed to be strictly negative. For an exact definition of MLE, see Kutoyants (1994). Nevertheless, this definition of MLE is widely used. It is well known that $\hat\theta_T$ is strongly consistent and $T^{1/2}(\hat\theta_T - \theta)$ asymptotically $\mathcal{N}(0, -2\theta)$ distributed as $T \to \infty$ (see Kutoyants (1984)). Bishwal (2000a) obtained weak convergence bound of the order $O(T^{-1/2})$ for the MLE. Bishwal (2000b) obtained weak convergence bound of the order $O(T^{-1/2})$ for the Bayes estimators, with smooth priors and loss functions. These two papers used parameter dependent nonrandom norming. Bishwal (2001) obtained weak convergence bound of the order $O(T^{-1/2})$ for the MLE and Bayes estimators using two different random normings which are useful for computation of a confidence interval. Bishwal and Bose (2001) obtained weak convergence bound for two approximate maximum likelihood estimators (AMLEs).

Note that in the stationary case where X_0 has $\mathcal{N}(0, -1/2\theta)$ distribution, the exact log-likelihood function is given by

$$l_T(\theta) := \frac{1}{2}\log\left(-\frac{\theta}{\pi}\right) - \frac{\theta^2}{2}\int_0^T X_t^2 dt + \left(\frac{\theta}{2}\right)[X_T^2 + X_0^2] + \frac{\theta T}{2}. \qquad (5.4)$$

In this case the likelihood equation has two solutions:

$$\check{\theta}_{T,1} = \frac{[X_T^2 + X_0^2 - T] + \{[X_T^2 + X_0^2]^2 - \pi \int_0^T X_t^2 dt\}^{1/2}}{4 \int_0^T X_t^2 dt}, \tag{5.5}$$

$$\check{\theta}_{T,2} = \frac{[X_T^2 + X_0^2 - T] - \{[X_T^2 + X_0^2]^2 - \pi \int_0^T X_t^2 dt\}^{1/2}}{4 \int_0^T X_t^2 dt}. \tag{5.6}$$

As an alternative to maximum likelihood method and to obtain estimators with higher order accuracy our aim is to use contrast functions. Using Itô formula, the score function $\gamma_T(\theta)$ can be written as

$$\gamma_T(\theta) = \frac{1}{2} X_T^2 - \int_0^T (\theta X_t^2 + \frac{1}{2}) dt. \tag{5.7}$$

Consider the estimating function

$$M_T(\theta) = - \int_0^T (\theta X_t^2 + \frac{1}{2}) dt \tag{5.8}$$

and the minimum contrast estimate (MCE)

$$\tilde{\theta}_T := -\frac{T}{2} \left\{ \int_0^T X_t^2 dt \right\}^{-1}. \tag{5.9}$$

M-estimator is reduced to the minimum contrast estimator. It is well known that $\tilde{\theta}_T$ is strongly consistent and asymptotically $\mathcal{N}(0, -2\theta)$ distributed as $T \to \infty$ (see Lanksa (1979)). The large deviations of $\hat{\theta}_T$ and $\tilde{\theta}_T$ were obtained in Florens-Landais and Pham (1999). In particular, it was shown the large deviation probabilities of $\hat{\theta}_T$ are identical to those of $\tilde{\theta}_T$ for $\theta \le \theta_0/3$ but weaker for $\theta > \theta_0/3$ where θ_0 is the true value of the parameter. However, as far as the rate of convergence to normality is concerned, Bishwal (2004a) showed that $\hat{\theta}_T$ and $\tilde{\theta}_T$ have the same Berry-Esseen bound of the order $O(T^{-1/2})$.

In this section we obtain weak convergence bounds for several approximate minimum contrast estimators (AMCEs) which are simpler and robust. In order to define the approximate minimum contrast estimators (AMCEs), we use various discrete approximations of the integral in the definition (5.9) of MCE.

Define a weighted sum of squares

$$M_{n,T} := \frac{T}{n} \left\{ \sum_{i=1}^{n} w_{t_i} X_{t_{i-1}}^2 + \sum_{i=2}^{n+1} w_{t_i} X_{t_{i-1}}^2 \right\}. \tag{5.10}$$

where $w_{t_i} \ge 0$ is a weight function.

Denote the discrete increasing functions

$$I_{n,T} := \frac{T}{n} \sum_{i=1}^{n} X_{t_{i-1}}^2, \tag{5.11}$$

$$J_{n,T} := \frac{T}{n} \sum_{i=2}^{n+1} X_{t_{i-1}}^2 = \frac{T}{n} \sum_{i=1}^{n} X_{t_i}^2. \tag{5.12}$$

General weighted AMCE is defined as

$$\tilde{\theta}_{n,T} := -\left\{ \frac{2}{n} M_{n,T} \right\}^{-1}. \tag{5.13}$$

With $w_{t_i} = 1$, we obtain the forward AMCE as

$$\tilde{\theta}_{n,T,F} := -\left\{ \frac{2}{n} I_{n,T} \right\}^{-1}. \tag{5.14}$$

With $w_{t_i} = 0$, we obtain the backward AMCE as

$$\tilde{\theta}_{n,T,B} := -\left\{ \frac{2}{n} J_{n,T} \right\}^{-1}. \tag{5.15}$$

Analogous to the estimators for the discrete AR (1) model, we define the simple symmetric and weighted symmetric estimators (see Fuller (1996)):
With $w_{t_i} = 0.5$, the simple symmetric AMCE is defined as

$$\tilde{\theta}_{n,T,z} := -\left\{ \frac{1}{n} [I_{n,T} + J_{n,T}] \right\}^{-1} = -\left\{ \frac{2}{n} \sum_{i=2}^{n} X_{t_{i-1}}^2 + 0.5(X_{t_0}^2 + X_{t_n}^2) \right\}^{-1}. \tag{5.16}$$

With the weight function

$$w_{t_i} = \begin{cases} 0 & : \quad i = 1 \\ \frac{i-1}{n} & : \quad i = 2, 3, \cdots, n \\ 1 & : \quad i = n+1 \end{cases}$$

the weighted symmetric AMCE is defined as

$$\tilde{\theta}_{n,T,w} := -\left\{ \frac{2}{n} \sum_{i=2}^{n} X_{t_{i-1}}^2 + \frac{1}{n} \sum_{i=1}^{n} X_{t_{i-1}}^2 \right\}^{-1}. \tag{5.17}$$

Note that estimator (1.13) is analogous to the trapezoidal rule in numerical analysis. One can instead use the midpoint rule to define another estimator

$$\tilde{\theta}_{n,T,A} := -\left\{ \frac{2}{n} \sum_{i=1}^{n} \left(\frac{X_{t_{i-1}} + X_{t_i}}{2} \right)^2 \right\}^{-1}. \tag{5.18}$$

One can use the Simpson's rule to define another estimator where the denominator is a convex combination of the denominators in (1.11) and (1.12)

$$\tilde{\theta}_{n,T,S} := -\left\{\frac{1}{3n}\sum_{i=1}^{n}\left\{X_{t_{i-1}}^2 + 4\left(\frac{X_{t_{i-1}} + X_{t_i}}{2}\right)^2 + X_{t_i}^2\right\}\right\}^{-1}. \qquad (5.19)$$

In general, one can generalize Simpson's rule as

$$\tilde{\theta}_{n,T,GS} := -\left\{\frac{2}{n}\sum_{i=1}^{n}\left\{\alpha\frac{X_{t_{i-1}}^2 + X_{t_i}^2}{2} + (1-\alpha)\left(\frac{X_{t_{i-1}} + X_{t_i}}{2}\right)^2\right\}\right\}^{-1} \qquad (5.20)$$

for any $0 \le \alpha \le 1$. The case $\alpha = 0$ produces the estimator (5.18). The case $\alpha = 1$ produces the estimator (5.17). The case $\alpha = \frac{1}{3}$ produces the estimator (5.19).

We propose a very general form of the quadrature based estimator as

$$\tilde{\theta}_{n,T,w} := -\left\{\frac{2}{n}\sum_{i=1}^{n}\sum_{j=1}^{m}\left[(1-s_j)X_{t_{i-1}} + s_j X_{t_i}\right]^2 p_j\right\}^{-1} \qquad (5.21)$$

where $p_j, j \in \{1, 2, \cdots, m\}$ is a probability mass function of a discrete random variable S on $0 \le s_1 < s_2 < \cdots < s_m \le 1$ with $P(S = s_j) := p_j, j \in \{1, 2, \cdots, m\}$.

Denote the k-th moment of the random variable S as $\mu_k := \sum_{j=1}^{m} s_j^k p_j, k = 1, 2, \cdots$.

If one chooses the probability distribution as uniform distribution for which the moments are a harmonic sequence $(\mu_1, \mu_2, \mu_3, \mu_4, \mu_5, \mu_6, \cdots) = (\frac{1}{2}, \frac{1}{3}, \frac{1}{4}, \frac{1}{5}, \frac{1}{7}, \cdots)$ then there is no change in rate of convergence than second order. If one can construct a probability distribution for which the harmonic sequence is truncated at a point, then there is a rate of convergence improvement at the point of truncation.

Given a positive integer m, construct a probability mass function $p_j, j \in \{1, 2, \cdots, m\}$ on $0 \le s_1 < s_2 < \cdots < s_m \le 1$ such that

$$\sum_{j=1}^{m} s_j^r p_j = \frac{1}{r+1}, \quad r \in \{0, \cdots, m-2\} \qquad (5.22)$$

$$\sum_{j=1}^{m} s_j^{m-1} p_j \ne \frac{1}{m}. \qquad (5.23)$$

Neither the probabilities p_j nor the atoms, s_j, of the distribution are specified in advance.

This problem is related to the truncated Hausdorff moment problem. I obtain examples of such probability distributions and use them to get higher order accurate (up to sixth order) AMCEs.

The order of approximation error (rate of convergence) of an estimator is $n^{-\nu}$ where

$$\nu := \inf\left\{ k : \mu_k \neq \frac{1}{1+k}, \ \mu_j = \frac{1}{1+j}, j = 1, 2, \cdots, k-1 \right\}. \qquad (5.24)$$

We construct probability distributions satisfying these moment conditions and obtain estimators of the rate of convergence up to order 6.

Probability $p_1 = 1$ at the point $s_1 = 0$ gives the estimator (5.11) for which $\mu_1 = 0$. Note that $\mu_1 \neq \frac{1}{2}$. Thus $\nu = 1$.

Probability $p_1 = 1$ at the point $s_1 = 1$ gives the estimator (5.12) for which $\mu_1 = 1$. Note that $\mu_1 \neq \frac{1}{2}$. Thus $\nu = 1$.

Probabilities $(p_1, p_2) = (\frac{1}{2}, \frac{1}{2})$ at the respective points $(s_1, s_2) = (0, 1)$ produces the estimator $\tilde{\theta}_{n,T,Z}$ for which $(\mu_1, \mu_2) = (\frac{1}{2}, \frac{1}{4})$. Thus $\nu = 2$.

Probability $p_j = 1$ at the point $s_j = \frac{1}{2}$ produce the estimator $\tilde{\theta}_{n,T,A}$ for which $(\mu_1, \mu_2) = (\frac{1}{2}, \frac{1}{2})$. Thus $\nu = 2$.

Probabilities $(p_1, p_2) = (\frac{1}{4}, \frac{3}{4})$ at the respective points $(s_1, s_2) = (0, \frac{2}{3})$ produce the asymmetric estimator

$$\tilde{\theta}_{n,T,3} := -\left\{ \frac{2}{n}\frac{1}{4}\sum_{i=1}^{n}\left[(X_{t_{i-1}})^2 + 3(\tfrac{X_{t_{i-1}} + 2X_{t_i}}{3})^2 \right]\right\}^{-1} \qquad (5.25)$$

for which $(\mu_1, \mu_2, \mu_3) = (\frac{1}{2}, \frac{1}{3}, \frac{2}{9})$. Thus $\nu = 3$.

Probabilities $(p_1, p_2) = \frac{3}{4}, \frac{1}{4}$ at the respective points $(s_1, s_2) = (\frac{1}{3}, 1)$ produce asymmetric estimator

$$\tilde{\theta}_{n,T,4} := -\left\{ \frac{2}{n}\frac{1}{4}\sum_{i=1}^{n}\left[3(\tfrac{2X_{t_{i-1}} + X_{t_i}}{3})^2 + (X_{t_i})^2 \right]\right\}^{-1} \qquad (5.26)$$

for which $(\mu_1, \mu_2, \mu_3) = (\frac{1}{2}, \frac{1}{3}, \frac{10}{36})$. Thus $\nu = 3$.

Probabilities $(p_1, p_2, p_3) = (\frac{1}{6}, \frac{2}{3}, \frac{1}{6})$ at the respective points $(s_1, s_2, s_3) = (0, \frac{1}{2}, 1)$ produce the estimator $\tilde{\theta}_{n,T,5}$ for which $(\mu_1, \mu_2, \mu_3, \mu_4) = (\frac{1}{2}, \frac{1}{3}, \frac{1}{4}, \frac{5}{25})$. Thus $\nu = 4$.

Probabilities $(p_1, p_2, p_3, p_4) = (\frac{1}{8}, \frac{3}{8}, \frac{3}{8}, \frac{1}{8})$ at the respective points $(s_1, s_2, s_3, s_4) = (0, \frac{1}{3}, \frac{2}{3}, 1)$ produce the symmetric estimator

$$\tilde{\theta}_{n,T,5} :=$$
$$-\left\{ \frac{2}{n}\frac{1}{8}\sum_{i=1}^{n}\left[(X_{t_{i-1}})^2 + 3(\tfrac{2X_{t_{i-1}} + X_{t_i}}{3})^2 + 3(\tfrac{X_{t_{i-1}} + 2X_{t_i}}{3})^2 + (X_{t_i})^2 \right]\right\}^{-1}$$
$$(5.27)$$

for which $(\mu_1, \mu_2, \mu_3, \mu_4) = (\frac{1}{2}, \frac{1}{3}, \frac{1}{4}, \frac{11}{54})$. Thus $\nu = 4$.

Probabilities $(p_1, p_2, p_3, p_4, p_5) = (\frac{1471}{24192}, \frac{6925}{24192}, \frac{1475}{12096}, \frac{2725}{12096}, \frac{5675}{24192}, \frac{1721}{24192})$ at the respective points $(s_1, s_2, s_3, s_4, s_5) = (0, \frac{1}{5}, \frac{2}{5}, \frac{3}{5}, \frac{4}{5}, 1)$ produce the asymmetric estimator

$$\tilde{\theta}_{n,T,7} := -\left\{\frac{2}{n}\frac{1}{24192}\sum_{i=1}^{n}\left[1471(X_{t_{i-1}})^2 + 6925(\frac{X_{t_{i-1}}+X_{t_i}}{5})^2\right.\right.$$

$$+2950(\frac{2X_{t_{i-1}}+2X_{t_i}}{5})^2 + 5450(\frac{3X_{t_{i-1}}+3X_{t_i}}{5})^2$$

$$\left.\left.+5675(\frac{4X_{t_{i-1}}+4X_{t_i}}{5})^2 + 1721(X_{t_i})^2\right]\right\}^{-1} \tag{5.28}$$

for which $(\mu_1, \mu_2, \mu_3, \mu_4, \mu_5, \mu_6) = (\frac{1}{2}, \frac{1}{3}, \frac{1}{4}, \frac{1}{5}, \frac{841}{5040})$. Thus $\nu = 5$.

Probabilities $(p_1, p_2, p_3, p_4, p_5) = (\frac{7}{90}, \frac{16}{45}, \frac{2}{15}, \frac{16}{45}, \frac{7}{90})$ at the respective points $(s_1, s_2, s_3, s_4, s_5) = (0, \frac{1}{4}, \frac{1}{2}, \frac{3}{4}, 1)$ produce the symmetric estimator $\tilde{\theta}_{n,T,8}$ given by

$$\tilde{\theta}_{n,T,8} := -\left\{\frac{2}{n}\frac{1}{90}\sum_{i=1}^{n}\left[7(X_{t_{i-1}})^2 + 32(\frac{3X_{t_{i-1}}+X_{t_i}}{4})^2 + 12(\frac{X_{t_{i-1}}+X_{t_i}}{2})^2\right.\right.$$

$$\left.\left.+ 32(\frac{X_{t_{i-1}}+3X_{t_i}}{4})^2 + 7(t_i, X_{t_i})^2\right]\right\}^{-1} \tag{5.29}$$

for which $(\mu_1, \mu_2, \mu_3, \mu_4, \mu_5, \mu_6) = (\frac{1}{2}, \frac{1}{3}, \frac{1}{4}, \frac{1}{5}, \frac{1}{6}, \frac{110}{768})$. Thus $\nu = 6$.

Probabilities $(p_1, p_2, p_3, p_4, p_5) = (\frac{19}{288}, \frac{75}{288}, \frac{50}{288}, \frac{50}{288}, \frac{75}{288}, \frac{19}{288})$ at the respective points
$(s_1, s_2, s_3, s_4, s_5) = (0, \frac{1}{5}, \frac{2}{5}, \frac{3}{5}, \frac{4}{5}, 1)$ produce symmetric estimator

$$\tilde{\theta}_{n,T,9} := -\left\{\frac{2}{n}\frac{1}{288}\sum_{i=1}^{n}\left[19(X_{t_{i-1}})^2 + 75(\frac{4X_{t_{i-1}}+X_{t_i}}{5})^2 + 50(\frac{3X_{t_{i-1}}+2X_{t_i}}{5})^2\right.\right.$$

$$\left.\left.+ 50(\frac{2X_{t_{i-1}}+3X_{t_i}}{5})^2 + 75(\frac{X_{t_{i-1}}+4X_{t_i}}{5})^2 + 19(X_{t_i})^2\right]\right\} \tag{5.30}$$

for which $(\mu_1, \mu_2, \mu_3, \mu_4, \mu_5, \mu_6) = (\frac{1}{2}, \frac{1}{3}, \frac{1}{4}, \frac{1}{5}, \frac{1}{6}, \frac{3219}{22500})$. Thus $\nu = 6$.

The estimator $\tilde{\theta}_{n,T,S}$ is based on the arithmetic mean of $I_{n,T}$ and $J_{n,T}$. One can use geometric mean and harmonic mean instead. The geometric mean based symmetric AMCE (which is based on the ideas of partial autocorrelation) is defined as

$$\tilde{\theta}_{n,T,G} := \frac{-T/2}{\sqrt{I_{n,T}\,J_{n,T}}} \tag{5.31}$$

The harmonic mean based symmetric AMCE is defined as

$$\tilde{\theta}_{n,T,H} := \frac{-T}{\frac{1}{I_{n,T}} + \frac{1}{J_{n,T}}} \tag{5.32}$$

Note that

$$\tilde{\theta}_{n,T,H} \leq \tilde{\theta}_{n,T,G} \leq \tilde{\theta}_{n,T,S}. \tag{5.33}$$

It is a desirable property of an estimator to lie inside the parameter space. The estimators $\tilde{\theta}_{n,T,H}$ and $\tilde{\theta}_{n,T,G}$ may lie inside the parameter space even if the estimator $\tilde{\theta}_{n,T,F}$ or $\tilde{\theta}_{n,T,B}$ may not.

Note that the symmetric estimators use both the end points of observations. These estimators are sufficient statistics. So one can take advantage of Rao-Blackwell theorem. If one excludes the end points of observations (as in the forward and the backward estimators), then the estimators will not be sufficient statistics.

From a time series viewpoint, a discretely observed stationary Ornstein-Uhlenbeck model is a nonlinear (in the parameter) autoregressive model given by

$$X_{t_i} = \exp(\theta(t_i - t_{i-1})) + \epsilon_i, \quad i \geq 1 \tag{5.34}$$

where

$$\epsilon_i := \int_{t_{i-1}}^{t_i} \exp\{\theta(t_i - u)\}dW_u \tag{5.36}$$

which are independent $\mathcal{N}\left(0, \frac{e^{2\theta(t_i - t_{i-1})} - 1}{2\theta}\right)$ random variables. The errors are heteroscedastic if the data points are unequally spaced, but as we assume equally spaced data points, the errors are homoscedastic and i.i.d. $\mathcal{N}(0, V_{n,T}(\theta))$ where

$$V_{n,T}(\theta) := \frac{e^{2\theta \frac{T}{n}} - 1}{2\theta}. \tag{5.37}$$

The log-likelihood is

$$l_{n,T}(\theta) := -\frac{n}{2}\log 2\pi - \frac{n}{2}V_{n,T}(\theta) - \frac{1}{2V_{n,T}(\theta)}\sum_{i=1}^{n}\left(X_{t_i} - e^{\frac{T}{n}\theta}X_{t_{i-1}}\right)^2. \tag{5.38}$$

The exact maximum likelihood estimator based on equally spaced discrete observations is explicit and is given by

$$\hat{\theta}_{n,T} := \frac{n}{T}\log\frac{\sum_{i=1}^{n} X_{t_{i-1}}^2}{\sum_{i=1}^{n} X_{t_i}X_{t_{i-1}}}. \tag{5.39}$$

Note that

$$\check{\theta}_{n,T} = 1 - \frac{n}{T}\exp\{\frac{T}{n}\hat{\theta}_{n,T}\}.$$

As far as the rate of convergence of this estimator to the continuous conditional MLE is concerned, it is of first order.

The Euler estimator which is the conditional least squares estimator is given by

$$\check{\theta}_{n,T} := \frac{n}{T}\frac{\sum_{i=1}^{n} X_{t_{i-1}}(X_{t_i} - X_{t_{i-1}})}{\sum_{i=1}^{n} X_{t_{i-1}}^2}. \tag{5.40}$$

Another contrast function, which is the least squares quadratic, is

$$\kappa_{n,T}(\theta) := \frac{2\theta}{e^{2\theta\frac{T}{n}} - 1} \sum_{i=1}^{n} \left(X_{t_i} - e^{\frac{T}{n}\theta} X_{t_{i-1}} \right)^2. \tag{5.41}$$

The solution of $\kappa'_{n,T}(\theta) = 0$ provides an AMCE but does not have an explicit form.

The simple symmetric approximate MLE was introduced and studied in Mishra and Bishwal (1995) for a more general diffusion process. Stochastic bound between the difference the AMLE and the continuous MLE was obtained.

In this section, we obtain rate of weak convergence to normal distribution of some AMCEs using different normings. We also obtain stochastic bound on the difference of the AMCEs and their continuous counterpart MCE when T is fixed.

We need the following lemmas in the sequel.

Lemma 5.1 *(Bishwal (2000a)) For every $\delta > 0$,*

$$P\left\{ \left| \frac{2\theta}{T} I_T - 1 \right| \geq \delta \right\} \leq CT^{-1}\delta^{-2}.$$

Lemma 5.2 *(Bishwal (2000a))*

$$\sup_{x\in\mathbb{R}} \left| P\left\{ \left(-\frac{2\theta}{T} \right)^{1/2} \left(\theta I_T - \frac{T}{2} \right) \leq x \right\} - \Phi(x) \right| \leq CT^{-1/2}.$$

Lemma 5.3

$$\text{(a)} \quad E|I_{n,T} - I_T|^2 = O\left(\frac{T^4}{n^2} \right).$$

$$\text{(b)} \quad E\left| \frac{I_{n,T} + J_{n,T}}{2} - I_T \right|^2 = O\left(\frac{T^4}{n^2} \right).$$

Part (a) is from Bishwal and Bose (2001). The proof of part (b) is analogous to part (a). We omit the details.

The following theorem gives the bound on the error of normal approximation of the AMCEs. Note that part (a) uses parameter dependent nonrandom norming. While this is useful for testing hypotheses about θ, it may not necessarily give a confidence interval. The normings in parts (b) and (c) are sample dependent which can be used for obtaining a confidence interval. Following theorem shows that asymptotic normality of the AMCEs need $T \to \infty$ and $\frac{T}{\sqrt{n}} \to 0$.

Theorem 5.1 *Denote $b_{n,T} := O(\max(T^{-1/2}(\log T)^{1/2}, (\frac{T^4}{n^2})(\log T)^{-1}))$.*

$$\text{(a)} \sup_{x\in\mathbb{R}} \left| P\left\{ \left(-\frac{T}{2\theta} \right)^{1/2} (\tilde{\theta}_{n,T,F} - \theta) \leq x \right\} - \Phi(x) \right| = O(b_{n,T}),$$

(b) $\sup\limits_{x \in \mathbb{R}} \left| P\left\{ I_{n,T}^{1/2}(\tilde{\theta}_{n,T,F} - \theta) \le x \right\} - \Phi(x) \right| = O(b_{n,T})$,

(c) $\sup\limits_{x \in \mathbb{R}} \left| P\left\{ \left(\dfrac{T}{2|\tilde{\theta}_{n,T,F}|} \right)^{1/2} (\tilde{\theta}_{n,T,F} - \theta) \le x \right\} - \Phi(x) \right| = O(b_{n,T})$.

Proof (a) Observe that

$$\left(-\frac{T}{2\theta} \right)^{1/2} (\tilde{\theta}_T - \theta) = \frac{\left(-\frac{2\theta}{T}\right)^{1/2} Y_T}{\left(-\frac{2\theta}{T}\right) I_T} \tag{5.42}$$

where

$$Y_T := -\theta I_T - \frac{T}{2} \quad \text{and} \quad I_T := \int_0^T X_t^2 \, dt.$$

Thus, we have

$$I_{n,T} \tilde{\theta}_{n,T,F} = -\frac{T}{2} = Y_T + \theta I_T.$$

Hence,

$$\left(-\frac{T}{2\theta} \right)^{1/2} (\tilde{\theta}_{n,T,F} - \theta)$$

$$= \frac{\left(-\frac{T}{2\theta}\right)^{1/2} Y_T + \theta \left(-\frac{T}{2\theta}\right)^{1/2} (I_T - I_{n,T})}{I_{n,T}}$$

$$= \frac{\left(-\frac{2\theta}{T}\right)^{1/2} Y_T + \left(-\frac{2\theta}{T}\right)^{1/2} (I_T - I_{n,T})}{\left(-\frac{2\theta}{T}\right) I_{n,T}} \tag{5.43}$$

Further,

$$P\left\{ \left| \left(\frac{-2\theta}{T} \right) (I_{n,T} - 1) \right| > \epsilon \right\}$$

$$= \left\{ \left| \left(\frac{-2\theta}{T} \right) (I_{n,T} - I_T + I_T) - 1 \right| > \epsilon \right\}$$

$$\le P\left\{ \left| \left(\frac{-2\theta}{T} \right) I_T - 1 \right| > \frac{\epsilon}{2} \right\} + P\left\{ \left(-\frac{2\theta}{T} \right) |I_{n,T} - I_T| > \frac{\epsilon}{2} \right\}$$

$$\le C \exp\left(\frac{T\theta}{16} \epsilon^2 \right) + \frac{16\theta^2}{T^2} \frac{E|I_{n,T} - I_T|^2}{\epsilon^2}$$

$$\le C \exp\left(\frac{T\theta}{16} \epsilon^2 \right) + C \frac{T^2/n^2}{\epsilon^2}. \tag{5.44}$$

Next, observe that

$$\sup\limits_{x \in \mathbb{R}} \left| P\left\{ \left(-\frac{T}{2\theta} \right)^{1/2} (\tilde{\theta}_{n,T,F} - \theta) \le x \right\} - \Phi(x) \right|$$

$$= \sup_{x \in \mathbb{R}} \left| P \left\{ \frac{\left(-\frac{2\theta}{T}\right)^{1/2} Y_T + \left(-\frac{2\theta}{T}\right)^{1/2} (I_T - I_{n,T})}{\left(-\frac{2\theta}{T}\right) I_{n,T}} \leq x \right\} - \Phi(x) \right|$$

$$\leq \sup_{x \in \mathbb{R}} \left| P \left\{ \left(-\frac{2\theta}{T}\right)^{1/2} Y_T \leq x \right\} - \Phi(x) \right|$$

$$+ P \left\{ \left| \theta \left(-\frac{2\theta}{T}\right)^{1/2} (I_{n,T} - I_T) \right| > \epsilon \right\} + P \left\{ \left| \left(-\frac{2\theta}{T}\right) I_{n,T} - 1 \right| > \epsilon \right\} + 2\epsilon$$

$$\leq CT^{-1/2} + \theta^2 \frac{\left(-\frac{2\theta}{T}\right) E|I_{n,T} - I_T|^2}{\epsilon^2} + C \exp \left(\frac{T\theta}{4} \epsilon^2 \right) + C \frac{T^2}{n^2 \epsilon^2} + 2\epsilon, \quad (5.45)$$

(the bound for the 3rd term in the right hand side of (2.4) is obtained from (5.44))

$$\leq CT^{-1/2} + C \frac{T^2}{n^2 \epsilon^2} + C \exp \left(\frac{T\theta}{4} \epsilon^2 \right) + C \frac{T}{n^2 \epsilon^2} + \epsilon \quad (5.46)$$

(by Lemma 3.2(c)).

Choosing $\epsilon = CT^{-1/2} (\log T)^{1/2}$, the terms in the right hand side of (5.46) are of the order $O(\max(T^{-1/2}(\log T)^{1/2}, (\frac{T^4}{n^2})(\log T)^{-1}))$.   □

(b)   From (5.42), we have

$$I_{n,T}^{1/2}(\tilde{\theta}_{n,T,F} - \theta) = \frac{Y_T + \theta(I_T - I_{n,T})}{I_{n,T}^{1/2}}.$$

Then,

$$\sup_{x \in \mathbb{R}} \left| P \left\{ I_{n,T}^{1/2}(\tilde{\theta}_{n,T,F} - \theta) \leq x \right\} - \Phi(x) \right|$$

$$= \sup_{x \in \mathbb{R}} \left| P \left\{ \frac{Y_T}{I_{n,T}^{1/2}} + \theta \frac{I_T - I_{n,T}}{I_{n,T}^{1/2}} \leq x \right\} - \Phi(x) \right|$$

$$\leq \sup_{x \in \mathbb{R}} \left| P \left\{ \frac{Y_T}{I_{n,T}^{1/2}} \leq x \right\} - \Phi(x) \right| + P \left\{ \left| \frac{\theta(I_T - I_{n,T})}{I_{n,T}^{1/2}} \right| > \epsilon \right\} + \epsilon$$

$$=: U_1 + U_2 + \epsilon. \quad (5.47)$$

We have from (5.44),

$$U_1 \leq CT^{-1/2} + C \exp \left(\frac{T\theta}{16} \epsilon^2 \right) + C \frac{T^2}{n^2 \epsilon^2} + \epsilon. \quad (5.48)$$

Now,

$$U_2 = P \left\{ |\theta| \left| \frac{I_{n,T} - I_T}{I_{n,T}^{1/2}} \right| > \epsilon \right\}$$

$$= P\left\{ |\theta| \frac{\left|\left(-\frac{2\theta}{T}\right)^{1/2}(I_{n,T} - I_T)\right|}{\left|\left(-\frac{2\theta}{T}\right)^{1/2} I_{n,T}^{1/2}\right|} > \epsilon \right\}$$

$$\leq P\left\{ \left|\left(-\frac{2\theta}{T}\right)^{1/2}\right| |I_{n,T} - I_T| > \delta \right\}$$

$$+ P\left\{ \left|\left(-\frac{2\theta}{T}\right)^{1/2} I_{n,T}^{1/2} - 1\right| > \delta_1 \right\} \tag{5.49}$$

(where $\delta = \epsilon - C\epsilon^2$ and $\delta_1 = (\epsilon - \delta)/\epsilon > 0$)

$$\leq \left(-\frac{2\theta}{T}\right) \frac{E|I_{n,T} - I_T|^2}{\delta^2} + P\left\{ \left|\left(-\frac{2\theta}{T}\right) I_{n,T} - 1\right| > \delta_1 \right\}$$

$$\leq C\frac{T^3}{n^2\delta^2} + C\exp\left(\frac{T\theta}{16}\delta_1^2\right) + C\frac{T^2}{n^2\delta_1^2}. \tag{5.50}$$

Here, the bound for the first term in the right hand side of (5.48) comes from Lemma 3.2(c) and that for the second term is obtained from (5.44).

Now, using the bounds (5.48) and (5.50) in (5.47) with $\epsilon = CT^{-1/2}(\log T)^{1/2}$, we obtain that the terms in (5.47) are of the order $O(\max(T^{-1/2}(\log T)^{1/2}, (\frac{T^4}{n^2})(\log T)^{-1}))$. $\square$

(c) Let

$$G_T := \left\{ |\tilde{\theta}_{n,T,F} - \theta| \leq d_T \right\}, \qquad \text{and} \qquad d_T := CT^{-1/2}(\log T)^{1/2}.$$

On the set G_T, expanding $(2|\tilde{\theta}_{n,T}|)^{1/2}$, we obtain

$$(-2\tilde{\theta}_{n,T})^{-1/2}$$

$$= (-2\theta)^{1/2}\left[1 - \frac{\theta - \tilde{\theta}_{n,T,F}}{\theta} \right]^{-1/2}$$

$$= (-2\theta)^{1/2}\left[1 + \frac{1}{2}\left(\frac{\theta - \tilde{\theta}_{n,T,F}}{\theta}\right) \right] + O(d_T^2).$$

Then,

$$\sup_{x\in\mathbb{R}}\left| P\left\{ \left(\frac{T}{2|\tilde{\theta}_{n,T,F}|}\right)^{1/2}(\tilde{\theta}_{n,T,F} - \theta) \leq x \right\} - \Phi(x) \right|$$

$$\leq \sup_{x\in\mathbb{R}}\left\{ P\left(\frac{T}{2|\tilde{\theta}_{n,T}|}\right)^{1/2}(\tilde{\theta}_{n,T,F} - \theta) \leq x, G_T \right\} + P(G_T^c).$$

Now,

$$P(G_T^c) = P\left\{|\tilde{\theta}_{n,T,F} - \theta| > CT^{-1/2}(\log T)^{1/2}\right\}$$

$$= P\left\{\left(-\frac{T}{2\theta}\right)^{1/2}|\tilde{\theta}_{n,T,F} - \theta| > C(\log T)^{1/2}(-2\theta)^{-1/2}\right\}$$

$$\leq C\max\left(T^{-1/2}(\log T)^{1/2}, \frac{T^3}{n^2}(\log T)^{-1}\right)$$

$$+2(1 - \Phi\log T^{1/2}(-2\theta)^{-1/2})$$

(by Theorem 5.1(a))

$$\leq C\max\left(T^{-1/2}(\log T)^{1/2}, \frac{T^3}{n^2}(\log T)^{-1}\right).$$

On the set G_T,

$$\left|\left(\frac{\tilde{\theta}_{n,T,F}}{\theta}\right)^{1/2} - 1\right| \leq CT^{-1/2}(\log T)^{1/2}.$$

Hence, upon choosing $\epsilon = CT^{-1/2}(\log T)^{1/2}, C$ large, we obtain

$$\left|P\left\{\left(\frac{T}{-2\tilde{\theta}_{n,T,F}}\right)^{1/2}(\tilde{\theta}_{n,T,F} - \theta) \leq x, G_T\right\} - \Phi(x)\right|$$

$$\leq \left|P\left\{\left(\frac{T}{-2\theta}\right)^{1/2}(\tilde{\theta}_{n,T,F} - \theta) \leq x, G_T\right\}\right|$$

$$+P\left\{\left|\left(\frac{\tilde{\theta}_{n,T,F}}{\theta}\right)^{1/2} - 1\right| > \epsilon, G_T\right\} + \epsilon$$

(by Lemma 1.1(b))

$$\leq C\max\left(T^{-1/2}(\log T)^{1/2}, \frac{T^4}{n^2}(\log T)^{-1}\right)$$

(by Theorem 2.1(a)).   □

In the following theorem, we improve the bound on the error of normal approximation using a mixture of random and nonrandom normings. Thus asymptotic normality of the AMCEs need $T \to \infty$ and $\frac{T}{n^{2/3}} \to 0$ which are sharper than the bound in Theorem 5.1.

Theorem 5.2 Let $\rho_{n,T} = O\left(\max\left(T^{-1/2}, \left(\frac{T^3}{n^2}\right)^{1/3}\right)\right)$.

$$\sup_{x\in\mathbb{R}}\left|P\left\{I_{n,T}\left(-\frac{2\theta}{T}\right)^{1/2}(\tilde{\theta}_{n,T,F} - \theta) \leq x\right\} - \Phi(x)\right| = O(\rho_{n,T}).$$

Proof From (2.2), we have

$$I_{n,T}\left(-\frac{2\theta}{T}\right)^{1/2}(\tilde{\theta}_{n,T,F} - \theta) = \left(-\frac{2\theta}{T}\right)^{1/2} Y_T + \theta\left(-\frac{2\theta}{T}\right)^{1/2}(I_T - I_{n,T}).$$

Hence, by Lemma 5.1–5.3

$$\sup_{x\in\mathbb{R}}\left|P\left\{I_{n,T}\left(-\frac{2\theta}{T}\right)^{1/2}(\tilde{\theta}_{n,T,F} - \theta) \le x\right\} - \Phi(x)\right|$$

$$= \sup_{x\in\mathbb{R}}\left|P\left\{\left(-\frac{2\theta}{T}\right)^{1/2} Y_T + \theta\left(-\frac{2\theta}{T}\right)^{1/2}(I_T - I_{n,T}) \le x\right\} - \Phi(x)\right|$$

$$\le \sup_{x\in R}\left|P\left\{\left(-\frac{2\theta}{T}\right)^{1/2} Y_T \le x\right\} - \Phi(x)\right|$$

$$+ P\left\{\left|\theta\left(-\frac{2\theta}{T}\right)^{1/2}(I_T - I_{n,T})\right| > \epsilon\right\} + \epsilon$$

$$\le CT^{-1/2} + C\frac{E|I_T - I_{n,T}|^2}{T\epsilon^2} + \epsilon \le CT^{-1/2} + C\frac{T^3}{n^2\epsilon^2} + \epsilon.$$

Choosing $\epsilon = (\frac{T^3}{n^2})^{1/3}$, the theorem follows.   □

The following theorem gives stochastic bound on the error of approximation of the continuous MCE by AMCEs.

Theorem 5.3

$$(a)\ \ |\tilde{\theta}_{n,T} - \tilde{\theta}_T| = O_P\left(\frac{T}{n}\right)^{1/2}.$$

$$(b)\ \ |\tilde{\theta}_{n,T,z} - \tilde{\theta}_T| = O_P\left(\frac{T^2}{n^2}\right)^{1/2}.$$

Proof From (5.9) and (5.14), we have

$$\tilde{\theta}_T = -\frac{T}{2I_T}, \quad \tilde{\theta}_{n,T} = -\frac{T}{2I_{n,T}}.$$

Hence, applying Lemma 2.1 with the aid of Lemma 5.3(a) and noting that $|\frac{I_{n,T}}{T}| = O_P(1)$ and $|\frac{I_T}{T}| = O_P(1)$ the part (a) of theorem follows.
    From (5.9) and (5.16), we have

$$\tilde{\theta}_T = -\frac{T}{2I_T}, \quad \tilde{\theta}_{n,T,z} = -\frac{T}{I_{n,t} + J_{n,T}}.$$

Applying Lemma 2.1 with the aid of Lemma 5.3(b) and noting that $|\frac{J_{n,T}}{T}| = O_P(1)$ and $|\frac{I_T}{T}| = O_P(1)$ the part (b) of theorem follows.   □

Remarks

(1) The bounds in Theorems 5.1 and 5.2 are uniform over compact subsets of the parameter space.
(2) It remains to investigate the nonuniform rates of convergence to normality which are more useful.
(3) Weak convergence rate for AMCE remains open when $T \to \infty$ and $\frac{T}{n} \to 0$.
(4) Weighted estimators converge faster than the unweighted estimators if the weights are chosen according to (5.22) and (5.23).

8.6 Berry-Esseen Bounds for Approximate Bayes Estimators

Using random, nonrandom and mixed normings, this section obtains uniform rates of weak convergence to the standard normal distribution of the distribution of several approximate Bayes estimators and approximate maximum a posteriori estimators of the drift parameter in the discretely observed Ornstein-Uhlenbeck process from high frequency data.

The Ornstein-Uhlenbeck process, also known as the Vasicek model in finance, is widely used in modeling short term interest rates. Also this process is the building block in stochastic volatility modelling, see Barndorff-Neilsen and Shephard (2001). In view of this, it becomes necessary to estimate the unknown parameters in the model from discrete data. In this section, we assume constant volatility and without loss of generality assume it to be one. To estimate the drift parameter, we adopt approximate Bayes estimation method and approximate maximum a posteriori method. We study the accuracy of the distributional approximation of the estimators from high frequency data.

We are motivated by the advent of complete record of quotes or transaction prices for many financial assets. Although market microstructure effects (e.g., discreteness of prices, bid/ask bounce, irregular trading, etc.) means that there is a mismatch between asset pricing theory based on semimartingales and the data at every time intervals, it does suggest the desirability of establishing an asymptotic distribution theory of estimation as we use more and more high frequency observations where we have returns over increasingly finer time points. We concentrate on refined central limit theorem, i.e., Berry-Esseen theorem.

This section deals with computational Bayesian method for parameter estimation. In computational Bayesian method, one seeks approximation the posterior distribution. Ogunyemi, Hutton and Nelson (1993) studied approximate Bayes estimation for discrete stochastic processes which include birth and death process and branching process, where they used normal approximation of posterior, which is implied from the Bernstein-von Mises phenomenon. Our approximate Bayesian estimation method for diffusions uses Riemann type approximation of posterior distribution. An alternative way is to use particle filtering where the approximation of posterior distribution is done by sequential monte-carlo method. More precisely, in the particle filtering method, which is a sequential Bayes method, the posterior distribution is approximated by a large set of Dirac-delta masses (samples/particles) that evolve randomly in time according to the dynamics of the model and observations. Since particles are interacting, classical i.i.d. limit results are not applicable. See Doucet *et al.* (2001)for this exciting area of particle filtering.

Let the continuous realization $\{X_t, 0 \leq t \leq T\}$ be denoted by X_0^T. Let P_θ^T be the measure generated on the space (C_T, B_T) of continuous functions on $[0, T]$ with the associated Borel σ-algebra B_T under the supremum norm by the process X_0^T and let P_0^T be the standard Wiener measure. It is well known by Girsanov theorem that when θ is the true value of the parameter P_θ^T is absolutely continuous with respect to P_0^T and the Radon-Nikodym derivative (conditional likelihood) of P_θ^T with respect to P_0^T based on X_0^T is given by

$$\frac{dP_\theta^T}{dP_0^T}(X_0^T) = \exp\left\{\theta \int_0^T X_t dX_t - \frac{\theta^2}{2}\int_0^T X_t^2 dt\right\}. \tag{6.1}$$

Consider the score function, the derivative of the conditional log-likelihood function, which is given by

$$L_{T,1}(\theta) := \int_0^T X_t dX_t - \theta \int_0^T X_t^2 dt. \tag{6.2}$$

A solution of the conditional likelihood equation $L_{T,1}(\theta) = 0$ provides the (conditional) maximum likelihood estimate (MLE)

$$\theta_{T,1} := \frac{\int_0^T X_t dX_t}{\int_0^T X_t^2 dt}. \tag{6.3}$$

Using Itô formula, the score function $L_{T,1}(\theta)$ can be written as

$$L_{T,2}(\theta) := \frac{1}{2}X_T^2 - \int_0^T (\theta X_t^2 + \frac{1}{2})dt. \tag{6.4}$$

$$\theta_{T,2} := \frac{X_T^2 - T}{2\int_0^T X_t^2 dt}. \tag{6.5}$$

Consider the estimating function

$$L_{T,3}(\theta) := -\int_0^T (\theta X_t^2 + \frac{1}{2})dt \tag{6.6}$$

and the minimum contrast estimate (MCE)

$$\theta_{T,2} := -\frac{T}{2\int_0^T X_t^2 dt}. \tag{6.6}$$

Let us introduce the Bayes procedure. Let $\lambda(\cdot)$ be the prior density which is continuous and positive on the parameter space Θ.

The posterior density is given by

$$\lambda_T(\theta|X_0^T) := \frac{\lambda(\theta)L_T(\theta)}{\int_\Theta \lambda(\theta)L_T(\theta)d\theta}.$$

For squared error loss, the Bayes estimator, which is the posterior mean, is given by

$$\tilde{\tilde{\theta}}_T = \int_\Theta \theta\lambda_T(\theta|X_0^T)d\theta.$$

The maximum a posteriori (MAP) estimator, which is the posterior mode, is given by

$$\hat{\hat{\theta}}_T := \arg\max \lambda_T(\theta|X_0^T).$$

The maximum probability estimator (MPE) in the sense of Wolfowitz (1975) is defined as

$$\bar{\theta}_T = \arg\max_{\theta\in\Theta} \int_{\theta-T^{-1/2}}^{\theta+T^{-1/2}} \lambda_T(\phi|X_0^T)d\phi.$$

As it is a regular case, the Bernstein-von Mises theorem applies and the posterior is asymptotically normal (see Bishwal (2000b)). Bishwal (2000b) obtained weak convergence bound of the order $O(T^{-1/2})$ for the posterior distribution and Bayes estimator using nonrandom norming. Bishwal (2001) obtained weak convergence bound of the order $O(T^{-1/2})$ for the MLE and Bayes estimators using two different random normings which are useful for computation of a confidence interval.

Based one discrete observations, we consider the following three contrast functions which are discretizations of $L_{T,1}, L_{T,2}, L_{T,3}$ respectively:

$$L_{n,T,1}(\theta) := \theta\sum_{i=1}^{n} X_{t_{i-1}}(X_{t_i} - X_{t_{i-1}}) - \frac{\theta^2}{2}\sum_{i=1}^{n} X_{t_{i-1}}^2(t_i - t_{i-1}),$$

$$L_{n,T,2}(\theta) := \frac{\theta}{2}(X_T^2 - T) - \frac{\theta^2}{2}\sum_{i=1}^{n} X_{t_{i-1}}^2(t_i - t_{i-1}),$$

$$L_{n,T,3}(\theta) := -\frac{\theta^2}{2}\sum_{i=1}^{n} X_{t_{i-1}}^2(t_i - t_{i-1}).$$

Define the estimators

$$\theta_{n,T,i} := \arg\max_\theta L_{n,T,i}, \quad i = 1, 2, 3.$$

which are given by

$$\theta_{n,T,1} = \frac{\sum_{i=1}^{n} X_{t_{i-1}}(X_{t_i} - X_{t_{i-1}})}{\sum_{i=1}^{n} X_{t_{i-1}}^2(t_i - t_{i-1})},$$

$$\theta_{n,T,2} = \frac{X_T^2 - T}{2\sum_{i=1}^{n} X_{t_{i-1}}^2(t_i - t_{i-1})},$$

$$\theta_{n,T,3} = \frac{-T}{2\sum_{i=1}^{n} X_{t_{i-1}}^2(t_i - t_{i-1})}.$$

Bishwal and Bose (2001) obtained weak convergence bounds for the approximate maximum likelihood estimators (AMLEs) $\theta_{n,T,1}$ and $\theta_{n,T,2}$ where the later one is shown to have sharper bound than the former. The weak convergence bound for the approximate minimum contrast estimator $\theta_{n,T,3}$ was obtained in Bishwal (2006a) where it is shown that it has the same error bound as the estimator $\theta_{n,T,2}$, and moreover this estimator is robust and simple.

First order asymptotic theory of approximate Bayes estimators (ABEs) for discretely observed nonlinear diffusion processes was first introduced and studied in Bishwal (1999c, 2005a). In this section, beyond the first order asymptotic theory, we obtain the Berry-Esseen bounds of the ABEs for the Ornstein-Uhlenbeck process.

We also study approximate maximum a posteriori estimators (AMAPEs). The AML estimate can be seen as an approximate Bayes estimate when the loss function is not specified. The AMAP estimate incorporates prior information. The difference between AMAP and AML estimation lies in the assumption of an appropriate prior distribution of the parameters to be estimated.

The approximate posterior densities are defined as

$$\lambda_{n,T,i}(\theta|X_0^{n,T}) := \frac{\lambda(\theta)L_{n,T,i}(\theta)}{\int_{\Theta} \lambda(\theta)L_{n,T,i}(\theta)d\theta}, \quad i = 1, 2, 3.$$

We consider squared error loss function. The approximate Bayes estimators (ABEs), which are the approximate posterior means, are given by

$$\tilde{\theta}_{n,T,i} := \int_{\Theta} \theta\lambda_{n,T,i}(\theta|X_0^{n,T})d\theta, \quad i = 1, 2, 3.$$

The AMAPEi estimators are defined as

$$\hat{\theta}_{n,T,i} := \arg\max_{\theta} \lambda_{n,T,i}(\theta|X_0^{n,T}), \quad i = 1, 2, 3.$$

If the parameter is assumed to be fixed but unknown, then there is no knowledge about the parameter, which is equivalent to assuming a non-informative improper prior. The above case then reduces to the familiar ML formulation.

We obtain Berry-Esseen bounds for the estimators ABE$i := \tilde{\theta}_{n,T,i}$ and AMAPE$i := \hat{\theta}_{n,T,i}$, $i = 1, 2, 3$.

Note that $\theta_{n,T,3}$ is not an approximate Bayes estimator in the true sense since $L_{n,T,3}$ is not an approximate likelihood, it is a pseudo likelihood. Nevertheless we keep the terminology. It is an approximate generalised Bayes estimator in the terminology in Strasser (1976).

We obtain various rates of convergence to normality of the ABEs and AMAPEs using several normings. We also obtain stochastic bound on the difference between the ABEs and AMAPEs from their corresponding continuous counterparts.

Throughout the section $\tilde{\tilde{\theta}}$ refers to posterior mean and $\hat{\hat{\theta}}$ refers to posterior mode.

The following theorem gives the bound on the error of normal approximation of the ABEs and AMAPEs. Note that parts (a) and (d) use parameter dependent nonrandom norming. While this is useful for testing hypotheses about θ, it may not necessarily give a confidence interval. The normings in parts (b) (c), (e), and (f) are sample dependent which can be used for obtaining a confidence interval.

Theorem 6.1 Denote $\alpha_{n,T} := O(\max(T^{-1/2}(\log T)^{1/2},\ (\frac{T^3}{n^2})(\log T)^{-1}))$.

(a) $\displaystyle\sup_{x\in\mathbb{R}}\left|P\left\{\left(-\frac{T}{2\theta}\right)^{1/2}(\tilde{\tilde{\theta}}_{n,T,1}-\theta)\le x\right\}-\Phi(x)\right|=O(\alpha_{n,T})$,

(b) $\displaystyle\sup_{x\in\mathbb{R}}\left|P\left\{I_{n,T}^{1/2}(\tilde{\tilde{\theta}}_{n,T,1}-\theta)\le x\right\}-\Phi(x)\right|=O(\alpha_{n,T})$,

(c) $\displaystyle\sup_{x\in\mathbb{R}}\left|P\left\{\left(\frac{T}{2|\tilde{\tilde{\theta}}_{n,T,1}|}\right)^{1/2}(\tilde{\tilde{\theta}}_{n,T,1}-\theta)\le x\right\}-\Phi(x)\right|=O(\alpha_{n,T})$,

(d) $\displaystyle\sup_{x\in\mathbb{R}}\left|P\left\{\left(-\frac{T}{2\theta}\right)^{1/2}(\hat{\hat{\theta}}_{n,T,1}-\theta)\le x\right\}-\Phi(x)\right|=O(\alpha_{n,T})$,

(e) $\displaystyle\sup_{x\in\mathbb{R}}\left|P\left\{I_{n,T}^{1/2}(\hat{\hat{\theta}}_{n,T,1}-\theta)\le x\right\}-\Phi(x)\right|=O(\alpha_{n,T})$,

(f) $\displaystyle\sup_{x\in\mathbb{R}}\left|P\left\{\left(\frac{T}{2|\hat{\hat{\theta}}_{n,T,1}|}\right)^{1/2}(\hat{\hat{\theta}}_{n,T,1}-\theta)\le x\right\}-\Phi(x)\right|=O(\alpha_{n,T})$.

The following theorem improves the Berry-Esseen bound in Theorem 2.1 using mixed norming.

Theorem 6.2 Let $\rho_{n,T}=O\left(\max\left(T^{-1/2},\left(\frac{T^3}{n^2}\right)^{1/3}\right)\right)$.

(a) $\sup\limits_{x\in\mathbb{R}}\left|P\left\{I_{n,T}\left(-\dfrac{2\theta}{T}\right)^{1/2}(\tilde{\tilde{\theta}}_{n,T,1}-\theta)\le x\right\}-\Phi(x)\right|=O(\rho_{n,T}),$

(b) $\sup\limits_{x\in\mathbb{R}}\left|P\left\{I_{n,T}\left(-\dfrac{2\theta}{T}\right)^{1/2}(\hat{\tilde{\theta}}_{n,T,1}-\theta)\le x\right\}-\Phi(x)\right|=O(\rho_{n,T}).$

The following theorem shows that ABE2 and AMAPE2 have sharper Berry-Esseen bounds than ABE1 and AMAPE1.

Theorem 6.3 *Denote* $\beta_{n,T}:=O(\max(T^{-1/2}(\log T)^{1/2},\ (\frac{T^4}{n^2})(\log T)^{-1})).$

(a) $\sup\limits_{x\in\mathbb{R}}\left|P\left\{\left(-\dfrac{T}{2\theta}\right)^{1/2}(\tilde{\tilde{\theta}}_{n,T,2}-\theta)\le x\right\}-\Phi(x)\right|=O(\beta_{n,T}),$

(b) $\sup\limits_{x\in\mathbb{R}}\left|P\left\{I_{n,T}^{1/2}(\tilde{\tilde{\theta}}_{n,T,2}-\theta)\le x\right\}-\Phi(x)\right|=O(\beta_{n,T}),$

(c) $\sup\limits_{x\in\mathbb{R}}\left|P\left\{\left(\dfrac{T}{2|\tilde{\theta}_{n,T,2}|}\right)^{1/2}(\tilde{\tilde{\theta}}_{n,T,2}-\theta)\le x\right\}-\Phi(x)\right|=O(\beta_{n,T}).$

(d) $\sup\limits_{x\in\mathbb{R}}\left|P\left\{\left(-\dfrac{T}{2\theta}\right)^{1/2}(\hat{\tilde{\theta}}_{n,T,2}-\theta)\le x\right\}-\Phi(x)\right|=O(\beta_{n,T}),$

(e) $\sup\limits_{x\in\mathbb{R}}\left|P\left\{I_{n,T}^{1/2}(\hat{\tilde{\theta}}_{n,T,2}-\theta)\le x\right\}-\Phi(x)\right|=O(\beta_{n,T}),$

(f) $\sup\limits_{x\in\mathbb{R}}\left|P\left\{\left(\dfrac{T}{2|\hat{\theta}_{n,T,2}|}\right)^{1/2}(\hat{\theta}_{n,T,2}-\theta)\le x\right\}-\Phi(x)\right|=O(\beta_{n,T}).$

The following theorem improves the Berry-Esseen bound in Theorem 3.1 using mixed norming.

Theorem 6.4 *Let* $\rho_{n,T}=O\left(\max\left(T^{-1/2},\left(\frac{T^3}{n^2}\right)^{1/3}\right)\right).$

(a) $\sup\limits_{x\in\mathbb{R}}\left|P\left\{I_{n,T}\left(-\dfrac{2\theta}{T}\right)^{1/2}(\tilde{\tilde{\theta}}_{n,T,2}-\theta)\le x\right\}-\Phi(x)\right|=O(\rho_{n,T}).$

(b) $\sup\limits_{x\in\mathbb{R}}\left|P\left\{I_{n,T}\left(-\dfrac{2\theta}{T}\right)^{1/2}(\hat{\tilde{\theta}}_{n,T,2}-\theta)\le x\right\}-\Phi(x)\right|=O(\rho_{n,T}).$

The following theorem shows that ABE3 has a sharper Berry-Esseen bound than ABE1.

Theorem 6.5 *Denote* $\gamma_{n,T} := O(\max(T^{-1/2}(\log T)^{1/2}, (\frac{T^4}{n^2})(\log T)^{-1}))$.

(a) $\displaystyle\sup_{x\in\mathbb{R}} \left| P\left\{ \left(-\frac{T}{2\theta}\right)^{1/2} (\tilde{\tilde{\theta}}_{n,T,3} - \theta) \le x \right\} - \Phi(x) \right| = O(\gamma_{n,T})$,

(b) $\displaystyle\sup_{x\in\mathbb{R}} \left| P\left\{ I_{n,T}^{1/2}(\tilde{\tilde{\theta}}_{n,T,3} - \theta) \le x \right\} - \Phi(x) \right| = O(\gamma_{n,T})$,

(c) $\displaystyle\sup_{x\in\mathbb{R}} \left| P\left\{ \left(\frac{T}{2|\tilde{\tilde{\theta}}_{n,T,3}|}\right)^{1/2} (\tilde{\tilde{\theta}}_{n,T,3} - \theta) \le x \right\} - \Phi(x) \right| = O(\gamma_{n,T})$.

(d) $\displaystyle\sup_{x\in\mathbb{R}} \left| P\left\{ \left(-\frac{T}{2\theta}\right)^{1/2} (\hat{\tilde{\theta}}_{n,T,3} - \theta) \le x \right\} - \Phi(x) \right| = O(\gamma_{n,T})$,

(e) $\displaystyle\sup_{x\in\mathbb{R}} \left| P\left\{ I_{n,T}^{1/2}(\hat{\tilde{\theta}}_{n,T,3} - \theta) \le x \right\} - \Phi(x) \right| = O(\gamma_{n,T})$,

(f) $\displaystyle\sup_{x\in\mathbb{R}} \left| P\left\{ \left(\frac{T}{2|\hat{\tilde{\theta}}_{n,T,3}|}\right)^{1/2} (\hat{\tilde{\theta}}_{n,T,3} - \theta) \le x \right\} - \Phi(x) \right| = O(\gamma_{n,T})$.

In the following theorem, we improve the bound on the error of normal approximation using a mixture of random and nonrandom normings. Thus asymptotic normality of the ABE needs $T \to \infty$ and $\frac{T}{n^{2/3}} \to 0$ which are sharper than the bound in Theorem 4.1.

The following theorem improves the Berry-Esseen bounds in Theorem 6.5 using mixed norming.

Theorem 6.6 *Let* $\rho_{n,T} = O\left(\max\left(T^{-1/2}, \left(\frac{T^3}{n^2}\right)^{1/3} \right) \right)$.

(a) $\displaystyle\sup_{x\in\mathbb{R}} \left| P\left\{ I_{n,T} \left(-\frac{2\theta}{T}\right)^{1/2} (\tilde{\tilde{\theta}}_{n,T,3} - \theta) \le x \right\} - \Phi(x) \right| = O(\rho_{n,T})$.

(b) $\displaystyle\sup_{x\in\mathbb{R}} \left| P\left\{ I_{n,T} \left(-\frac{2\theta}{T}\right)^{1/2} (\hat{\tilde{\theta}}_{n,T,3} - \theta) \le x \right\} - \Phi(x) \right| = O(\rho_{n,T})$.

The following theorem gives bound on the error of approximation of the continuous BE by the discrete ABEs when T is fixed.

Theorem 6.7

(a) $|\tilde{\tilde{\theta}}_{n,T,1} - \tilde{\tilde{\theta}}_T| = O_P \left(\frac{T^2}{n}\right)^{1/2}$,

(b) $|\tilde{\tilde{\theta}}_{n,T,2} - \tilde{\tilde{\theta}}_T| = O_P \left(\frac{T^3}{n^2}\right)^{1/2}$,

$$(c) \quad |\tilde{\tilde{\theta}}_{n,T,3} - \tilde{\tilde{\theta}}_T| = O_P\left(\frac{T^4}{n^2}\right)^{1/2}.$$

Proof For $i = 1, 2, 3$, we have

$$\tilde{\tilde{\theta}}_{n,T,i} - \tilde{\tilde{\theta}}_T$$

$$= \int_\Theta \theta \lambda_{n,T,i}(\theta|X_0^{n,T}) d\theta - \int_\Theta \theta \lambda(\theta|X_0^T) d\theta$$

$$= \int_\Theta \theta \left[\lambda_{n,T,i}(\theta|X_0^{n,T}) - \lambda(\theta|X_0^T)\right] d\theta$$

$$= \int_\Theta \theta \left[\frac{\lambda(\theta)L_{n,T,i}(\theta)}{\int_\Theta \lambda(\theta)L_{n,T,i}(\theta)d\theta} - \frac{\lambda(\theta)L_T(\theta)}{\int_\Theta \lambda(\theta)L_T(\theta)d\theta}\right] d\theta$$

$$= \int_\Theta \theta\lambda(\theta) \left[\frac{L_{n,T,i}(\theta)}{\int_\Theta \lambda(\theta)L_{n,T,i}(\theta)d\theta} - \frac{L_T(\theta)}{\int_\Theta \lambda(\theta)L_T(\theta)d\theta}\right] d\theta.$$

We have

$$|L_{n,T,1}(\theta) - L_T(\theta)| = O\left(\frac{T^2}{n}\right)^{1/2},$$

$$|L_{n,T,2}(\theta) - L_T(\theta)| = O\left(\frac{T^3}{n^2}\right),$$

$$|L_{n,T,3}(\theta) - L_T(\theta)| = O\left(\frac{T^4}{n^2}\right)^{1/2},$$

and hence

$$\left|\int_\Theta \theta\lambda(\theta)L_{n,T,1}(\theta)d\theta - \int_\Theta \theta\lambda(\theta)L_T(\theta)d\theta\right| = O\left(\frac{T^2}{n}\right)^{1/2},$$

$$\left|\int_\Theta \theta\lambda(\theta)L_{n,T,2}(\theta)d\theta - \int_\Theta \theta\lambda(\theta)L_T(\theta)d\theta\right| = O\left(\frac{T^3}{n^2}\right)^{1/2},$$

$$\left|\int_\Theta \theta\lambda(\theta)L_{n,T,3}(\theta)d\theta - \int_\Theta \theta\lambda(\theta)L_T(\theta)d\theta\right| = O\left(\frac{T^4}{n^2}\right)^{1/2}.$$

An application of Lemma 2.1 gives the result. □

Theorem 6.8

$$(a) \quad |\hat{\tilde{\theta}}_{n,T,1} - \hat{\tilde{\theta}}_T| = O_P\left(\frac{T^2}{n}\right)^{1/2},$$

$$(b) \quad |\hat{\tilde{\theta}}_{n,T,2} - \hat{\tilde{\theta}}_T| = O_P\left(\frac{T^3}{n^2}\right)^{1/2},$$

$$(c) \ |\hat{\hat{\theta}}_{n,T,3} - \hat{\hat{\theta}}_T| = O_P \left(\frac{T^4}{n^2}\right)^{1/2}.$$

Proof. For $i = 1, 2, 3$, we have

$$\hat{\hat{\theta}}_{n,T,i} - \hat{\hat{\theta}}_T$$

$$= \min_{\theta \in \Theta} \lambda_{n,T,i}(\theta|X_0^{n,T}) - \min_{\theta \in \Theta} \lambda(\theta|X_0^T)$$

$$= \min_{\theta \in \Theta} \left[\lambda_{n,T,i}(\theta|X_0^{n,T}) - \lambda(\theta|X_0^T)\right]$$

$$= \min_{\theta \in \Theta} \left[\frac{\lambda(\theta)L_{n,T,i}(\theta)}{\int_\Theta \lambda(\theta)L_{n,T,i}(\theta)d\theta} - \frac{\lambda(\theta)L_T(\theta)}{\int_\Theta \lambda(\theta)L_T(\theta)d\theta}\right]$$

$$= \min_{\theta \in \Theta} \lambda(\theta) \left[\frac{L_{n,T,i}(\theta)}{\int_\Theta \lambda(\theta)L_{n,T,i}(\theta)d\theta} - \frac{L_T(\theta)}{\int_\Theta \lambda(\theta)L_T(\theta)d\theta}\right].$$

We have

$$|L_{n,T,1}(\theta) - L_T(\theta)| = O\left(\frac{T^2}{n}\right)^{1/2},$$

$$|L_{n,T,2}(\theta) - L_T(\theta)| = O\left(\frac{T^3}{n^2}\right),$$

$$|L_{n,T,3}(\theta) - L_T(\theta)| = O\left(\frac{T^4}{n^2}\right)^{1/2},$$

and hence

$$\left|\int_\Theta \lambda(\theta)L_{n,T,1}(\theta)d\theta - \int_\Theta \lambda(\theta)L_T(\theta)\right| = O\left(\frac{T^2}{n}\right)^{1/2},$$

$$\left|\int_\Theta \lambda(\theta)L_{n,T,2}(\theta)d\theta - \int_\Theta \lambda(\theta)L_T(\theta)d\theta\right| = O\left(\frac{T^3}{n^2}\right)^{1/2},$$

$$\left|\int_\Theta \lambda(\theta)L_{n,T,3}(\theta)d\theta - \int_\Theta \lambda(\theta)L_T(\theta)d\theta\right| = O\left(\frac{T^4}{n^2}\right)^{1/2}.$$

An application of Lemma 2.1 gives the result. $\qquad\square$

Local Asymptotic Normality for Discretely Observed Homogeneous Diffusions

9.1 Introduction

In Chapters 7 and 8 we studied the rates of convergence of approximate maximum likelihood estimators (AMLEs) of the drift parameter appearing linearly in the drift coefficient of linearly parametrized diffusion processes. In this Chapter we consider the nonlinear homogeneous SDE

$$dX_t = f(\theta, X_t)dt + dW_t, \ t \geq 0$$
$$X_0 = x_o$$

where $\{W_t, t \geq 0\}$ is a one dimensional standard Wiener process, $\theta \in \Theta$, Θ is a closed interval in $\mathbb{R}$, f is a known real valued function defined on $\Theta \times \mathbb{R}$, the unknown parameter θ is to be estimated on the basis of observation of the process $\{X_t, t \geq 0\}$. Let θ_0 be the true value of the parameter which is in the interior of Θ. As in Chapters 4 and 5, we assume that the process $\{X_t, t \geq 0\}$ is observed at $0 = t_0 < t_1 < \ldots < t_n = T$ with $t_i - t_{i-1} = \frac{T}{n} = h$. We estimate θ from the observations $\{X_{t_0}, X_{t_1}, \ldots, X_{t_n}\} \equiv X_0^{n,h}$. This model was first studied by Dorogovcev (1976) who obtained weak consistency of the conditional least squares estimator (CLSE) under some assumptions as $T \to \infty$ and $\frac{T}{n} \to 0$. We call it the *slowly increasing experimental design* (SIED) condition. Kasonga (1988) obtained the strong consistency of the CLSE under the SIED condition. Under some stronger regularity conditions Prakasa Rao (1983) obtained the asymptotic normality and asymptotic efficiency of the CLSE as $T \to \infty$ and $\frac{T}{\sqrt{n}} \to 0$. This condition, i.e., $T \to \infty$ and $\frac{T}{\sqrt{n}} \to 0$ is called the *rapidly increasing experimental design* (RIED) condition (see Prakasa Rao (1988)). Florens-Zmirou (1989) studied minimum contrast estimator, based on an Euler-Maruyama type first order approximate discrete time scheme of the SDE. She showed L_2-consistency of the estimator as $T \to \infty$ and $\frac{T}{n} \to 0$ and asymptotic normality of the estimator as $T \to \infty$ and $\frac{T}{n^{2/3}} \to 0$. We call the last condition as the *moderately increasing experimental design* (MIED) condition. Kessler (1997) studied the minimum contrast estimator based on the Euler scheme. He proved weak consistency of the estimator as $T \to \infty$ and $\frac{T}{n} \to 0$ and asymptotic normality of the estimator as $T \to \infty$ and $\frac{T^p}{n^{p-1}} \to 0$ where p is an arbitrary integer which stands for the order of differentiability of the drift function with respect to x. We call the last condition as the *generally increasing experimental design* (GIED) condition. Stronger design condition for asymptotic normality than for consistency implies that one needs larger number of observations for asymptotic normality than for consistency. The likelihood based estimators and the posterior densities are based on two types of approximations of the continuous likelihood of the process $\{X_t\}$ on an interval $[0, T]$. One approximate likelihood is based on the Itô type approximation of the stochastic integral and the rectangular rule type approximation of the ordinary integral in the continuous likelihood L_T. Using this approximate likelihood, Yoshida (1992) proved the consistency of an approximate maximum likelihood estimator (AMLE) as $T \to \infty$ and $\frac{T}{n} \to 0$ and asymptotic normality of the estimator as $T \to \infty$ and $\frac{T}{n^{2/3}} \to 0$. Our other approximate likelihood is based on the Fisk-Stratonovich approximation of the stochastic integral and rectangular rule type of approximation of the ordinary integrals in L_T. We study the asymptotic properties of approximate posterior density (APD), approximate Bayes estimator (ABE), approximate maximum probability estimator (AMPE) based on the two approximate likelihoods. The method used for the study of asymptotic behaviour of estimators here are through the weak convergence of the approximate likelihood ratio random field. We obtain asymptotic normality of the conditional least squares estimator (CLSE) under weaker regularity conditions and weaker design conditions than in Prakasa Rao (1983). Under some regularity conditions, all the above estimators are shown to be asymptotically normal and asymptotically efficient under the MIED condition. We also prove the in probability versions of *Bernstein-von Mises type theorems* concerning the convergence of suitably normalized approximate posterior densities to normal density under the MIED condition.

The organization of the Chapter is as follows: In Section 9.2 we prepare notations, assumptions and preliminaries. In Section 9.3 we study the weak convergence of the approximate likelihood ratio random field. In Section 9.4 we study the *Bernstein-von Mises type theorems* and the asymptotic behaviour of different estimators.

This Chapter is adapted from Bishwal (2005b).

9.2 Model, Assumptions and Preliminaries

Recall that our observation process $\{X_t\}$ satisfies the SDE

$$dX_t = f(\theta, X_t)dt + dW_t, \quad t \geq 0$$
$$X_0 = x_0$$
(2.1)

We will use the following notations throughout the Chapter: $\Delta X_i = X_{t_i} - X_{t_{i-1}}$, $\Delta W_i = W_{t_i} - W_{t_{i-1}}$, C is a generic constant independent of h, n and other variables (perhaps it may depend on θ). *Prime denotes derivative w.r.t. θ and dot denotes derivative w.r.t. x.* Suppose that θ_0 denote the true value of the parameter and $\theta_0 \in \Theta$.

If continuous observation of $\{X_t\}$ on the interval $[0, T]$ were given, then the likelihood function of θ would be (see Chapter 2)

$$L_T(\theta) = \exp\{ \int_0^T f(\theta, X_t)dX_t - \frac{1}{2} \int_0^T f^2(\theta, X_t)dt \}, \qquad (2.2)$$

and the maximum likelihood estimate (MLE) would be

$$\theta_T := \arg\max_{\theta \in \Theta} L_T(\theta).$$

In our case we have discrete data and we have to approximate the likelihood to compute the MLE. First, we take Itô type approximation of the stochastic integral and rectangle rule approximation of the ordinary integral in (2.2) and obtain the approximate likelihood function

$$L_{n,T}(\theta) := \exp\left\{ \sum_{i=1}^n f(\theta, X_{t_{i-1}})(X_{t_i} - X_{t_{i-1}}) - \frac{h}{2} \sum_{i=1}^n f^2(\theta, X_{t_{i-1}}) \right\}. \qquad (2.3)$$

Next, transforming the Itô integral in (2.2) to Fisk-Stratonovich (FS) integral (using (2.20) in Chapter 7), gives the continuous likelihood as

$$L_T(\theta) = \exp\{ \oint_0^T f(\theta, X_t)dX_t - \frac{1}{2} \int_0^T \dot{f}(\theta, X_t)dt - \frac{1}{2} \int_0^T f^2(\theta, X_t)dt \}, \qquad (2.4)$$

We apply RFS type approximation of the stochastic integral and rectangular approximation of the ordinary integrals in (2.4) and obtain the approximate likelihood

$$\tilde{L}_{n,T}(\theta) := \exp\left\{ \frac{1}{2} \sum_{i=1}^n [f(\theta, X_{t_{i-1}}) + f(\theta, X_{t_i})](X_{t_i} - X_{t_{i-1}}) - \frac{h}{2} \sum_{i=1}^n \dot{f}(\theta, X_{t_{i-1}}) - \frac{h}{2} \sum_{i=1}^n f^2(\theta, X_{t_{i-1}}) \right\}. \qquad (2.5)$$

The approximate maximum likelihood estimate (AMLE1) based on $L_{n,T}$ is defined as

$$\hat{\theta}_{n,T} := \arg\max_{\theta \in \Theta} L_{n,T}(\theta).$$

The approximate maximum likelihood estimator (AMLE2) based on $\tilde{L}_{n,T}$ is defined as

$$\hat{\tilde{\theta}}_{n,T} := \arg\max_{\theta \in \Theta} \tilde{L}_{n,T}(\theta).$$

Consider the interval $(\theta - T^{-1/2}, \theta + T^{-1/2})$ for sufficiently large T and define the integrated approximate likelihoods

$$D_{n,T}(\theta) = \int_{\theta-T^{-1/2}}^{\theta+T^{-1/2}} L_{n,T}(\phi)d\phi,$$

$$\tilde{D}_{n,T}(\theta) = \int_{\theta-T^{-1/2}}^{\theta+T^{-1/2}} \tilde{L}_{n,T}(\phi)d\phi.$$

The approximate maximum probability estimator (AMPE1) based on $L_{n,T}$ is defined as

$$\bar{\theta}_{n,T} := \arg\max_{\theta\in\Theta} D_{n,T}(\theta).$$

The approximate maximum probability estimator (AMPE2) based on $\tilde{L}_{n,T}$ is defined as

$$\bar{\bar{\theta}}_{n,T} := \arg\max_{\theta\in\Theta} \tilde{D}_{n,T}(\theta).$$

Supppose that Λ is a prior probability measure on $(\Theta, \mathcal{B})$ where $\mathcal{B}$ is the σ-algebra of Borel subsets of Θ. Suppose that Λ has a density $\lambda(\cdot)$ with respect to Lebesgue measure which is continuous and positive.

Let $M_{n,T} : \Theta \times \Theta \to \mathbb{R}, n \geq 1, T \geq 0$ be loss functions with $M_{n,T}(\theta, \phi) = |\theta - \phi|^{\alpha}, \alpha \geq 1$. An approximate posterior density (APD1) or *pseudo-posterior density* of θ given the observations $X_0^{n,h}$ with respect to the approximate likelihood $L_{n,T}$ and the prior density $\lambda(\cdot)$ is given by

$$p_{n,T}(\theta|X_0^{n,h}) = \frac{\lambda(\theta)L_{n,T}(\theta)}{\int_{\Theta} \lambda(\theta)L_{n,T}(\theta)d\theta}$$

Let $t = T^{1/2}(\theta - \hat{\theta}_{n,T})$. Then the approximate posterior density (APD1) of $T^{1/2}(\theta - \hat{\theta}_{n,T})$ given $X_0^{n,h}$ is given by

$$p_{n,T}^*(t|X_0^{n,h}) = T^{-1/2}p_{n,T}(\hat{\theta}_{n,T} + T^{-1/2}t|X_0^{n,h}).$$

This is the normalized pseudo-posterior density centered at the AMLE1.

Similarly another approximate posterior density (APD2) or *pseudo-posterior density* based on $\tilde{L}_{n,T}$ and the prior density $\lambda(\cdot)$ is given by

$$\tilde{p}_{n,T}(\theta|X_0^{n,h}) = \frac{\lambda(\theta)\tilde{L}_{n,T}(\theta)}{\int_{\Theta} \lambda(\theta)\tilde{L}_{n,T}(\theta)d\theta}.$$

Let $s = T^{1/2}(\theta - \hat{\hat{\theta}}_{n,T})$. Then the approximate posterior density (APD2) of $T^{1/2}(\theta - \hat{\hat{\theta}}_{n,T})$ given $X_0^{n,h}$ is given by

$$\tilde{p}_{n,T}^*(s|X_0^{n,h}) = T^{-1/2}p_{n,T}^*(\hat{\hat{\theta}}_{n,T} + T^{-1/2}s|X_0^{n,h}).$$

This is the normalized pseudo-posterior density centered at the AMLE2. Note that APD1 and APD2 are not the posterior densities given discrete data $X_0^{n,h}$ since these are based on approximate likelihoods.

An approximate Bayes estimator (ABE1) or *pseudo-Bayes estimator* with respect to the loss function $M_{n,T}$ and the prior density $\lambda(\cdot)$ is defined as

$$\widetilde{\theta}_{n,T} := \arg\min_{\phi \in \Theta} B_{n,T}(\phi)$$

where $B_{n,T}(\phi) = \int_{\Theta} M_{n,T}(\theta, \phi) p_{n,T}(\theta | X_0^{n,h}) d\theta$ is an pseudo-posterior risk.

Another approximate Bayes estimator (ABE2) or *pseudo-Bayes estimator* is defined as

$$\widetilde{\widetilde{\theta}}_{n,T} := \arg\min_{\phi \in \Theta} \widetilde{B}_{n,T}(\phi)$$

where $\widetilde{B}_{n,T}(\phi) = \int_{\Theta} M_{n,T}(\theta, \phi) \widetilde{p}_{n,T}(\theta | X_0^{n,h}) d\theta$. The above six estimators $\hat{\theta}_{n,T}, \hat{\hat{\theta}}_{n,T}, \bar{\theta}_{n,T}, \bar{\bar{\theta}}_{n,T}, \widetilde{\theta}_{n,T}$ and $\widetilde{\widetilde{\theta}}_{n,T}$ are based on approximate likelihood functions.

Consider the conditional least squares estimator (CLSE) (see Dorogocev (1976), Kasonga (1989)) of θ as

$$\theta_{n,T} := \arg\min_{\theta \in \Theta} Q_{n,T}(\theta)$$

where

$$Q_{n,T}(\theta) := \sum_{i=1}^{n} \left[X_{t_i} - X_{t_{i-1}} - f(\theta, X_{t_{i-1}})h \right]^2 .$$

Consider the minimum contrast estimator (see Florens-Zmirou (1989) or the so called Euler-Maruyama estimator (EME) (see Shoji (1997)) where the contrast is constructed by using the Euler-Maruyama discretization scheme (see Kloeden and Platen (1995)). Recall that the Euler-Maruyama discretization scheme for the SDE (2.1) is given by

$$Y_{t_i} - Y_{t_{i-1}} = f(\theta, Y_{t_{i-1}})h + W_{t_i} - W_{t_{i-1}}, i \geq 1, Y_0 = x_0. \qquad (2.5)$$

The log-likelihood function of $\{Y_{t_i}, 0 \leq i \leq n\}$ is given by

$$-\frac{1}{2} \sum_{i=1}^{n} \left\{ \frac{\left[Y_{t_i} - Y_{t_{i-1}} - f(\theta, Y_{t_{i-1}})h \right]^2}{h} + \log(2\pi h) \right\}. \qquad (2.6)$$

A contrast is derived from (2.6) by substituting $\{Y_{t_i}, 0 \leq i \leq n\}$ with $\{X_{t_i}, 0 \leq i \leq n\}$. The resulting contrast is given by

$$\kappa_{n,T} := -\frac{1}{2} \sum_{i=1}^{n} \left\{ \frac{\left[X_{t_i} - X_{t_{i-1}} - f(\theta, X_{t_{i-1}})h \right]^2}{h} + \log(2\pi h) \right\}. \qquad (2.7)$$

Note that the AMLE1, CLSE and the EME are the same estimators as seen from the following simple relationship:

$$AMLE1 = \arg\max_{\theta \in \Theta} L_{n,T}(\theta)$$

$$= \arg\max_{\theta \in \Theta} L_{n,T}(\theta)h$$

$$= \arg\max_{\theta \in \Theta} \left\{ L_{n,T}(\theta)h - \frac{1}{2} \sum_{i=1}^{n} \left[X_{t_i} - X_{t_{i-1}} \right]^2 \right\}$$

$$= \arg\max_{\theta \in \Theta} \left\{ -\frac{1}{2} Q_{n,T}(\theta) \right\}$$

$$= \arg\min_{\theta \in \Theta} Q_{n,T}(\theta)$$

$$= CLSE.$$

On the otherhand,

$$CLSE = \arg\min_{\theta \in \Theta} Q_{n,T}(\theta)$$

$$= \arg\max_{\theta \in \Theta} \left\{ -\frac{1}{2} Q_{n,T}(\theta) \right\}$$

$$= \arg\max_{\theta \in \Theta} - \left\{ \frac{1}{2h} Q_{n,T}(\theta) - \frac{n}{2} \log \frac{2\pi}{h} \right\}$$

$$= EME.$$

We assume that measurable versions of the above estimators exist. We assume that the following conditions are satisfied:

(A1) There exists a constant K such that
$$|f(\theta, x)| \leq K(1 + |x|),$$
$$|f(\theta, x) - f(\theta, y)| \leq K|x - y|.$$

(A2) The diffusion process X is ergodic with invariant measure ν, i.e., for any g with $E[g(\cdot)] < \infty$

$$\frac{1}{n} \sum_{i=1}^{n} g(X_{t_i}) \overset{P}{\to} \int_{\mathbb{R}} g(x)\nu(dx) \text{ as } T \to \infty \text{ and } h \to 0.$$

(A3) For each $p > 0, \sup_{t} E|X_t|^p < \infty.$

(A4) $f(\theta, x)$ is twice differentiable in $\theta \in \Theta$ and

$$|f'(\theta, x)| + |f''(\theta, x)| \leq C(1 + |x|^c),$$

$$|f'(\theta, x) - f'(\theta, y)| \leq C|x - y|.$$

(A5) The function f and f' are smooth in x and their derivatives are of polynomial growth order in x uniformly in θ.

(A6) $\Gamma \equiv \Gamma(\theta_0) := \int_{\mathbb{R}} f'^2(\theta_0, x)\nu(dx) > 0.$

(A7) $l(\theta)$ has its unique maximum at $\theta = \theta_0$ in Θ
where $l(\theta) = \int_{\mathbb{R}} f(\theta, x)\{f(\theta_0, x) - \frac{1}{2}f(\theta, x)\}\nu(dx).$

(A8) f is twice continuously differentiable function in x with

$$\sup_t E|\, \dot{f}\,(X_t)|^2 < \infty, \qquad \sup_t E|\ddot{f}(X_t)|^2 < \infty.$$

We shall use the following lemma in the sequel.

Lemma 2.1 *(Yoshida (1990)): Let* $\{S_T(\theta), \theta \in \Theta\}, T \geq 0$ *be a family of random fields on* Θ, *a convex compact in* $\mathbb{R}^k$. *Suppose that there exists constants* p *and* l *such that* $p \geq l > k$, *and for any* $\theta, \theta_1, \theta_2$,
(1) $E|S_T(\theta_2) - S_T(\theta_1)|^p \leq C|\theta_2 - \theta_1|^l$,
(2) $E|S_T(\theta)|^p \leq C$,
(3) $S_T(\theta) \to 0$ *in probability,*
where C *is independent of* $\theta, \theta_1, \theta_2$ *and* T. *Then*

$$\sup_{\theta \in \Theta} |S_T(\theta)| \to 0 \quad \text{in probability.}$$

9.3 Weak Convergence of the Approximate Likelihood Ratio Random Fields

Let $\theta = \theta_0 + T^{-1/2}u$, $u \in \mathbb{R}$. Consider the approximate likelihood ratio (ALR) random fields

$$Z_{n,T}(u) = \frac{L_{n,T}(\theta)}{L_{n,T}(\theta_0)}, \quad \widetilde{Z}_{n,T}(u) = \frac{\widetilde{L}_{n,T}(\theta)}{\widetilde{L}_{n,T}(\theta_0)}. \tag{3.1}$$

Let

$$A_{\alpha,T} := \{u \in \mathbb{R} : |u| \leq \alpha, \theta_0 + T^{-1/2}u \in \Theta\}, \ \alpha > 0,$$

$$l_{n,T}(\theta) := \frac{1}{T} \log L_{n,T}(\theta), \quad \widetilde{l}_{n,T}(\theta) = \frac{1}{T} \log \widetilde{L}_{n,T}(\theta),$$

and

$$Df(\theta_0, X_{t_{i-1}}, u) := f(\theta, X_{t_{i-1}}) - f(\theta_0, X_{t_{i-1}}) - T^{-1/2}uf'(\theta_0, X_{t_{i-1}})$$

Yoshida (1992) proved the weak convergence of the ALR random field $Z_{n,T}(u)$ through the following four lemmas. The proofs can be obtained from from Yoshida (1992) when the diffusion coefficient is known. Lemma 3.2 is an analogue of local asymptotic normality (LAN) for ergodic models.

Lemma 3.1 *Under the assumptions (A1) - (A7), we have*

$$\sup_{\theta \in \Theta} |l_{n,T}(\theta) - l(\theta)| \xrightarrow{P} 0 \ \text{as } T \to \infty, n \to \infty \ \text{and} \ \frac{T}{n} \to 0.$$

Lemma 3.2 *Under the assumptions (A1) - (A7), for all* $u \in \mathbb{R}$, *we have*

$$\log Z_{n,T}(u) = u\Delta_{n,T} - \frac{1}{2}u^2\Gamma + r_{n,T}(u)$$

where $\Delta_{n,T} \xrightarrow{\mathcal{D}} \Delta$, $\Delta \sim \mathcal{N}(0,\Gamma)$ and $r_{n,T}(u) \xrightarrow{P} 0$ as $T \to \infty$ and $\frac{T}{n^{2/3}} \to 0$.

Lemma 3.3 *Under the assumptions (A1) - (A7), for each $\epsilon > 0$, we have*

$$\lim_{\delta \to 0} \limsup_{T \to \infty, \frac{T}{n^{2/3}} \to 0} P \left\{ \sup_{u_1, u_2 \in A_{\alpha,T}, |u_2 - u_1| \leq \delta} |\log Z_{n,T}(u_2) - \log Z_{n,T}(u_1)| > \epsilon \right\} = 0.$$

Lemma 3.4 *Under the assumption (A1) - (A7), for each $\epsilon > 0$, we have*

$$\lim_{\alpha \to \infty} \limsup_{T \to \infty, \frac{T}{n^{2/3}} \to 0} P \left\{ \sup_{|u| \geq \alpha} Z_{n,T}(u) > \epsilon \right\} = 0.$$

We prove the weak convergence of the ALR random field $\tilde{Z}_{n,T}(u)$ through the following lemmas.

Lemma 3.5 *Under the assumptions (A1) - (A8),*

$$\sup_{\theta \in \Theta} |\tilde{l}_{n,T}(\theta) - l(\theta)| \xrightarrow{P} 0 \quad as \ T \to \infty \ \ and \ \frac{T}{n} \to 0.$$

Proof: We have

$$\tilde{l}_{n,T}(\theta) = T^{-1} \sum_{i=1}^{n} \frac{f(\theta, X_{t_{i-1}}) + f(\theta, X_{t_i})}{2} (X_{t_i} - X_{t_{i-1}})$$

$$- \frac{1}{2} T^{-1} h \sum_{i=1}^{n} \dot{f}(\theta, X_{t_{i-1}}) - \frac{1}{2} T^{-1} h \sum_{i=1}^{n} f^2(\theta, X_{t_{i-1}})$$

$$= \left\{ T^{-1} \sum_{i=1}^{n} \frac{f(\theta, X_{t_{i-1}}) + f(\theta, X_{t_i})}{2} \Delta W_i - \frac{1}{2} T^{-1} h \sum_{i=1}^{n} \dot{f}(\theta, X_{t_{i-1}}) \right\}$$

$$+ \frac{1}{2} T^{-1} \sum_{i=1}^{n} \int_{t_{i-1}}^{t_i} [f(\theta, X_{t_{i-1}}) + f(\theta, X_{t_i})] f(\theta, X_t) dt$$

$$- \frac{1}{2} T^{-1} h \sum_{i=1}^{n} f^2(\theta, X_{t_{i-1}})$$

$$=: F_1 + F_2 + F_3.$$

Note that

$$F_1 = \left[\left\{ T^{-1} \sum_{i=1}^{n} \frac{f(\theta, X_{t_{i-1}}) + f(\theta, X_{t_i})}{2} \Delta W_i - \frac{1}{2} T^{-1} h \sum_{i=1}^{n} f(\theta, X_{t_{i-1}}) \right\} \right.$$

$$\left. - T^{-1} \sum_{i=1}^{n} f(\theta, X_{t_{i-1}}) \Delta W_i \right] + T^{-1} \sum_{i=1}^{n} f(\theta, X_{t_{i-1}}) \Delta W_i$$

$$=: G_1 + G_2.$$

Further note that

$$
G_1 = T^{-1}\left\{\sum_{i=1}^{n}\frac{f(\theta, X_{t_{i-1}}) + f(\theta, X_{t_i})}{2}\Delta W_i - \oint_0^T f(\theta, X_t)dW_t\right\}
$$

$$
-\frac{1}{2}T^{-1}\left\{h\sum_{i=1}^{n}\dot{f}(\theta, X_{t_{i-1}}) - \int_0^T \dot{f}(\theta, X_t)dt\right\}
$$

$$
-T^{-1}\left\{\sum_{i=1}^{n}f(\theta, X_{t_{i-1}})\Delta W_i - \int_0^T f(\theta, X_t)dW_t\right\}
$$

$$
+T^{-1}\left\{\oint_0^T f(\theta, X_t)dW_t - \frac{1}{2}\int_0^T \dot{f}(\theta, X_t)dt - \int_0^T f(\theta, X_t)dW_t\right\}
$$

$$
=: J_1 + J_2 + J_3 + 0.
$$

The last term here is zero from the relation of Itô and RFS integrals (see (2.20)) of Chapter 7. By Theorem 7.3.1 (b), we have $E|J_1|^2 \leq C\frac{T}{n^2}$. By Theorem 7.3.1 (a), we have $E|J_3|^2 \leq \frac{C}{n}$. By Theorem 7.4.1, we have $E|J_2|^2 \leq C(\frac{T}{n})^2$. Thus $E|G_1|^2 \leq C(\frac{T}{n})^2$. By Burkholder-Davis-Guindy inequality

$$
E|G_2|^2 \leq CT^{-1}, E|F_2|^2 \leq C, E|F_3|^2 \leq C.
$$

Thus $E|\tilde{l}_{n,T}(\theta)|^2 \leq C$. By (A5) we have, $E|\tilde{l}_{n,T}(\theta_2) - \tilde{l}_{n,T}(\theta_1)|^2 \leq C|\theta_2 - \theta_1|$ for $\theta, \theta_1, \theta_2 \in \Theta$. Now use Lemma 2.1. The family of distributions of $\tilde{l}_{n,T}(\cdot)$ on the Banach space $C(\Theta)$ with sup-norm is tight. Since $l(\cdot)$ is a point of $C(\Theta)$ and since by the ergodic property $\tilde{l}_{n,T}(\theta) \xrightarrow{P} l(\theta)$ as $T \to \infty$ and $\frac{T}{n} \to 0$, hence this completes the proof of the lemma. □

The next lemma is an analogue of local asymptotic normality (LAN) for ergodic models for the ALR random field $\tilde{Z}_{n,T}(u)$.

Lemma 3.6 *Under the assumptions (A1) - (A8), for all $u \in \mathbb{R}$,*

$$
\log \tilde{Z}_{n,T}(u) = u\tilde{\Delta}_{n,T} - \frac{1}{2}u^2\Gamma + \tilde{\gamma}_{n,T}(u)
$$

where $\tilde{\Delta}_{n,T} \xrightarrow{D} \Delta$, $\Delta \sim \mathcal{N}(0, \Gamma)$ and $\tilde{\gamma}_{n,T}(u) \xrightarrow{P} 0$ as $T \to \infty$ and $\frac{T}{n^{2/3}} \to 0$.
Proof: We have for $\theta = \theta_0 + T^{-1/2}u$,

$$
\log \tilde{Z}_{n,T}(u) = \log \frac{\tilde{L}_{n,T}(\theta)}{\tilde{L}_{n,T}(\theta_0)}
$$

$$
= \sum_{i=1}^{n}\left[\frac{f(\theta, X_{t_i}) + f(\theta, X_{t_{i-1}})}{2} - \frac{f(\theta_0, X_{t_i}) + f(\theta_0, X_{t_{i-1}})}{2}\right]\Delta X_i
$$

$$-\frac{h}{2}\sum_{i=1}^{n}\left[\dot{f}\left(\theta,X_{t_{i-1}}\right)-\dot{f}\left(\theta_0,X_{t_{i-1}}\right)\right]$$

$$-\frac{h}{2}\sum_{i=1}^{n}\left[f^2(\theta,X_{t_{i-1}})-f^2(\theta_0,X_{t_{i-1}})\right]$$

$$=\frac{1}{2}\sum_{i=1}^{n}\left[\{f(\theta,X_{t_i})-f(\theta_0,X_{t_i})\}+\{f(\theta,X_{t_{i-1}})-f(\theta_0,X_{t_{i-1}})\}\right]\Delta W_i$$

$$+\frac{1}{2}\sum_{i=1}^{n}\int_{t_{i-1}}^{t_i}\left[\{f(\theta,X_{t_i})-f(\theta_0,X_{t_i})\}\right.$$
$$\left.+\{f(\theta,X_{t_{i-1}})-f(\theta_0,X_{t_{i-1}})\}\right]f(\theta_0,X_t)dt$$

$$-\frac{h}{2}\sum_{i=1}^{n}\left[\dot{f}\left(\theta,X_{t_{i-1}}\right)-\dot{f}\left(\theta_0,X_{t_{i-1}}\right)\right]-\frac{h}{2}\sum_{i=1}^{n}\left[f^2(\theta,X_{t_{i-1}})-f^2(\theta_0,X_{t_{i-1}})\right]$$

$$=T^{-1/2}u\sum_{i=1}^{n}\frac{f(\theta_0,X_{t_{i-1}})+f(\theta_0,X_{t_i})}{2}\Delta W_i$$

$$+\frac{1}{2}\sum_{i=1}^{n}\left[Df(\theta_0,X_{t_{i-1}},u)+Df(\theta_0,X_{t_i},u)\right]\Delta W_i$$

$$-\frac{h}{2}T^{-1/2}u\sum_{i=1}^{n}\dot{f}'\left(\theta_0,X_{t_{i-1}}\right)-\frac{h}{2}\sum_{i=1}^{n}D\dot{f}\left(\theta_0,X_{t_{i-1}},u\right)$$

$$+\frac{1}{2}\sum_{i=1}^{n}\int_{t_{i-1}}^{t_i}\left[\{f(\theta,X_{t_i})-f(\theta_0,X_{t_i})\}\right.$$
$$\left.+\{f(\theta,X_{t_{i-1}})-f(\theta_0,X_{t_{i-1}})\}\right]f(\theta_0,X_t)dt$$

$$-\frac{h}{2}\sum_{i=1}^{n}\left[f(\theta,X_{t_{i-1}})-f(\theta_0,X_{t_{i-1}})\right]^2$$

$$-\sum_{i=1}^{n}\int_{t_{i-1}}^{t_i}\left[f(\theta,X_{t_{i-1}})-f(\theta_0,X_{t_{i-1}})\right]f(\theta_0,X_{t_{i-1}})dt$$

$$=:u\widetilde{\Delta}_{n,T}-\frac{1}{2}\Gamma_{n,T}+\widetilde{\rho}_{n,T}(u)$$

where

$$\widetilde{\Delta}_{n,T}=T^{-1/2}\left[\sum_{i=1}^{n}\frac{f'(\theta_0,X_{t_{i-1}})+f'(\theta_0,X_{t_i})}{2}\Delta W_i-\frac{h}{2}\sum_{i=1}^{n}\dot{f}'\left(\theta_0,X_{t_{i-1}}\right)\right],$$

$$\Gamma_{n,T}=h\sum_{i=1}^{n}\left[f(\theta,X_{t_{i-1}})-f(\theta_0,X_{t_{i-1}})\right]^2,$$

and

$$\tilde{\rho}_{n,T}(u)$$

$$= \left\{ \sum_{i=1}^{n} \frac{D\dot{f}(\theta_0, X_{t_{i-1}}, u) + Df(\theta_0, X_{t_i}, u)}{2} \Delta W_i - \frac{h}{2} \sum_{i=1}^{n} Df(\theta_0, X_{t_{i-1}}, u) \right\}$$

$$+ T^{-1/2} u \sum_{i=1}^{n} \int_{t_{i-1}}^{t_i} f'(\theta_0, X_{t_{i-1}}) \left[f(\theta_0, X_t) - f(\theta_0, X_{t_{i-1}}) \right] dt$$

$$+ \sum_{i=1}^{n} \int_{t_{i-1}}^{t_i} Df(\theta_0, X_{t_{i-1}}, u) \left[f(\theta_0, X_t) - f(\theta_0, X_{t_{i-1}}) \right] dt$$

$$+ \frac{1}{2} \sum_{i=1}^{n} \int_{t_{i-1}}^{t_i} \{ [f(\theta, X_{t_i}) - f(\theta_0, X_{t_i})] $$

$$- \left[f(\theta, X_{t_{i-1}}) - f(\theta_0, X_{t_{i-1}}) \right] \} f(\theta_0, X_t) dt$$

$$=: S_1(u) + S_2(u) + S_3(u) + S_4(u).$$

Thus

$$\log \tilde{Z}_{n,T}(u) = u\tilde{\Delta}_{n,T} - \frac{1}{2}\Gamma_{n,T} + \tilde{\rho}_{n,T}(u)$$

$$= u\tilde{\Delta}_{n,T} - \frac{1}{2}u^2\Gamma - \frac{1}{2}(\Gamma_{n,T} - u^2\Gamma) + \tilde{\rho}_{n,T}(u)$$

$$=: u\tilde{\Delta}_{n,T} - \frac{1}{2}u^2\Gamma + \tilde{\gamma}_{n,T}(u).$$

Observe that

$$\Gamma_{n,T} - u^2\Gamma$$

$$= h \sum_{i=1}^{n} [f(\theta, X_{t_{i-1}}) - f(\theta_0, X_{t_{i-1}})]^2 - u^2\Gamma$$

$$= h \sum_{i=1}^{n} [T^{-1/2}uf'(\theta_0, X_{t_{i-1}}) + hDf(\theta_0, X_{t_{i-1}}, u)]^2 - u^2\Gamma$$

$$= \left\{ hT^{-1}u^2 \sum_{i=1}^{n} f'^2(\theta_0, X_{t_{i-1}}) - u^2\Gamma \right\}$$

$$+ h \sum_{i=1}^{n} [Df(\theta_0, X_{t_{i-1}}, u]^2 + 2hT^{-1}u \sum_{i=1}^{n} f'(\theta_0, X_{t_{i-1}})Df(\theta_0, X_{t_{i-1}}, u)$$

$$=: I_1 + I_2 + I_3.$$

By ergodicity, we have $I_1 \to 0$ as $T \to \infty$ and $\frac{T}{n} \to 0$.
Note that by (3.10)

$$hE| \sum_{i=1}^{n} [Df(\theta_0, X_{t_{i-1}}]^2| \leq CnhT^{-2}u^4 = CT^{-1}u^4 \to 0 \text{ as } T \to 0.$$

Hence $I_2 \xrightarrow{P} 0$ as $T \to \infty$.
Further,

$$E|hT^{-1/2}u\sum_{i=1}^{n} f'(\theta_0, X_{t_{i-1}})Df(\theta_0, X_{t_{i-1}}, u)|$$

$$\leq hT^{-1/2}u\sum_{i=1}^{n} E|f'(\theta_0, X_{t_{i-1}})Df(\theta_0, X_{t_{i-1}}, u)|$$

$$\leq hT^{-1/2}u\sum_{i=1}^{n} \left\{E[f'(\theta_0, X_{t_{i-1}})]^2 E[Df(\theta_0, X_{t_{i-1}}, u)]^2\right\}^{1/2}$$

$$\leq hT^{-1/2}u\sum_{i=1}^{n} \left\{E[1 + |X_{t_{i-1}}|^c]^2 CT^{-2}u^4\right\}^{1/2} \quad \text{(by (A4))}$$

$$\leq CnhT^{-3/2}u^3$$

$$= CT^{-1/2}u^3.$$

Hence $I_3 \xrightarrow{P} 0$ as $T \to \infty$.
Now combining I_1, I_2 and I_3, we have

$$\Gamma_{n,T} - u^2\Gamma \xrightarrow{P} 0 \quad \text{as } T \to \infty \text{ and } \frac{T}{n} \to 0.$$

Next

$$\widetilde{\Delta}_{n,T}$$

$$= \left\{T^{-1/2}\left[\sum_{i=1}^{n} \frac{f'(\theta_0, X_{t_{i-1}}) + f'(\theta_0, X_{t_i})}{2}\Delta W_i - \frac{h}{2}\sum_{i=1}^{n} \dot{f}'(\theta_0, X_{t_{i-1}})\right]\right.$$

$$\left. -T^{-1/2}\sum_{i=1}^{n} f'(\theta_0, X_{t_{i-1}})\Delta W_i\right\} + T^{-1/2}\sum_{i=1}^{n} f'(\theta_0, X_{t_{i-1}})\Delta W_i$$

$$=: H_3 + \Delta_{n,T}.$$

Using arguments similar to obtain bounds for G_1, we have $H_3 \xrightarrow{P} 0$ as $\frac{T}{n^{2/3}} \to 0$.
Notice that with $g_i(t) := f'(\theta_0, X_{t_{i-1}}) - f'(\theta_0, X_t)$ for $t_{i-1} \leq t < t_i, i = 1, 2, \ldots, n$,

$$E|T^{-1/2}\sum_{i=1}^{n} f'(\theta_0, X_{t_{i-1}})\Delta W_i - T^{-1/2}\int_0^T f'(\theta_0, X_t)dW_t|$$

$$= T^{-1/2}E|\sum_{i=1}^{n} \int_{t_{i-1}}^{t_i} f'(\theta_0, X_{t_{i-1}})dW_t - \sum_{i=1}^{n} \int_{t_{i-1}}^{t_i} f'(\theta_0, X_t)dW_t|$$

$$= T^{-1/2}E|\sum_{i=1}^{n} \int_{t_{i-1}}^{t_i} \left[f'(\theta_0, X_{t_{i-1}}) - f'(\theta_0, X_t)\right]dW_t|$$

$$= T^{-1/2} E| \int_0^T g_i(t)dW_t|$$

$$\leq T^{-1/2} \left\{ E| \int_0^T g_i(t)dW_t|^2 \right\}^{1/2}$$

$$= T^{-1/2} \left\{ \int_0^T E|g_i(t)|^2 dt \right\}^{1/2}$$

$$\leq T^{-1/2} \left\{ \int_0^T CE|X_{t_{i-1}} - X_t|^2 dt \right\}^{1/2} \quad \text{(by (A4))}$$

$$\leq CT^{-1/2} \left\{ \sum_{i=1}^n \int_{t_{i-1}}^{t_i} (t - t_{i-1})dt \right\}^{1/2} \quad \text{(by (3.11) of Chapter 7)}$$

$$\leq CT^{-1/2} \left\{ n \left(\frac{T}{n} \right)^2 \right\}^{1/2}$$

$$\leq C \left(\frac{T}{n} \right)^{1/2}.$$

Thus

$$T^{-1/2} \sum_{i=1}^n f'(\theta_0, X_{t_{i-1}})\Delta W_i - T^{-1/2} \int_0^T f'(\theta_0, X_t)dW_t \xrightarrow{P} 0 \quad \text{as } \frac{T}{n} \to 0.$$

On the other hand, using the condition (A2), by the central limit theorem for stochastic integrals (see Basawa and Prakasa Rao (1980, Theorem 2.1, Appendix 2)),

$$T^{-1/2} \int_0^T f'(\theta_0, X_t)dW_t \xrightarrow{D} \mathcal{N}(0, \Gamma) \quad \text{as } T \to \infty.$$

Hence $\quad \Delta_{n,T} = T^{-1/2} \sum_{i=1}^n f'(\theta_0, X_{t_{i-1}})\Delta W_i \xrightarrow{D} \mathcal{N}(0, \Gamma)$ as $T \to \infty$ and $\frac{T}{n} \to 0$.

Thus to complete the proof of the lemma we have to show that $\tilde{\rho}_{n,T}(u) \xrightarrow{P} 0$.
Let us first estimate $S_1(u)$. Observe that

$$S_1(u)$$
$$= \left\{ \left[\sum_{i=1}^n \frac{Df(\theta_0, X_{t_{i-1}}, u) + Df(\theta_0, X_{t_i}, u)}{2} \Delta W_i - \frac{h}{2} \sum_{i=1}^n D\dot{f}(\theta_0, X_{t_{i-1}}, u) \right] \right.$$
$$\left. - \sum_{i=1}^n Df(\theta_0, X_{t_{i-1}}, u)\Delta W_i \right\} + \sum_{i=1}^n Df(\theta_0, X_{t_{i-1}}, u)\Delta W_i$$

$=: H_4 + r_1(u).$

Using arguments similar to obtain bounds for G_1, we have $H_4 \overset{P}{\to} 0$ as $\frac{T}{n^{2/3}} \to 0$. Next

$$
\begin{aligned}
E(r_1^2(u)) \\
= E \left[\sum_{i=1}^{n} Df(\theta_0, X_{t_{i-1}}, u) \Delta W_i \right]^2 \\
= \sum_{i=1}^{n} E|Df(\theta_0, X_{t_{i-1}}, u)|^2 E|\Delta W_i|^2 \\
= h \sum_{i=1}^{n} E|Df(\theta_0, X_{t_{i-1}}, u)|^2
\end{aligned}
\tag{3.9}
$$

But

$$
\begin{aligned}
Df(\theta_0, X_t, u) &= (\theta - \theta_0) f'(\theta^*, X_t) - T^{-1/2} u f'(\theta_0, X_{t_{i-1}}) \\
&\quad (\text{where } |\theta^* - \theta_0| < T^{-1/2} u) \\
&= T^{-1/2} u \left[f'(\theta^*, X_t) - f'(\theta_0, X_t) \right].
\end{aligned}
$$

Hence

$$
\begin{aligned}
E|Df(\theta_0, X_{t_{i-1}}, u)|^2 \\
= T^{-1} u^2 E|f'(\theta^*, X_{t_{i-1}}) - f'(\theta_0, X_{t_{i-1}})|^2 \\
\le T^{-1} u^2 E|J(X_{t_{j-1}})(\theta^* - \theta_0)|^2 \\
\le T^{-2} u^4 E[J^2(X_0)] \\
\le C T^{-2} u^4.
\end{aligned}
\tag{3.10}
$$

Substituting (3.10) into (3.9), we obtain

$$
\begin{aligned}
E(r_1^2(u)) &\le C T^{-2} u^4 n h \\
&\le C T^{-1}.
\end{aligned}
$$

Thus $r_1(u) \overset{P}{\to} 0$ as $T \to \infty$. Hence $S_1(u) \overset{P}{\to} 0$ as $T \to \infty$ and $\frac{T}{n^{2/3}} \to 0$. Next let us estimate $S_2(u)$. We have by Itô formula,

$$
\begin{aligned}
f(\theta_0, X_t) - f(\theta_0, X_{t_{i-1}}) \\
= \int_{t_{i-1}}^{t} \dot{f}(\theta_0, X_u) dX_u + \frac{1}{2} \int_{t_{i-1}}^{t} \ddot{f}(\theta_0, X_u) du \\
= \int_{t_{i-1}}^{t} \dot{f}(\theta_0, X_u) dW_u + \int_{t_{i-1}}^{t} [f(\theta_0, X_u) \dot{f}(\theta_0, X_u) + \frac{1}{2} \ddot{f}(\theta_0, X_u)] du \\
=: \int_{t_{i-1}}^{t} \dot{f}(\theta_0, X_u) dW_u + \int_{t_{i-1}}^{t} F(\theta_0, X_u) du.
\end{aligned}
$$

Thus

$$
\begin{aligned}
E|S_2(u)|^2 \\
E|T^{-1/2} u \sum_{i=1}^{n} \int_{t_{i-1}}^{t_i} f'(\theta_0, X_{t_{i-1}}) \left[f(\theta_0, X_t) - f(\theta_0, X_{t_{i-1}}) \right] dt|^2
\end{aligned}
$$

$$= E|T^{-1/2}u\sum_{i=1}^{n}\int_{t_{i-1}}^{t_i}\left[f'(\theta_0,X_{t_{i-1}})\,\dot{f}\,(\theta_0,X_u)dW_u\right.$$

$$\left.+\int_{t_{i-1}}^{t}f'(\theta_0,X_{t_{i-1}})F(\theta_0,X_u)du\right]dt|^2$$

$$\leq 2T^{-1}u^2\left\{E|\sum_{i=1}^{n}\int_{t_{i-1}}^{t_i}\int_{t_{i-1}}^{t}f'(\theta_0,X_{t_{i-1}})\,\dot{f}\,(\theta_0,X_u)dW_udt|^2\right.$$

$$\left.+E|\sum_{i=1}^{n}\int_{t_{i-1}}^{t_i}\int_{t_{i-1}}^{t}f'(\theta_0,X_{t_{i-1}})F(\theta_0,X_u)dudt|^2\right\}$$

$$=: 2T^{-1}u^2(M_1+M_2).$$

Observe that with $B_{i,t}:=\int_{t_{i-1}}^{t}f'(\theta_0,X_{t_{i-1}})\,\dot{f}\,(\theta_0,X_u)dW_u, t_{i-1}\leq t<t_i,$
and $B_{j,s}:=\int_{t_{j-1}}^{s}f'(\theta_0,X_{t_{j-1}})\,\dot{f}\,(\theta_0,X_u)dW_u, t_{j-1}\leq s<t_j, 1\leq i<j\leq n,$
we have

$$M_1=\sum_{i=1}^{n}E\left(\int_{t_{i-1}}^{t_i}B_{i,t}dt\right)^2+\sum_{j\neq i=1}^{n}E\left(\int_{t_{i-1}}^{t_i}B_{i,t}dt\right)\left(\int_{t_{j-1}}^{t_j}B_{j,s}ds\right)$$

$$\leq\sum_{i=1}^{n}(t_i-t_{i-1})\int_{t_{i-1}}^{t_i}E(B_{i,t}^2)dt$$

(the second term is zero due to orthogonality of the integrals $B_{i,t}$ and $B_{j,s}$)

$$\leq\sum_{i=1}^{n}(t_i-t_{i-1})\int_{t_{i-1}}^{t_i}\left\{\int_{t_{i-1}}^{t}E\left[f'(\theta_0,X_{t_{i-1}})\,\dot{f}\,(\theta_0,X_u)\right]^2du\right\}dt$$

$$\leq C\frac{T}{n}\sum_{i=1}^{n}\int_{t_{i-1}}^{t_i}(t-t_{i-1})dt \quad\text{(by (A4) and (A3))}$$

$$\leq C\frac{T}{n}\sum_{i=1}^{n}(t_i-t_{i-1})^2$$

$$= C\frac{T^3}{n^2}.$$

On the other hand, with $N_{i,t}:=\int_{t_{i-1}}^{t}f'(\theta_0,X_{t_{i-1}})F(\theta_0,X_u)du, t_{i-1}\leq t<t_i,$
and $N_{j,s}:=\int_{t_{j-1}}^{s}f'(\theta_0,X_{t_{i-1}})F(\theta_0,X_u)du, t_{j-1}\leq s<t_j, 1\leq i<j\leq n,$ we
have

$$M_2=E|\sum_{i=1}^{n}\int_{t_{i-1}}^{t_i}\int_{t_{i-1}}^{t}f'(\theta_0,X_{t_{i-1}})F(\theta_0,X_u)dudt|^2$$

$$=E|\sum_{i=1}^{n}\int_{t_{i-1}}^{t_i}N_{i,t}dt|^2$$

$$= \sum_{i=1}^{n} E\left(\int_{t_{i-1}}^{t_i} N_{i,t}dt\right)^2 + \sum_{j\neq i=1}^{n} E\left(\int_{t_{i-1}}^{t_i} N_{i,t}dt\right)\left(\int_{t_{j-1}}^{t_j} N_{j,s}ds\right)$$

$$\leq \sum_{i=1}^{n} (t_i - t_{i-1})E\left(\int_{t_{i-1}}^{t_i} N_{i,t}dt\right)^2$$

$$+ \sum_{j\neq i=1}^{n} \left\{E(\int_{t_{i-1}}^{t_i} N_{i,t}dt)^2 E(\int_{t_{j-1}}^{t_j} N_{j,s}ds)^2\right\}^{1/2}$$

$$\leq \sum_{i=1}^{n} (t_i - t_{i-1})^{3/2}\left(\int_{t_{i-1}}^{t_i} E(N_{i,t}^2)dt\right)^{1/2}$$

$$+ \sum_{j\neq i=1}^{n} \left\{(t_i - t_{i-1})\int_{t_{i-1}}^{t_i} E(N_{i,t}^2)dt\,(t_j - t_{j-1})\int_{t_{j-1}}^{t_j} E(N_{j,s}^2)ds\right\}^{1/2}$$

But $E(N_{i,t}^2) \leq C(t - t_{i-1})^2, 1 \leq i \leq n$ using (A4) and (A3). On substitution, the last term is bounded by

$$C\sum_{i=1}^{n} (t_i - t_{i-1})^3 + C\sum_{j\neq i=1}^{n} (t_i - t_{i-1})^2(t_j - t_{j-1})^2$$
$$= C\frac{T^3}{n^2} + C\frac{n(n-1)T^4}{2n^4} \leq C\frac{T^4}{n^2}$$

Hence

$$E|S_2(u)|^2 \leq CT^{-1}u^2\frac{T^4}{n^2} \leq C(\frac{T}{n^{2/3}})^3 u^2.$$

Combining bounds for M_1 and M_2, we have $S_2(u)\overset{P}{\to}0$ as $\frac{T}{n^{2/3}} \to 0$. Next let us estimate $S_3(u)$. Note that

$$E|S_3(u)|$$

$$= E|\sum_{i=1}^{n} \int_{t_{i-1}}^{t_i} Df(\theta_0, X_{t_{i-1}}, u)\left[f(\theta_0, X_t) - f(\theta_0, X_{t_{i-1}})\right]dt|$$

$$\leq \sum_{i=1}^{n} \int_{t_{i-1}}^{t_i} E|Df(\theta_0, X_{t_{i-1}}, u)||f(\theta_0, X_t) - f(\theta_0, X_{t_{i-1}})|dt$$

$$\leq \sum_{i=1}^{n} \int_{t_{i-1}}^{t_i} \left\{E|Df(\theta_0, X_{t_{i-1}}, u)|^2 E|f(\theta_0, X_t) - f(\theta_0, X_{t_{i-1}})|^2\right\}^{1/2} dt$$

$$\leq \sum_{i=1}^{n} \int_{t_{i-1}}^{t_i} \left\{CT^{-2}u^4 E|X_t - X_{t_{i-1}}|^2\right\}^{1/2} dt \quad \text{(by (3.10) and (A1))}$$

$$\leq CT^{-1}u^2 \sum_{i=1}^{n} \int_{t_{i-1}}^{t_i} (t - t_{i-1})^{1/2}dt \quad \text{(by (3.11) of Chapter 7)}$$

$$\leq CT^{-1}u^2 \sum_{i=1}^{n} \int_{t_{i-1}}^{t_i} (t_i - t_{i-1})^{1/2} dt$$

$$\leq CT^{-1}u^2 n \left(\frac{T}{n}\right)^{3/2} \leq C \left(\frac{T}{n}\right)^{1/2} u^2.$$

Thus $S_3(u) \xrightarrow{P} 0$ as $\frac{T}{n} \to 0$. Next let us estimate $S_4(u)$. Note that

$$2S_4(u)$$

$$= \sum_{i=1}^{n} \int_{t_{i-1}}^{t_i} \left\{ [f(\theta, X_{t_i}) - f(\theta_0, X_{t_i})] - [f(\theta, X_{t_{i-1}}) - f(\theta_0, X_{t_{i-1}})] \right\} f(\theta_0, X_t) dt$$

$$= T^{-1/2} u \sum_{i=1}^{n} \int_{t_{i-1}}^{t_i} \left[f'(\theta^*, X_{t_i}) - f'(\theta^{**}, X_{t_{i-1}}) \right] f(\theta_0, X_t) dt$$

where $|\theta^* - \theta_0| < T^{-1/2}u, |\theta^{**} - \theta_0| < T^{-1/2}u$.

Now proceeding similarly as in the proof of convergence of $S_2(u)$ to zero in probability, it can be shown that $S_4(u) \xrightarrow{P} 0$ as $\frac{T}{n^{2/3}} \to 0$. This completes the proof of the lemma. $\qquad\square$

Lemma 3.7 *Under the assumptions (A1) - (A8), for each $\epsilon > 0$,*

$$\lim_{\delta \to 0} \limsup_{T \to \infty, \frac{T}{n^{2/3}} \to 0} P \left\{ \sup_{u_1, u_2 \in A_{\alpha,T}, |u_2 - u_1| \leq \delta} |\log \tilde{Z}_{n,T}(u_2) - \log \tilde{Z}_{n,T}(u_1)| > \epsilon \right\} = 0.$$

Proof: From Lemma 3.6, we have

$$\left| \log \tilde{Z}_{n,T}(u_2) - \log \tilde{Z}_{n,T}(u_1) \right|$$

$$= \left| (u_2 - u_1) \tilde{\Delta}_{n,T} - \frac{1}{2}(u_2^2 - u_1^2)\Gamma + \tilde{\gamma}_{n,T}(u_2) - \tilde{\gamma}_{n,T}(u_1) \right|$$

$$\leq |u_2 - u_1||\tilde{\Delta}_{n,T}| + K|u_2 - u_1| + |\tilde{\gamma}_{n,T}(u_2)| + |\tilde{\gamma}_{n,T}(u_1)|$$

where K is a positive constant. Therefore

$$P \left\{ \sup_{u_1, u_2, \in A_{\alpha,T}, |u_2 - u_1| \leq \delta} |\log \tilde{Z}_{n,T}(u_2) - \log \tilde{Z}_{n,T}(u_1)| > \epsilon \right\}$$

$$\leq P \left\{ |\tilde{\Delta}_{n,T}| + K > \frac{\epsilon}{3\delta} \right\} + 2P \left\{ \sup_{u \in A_{\alpha,T}} |\tilde{\gamma}_{n,T}(u_1)| > \frac{\epsilon}{3} \right\}$$

Note that

$$P \left\{ \sup_{u \in A_{\alpha,T}} |\tilde{\gamma}_{n,T}(u)| > \frac{\epsilon}{3} \right\}$$

$$= P\left\{\sup_{|u|\leq\alpha} |S_1(u) + S_2(u) + S_3(u) + S_4(u)| > \frac{\epsilon}{6}\right\}$$

$$+ P\left\{\sup_{u\in A_{\alpha,T}} |\Gamma_{n,T} - u^2\Gamma| > \frac{\epsilon}{6}\right\}$$

$$\leq P\left\{\sup_{|u|\leq\alpha} |S_1(u)| > \frac{\epsilon}{24}\right\} + P\left\{\sup_{|u|\leq\alpha} |S_2(u)| > \frac{\epsilon}{24}\right\}$$

$$+ P\left\{\sup_{|u|\leq\alpha} |S_3(u)| > \frac{\epsilon}{24}\right\} + P\left\{\sup_{|u|\leq\alpha} |S_4(u)| > \frac{\epsilon}{24}\right\}$$

$$+ P\left\{\sup_{u\in A_{\alpha,T}} |\Gamma_{n,T} - u^2\Gamma| > \frac{\epsilon}{6}\right\}.$$

By Burkholder-Davis-Gundy inequality and Novikov's moment inequality it can be proved that

$$E|S_i(u)|^{2p} \leq C,$$
$$E|S_i(u_2) - S_i(u_1)|^{2p} \leq C|u_2 - u_1|^{2p}, \quad p > 1, \quad i = 1, 2, 3, 4.$$

Since $S_i(u) \to^P 0$ as $T \to \infty$ and $\frac{T}{n^{2/3}} \to 0, i = 1, 2, 3, 4$ from the proof of Lemma 3.6, hence upon using Lemma 2.1, we have for $i = 1, 2, 3, 4$

$$\sup_{u\in A_{\alpha,T}} |S_i(u)| \to^P 0 \text{ as } T \to \infty \text{ and } \frac{T}{n^{2/3}} \to 0.$$

Similarly it can be shown that

$$\sup_{u\in A_{\alpha,T}} |\Gamma_{n,T} - u^2\Gamma| \to^P 0 \text{ as } T \to \infty \text{ and } \frac{T}{n} \to 0.$$

Since $\widetilde{\Delta}_{n,T}$ converges in distribution to $\mathcal{N}(0, \Gamma)$, hence

$$\lim_{\delta\to 0} \lim_{T\to\infty, \frac{T}{n^{2/3}}\to 0} P\left\{\sup_{u_1,u_2\in A_{\alpha,T}, |u_2-u_1|\leq\delta} |\log\widetilde{Z}_{n,T}(u_2) - \log\widetilde{Z}_{n,T}(u_1)| > \epsilon\right\} = 0.$$

$\square$

Lemma 3.8. *Under the assumptions (A1) - (A8), we have for each $\epsilon > 0$,*

$$\lim_{\alpha\to\infty} \lim_{T\to\infty, \frac{T}{n^{2/3}}\to 0} \sup P\left\{\sup_{|u|\geq\alpha} \widetilde{Z}_{n,T}(u) > \epsilon\right\} = 0.$$

Proof: Since Γ is positive, there exists a number η such that $\eta u^2 \leq \frac{1}{4}\Gamma u^2$, $u \in \mathbb{R}$. By Lemma 3.6 $\log\widetilde{Z}_{n,T}(u) = u\widetilde{\Delta}_{n,T} - \frac{1}{2}u^2\Gamma + \widetilde{\gamma}_{n,T}(u)$. Let $\bar{S}_i(u) := \frac{1}{1+u^2}S_i(u)$, $i = 1, 2, 3, 4$.

and $U_1 := \{u : |u| \leq \delta T^{1/2}\}$ for any $\delta > 0$. For $p > 1$,

$$E|\bar{S}_1(u)|^{2p} \leq CT^{-p},$$
$$E|\bar{S}_1(u_2) - \bar{S}_1(u_1)|^{2p} \leq CT^{-p}|u_2 - u_1|^{2p}.$$

Therefore, using Lemma 2.1, $\sup_{u \in U_1} |\bar{S}_1(u)| \overset{P}{\to} 0$ as $T \to \infty$. Similarly it can be shown that $\sup_{u \in U_1} |\bar{S}_3(u)| \overset{P}{\to} 0$ as $\dfrac{T}{n} \to 0$. Next

$$S_2(u) = T^{-1/2}u \sum_{i=1}^{n} \int_{t_{i-1}}^{t_i} f'(\theta_0, X_{t_{i-1}}) \left[f(\theta_0, X_t) - f(\theta_0, X_{t_{i-1}}) \right] dt$$

$$= T^{-1/2}u \sum_{i=1}^{n} f'(\theta_0, X_{t_{i-1}}) \int_{t_{i-1}}^{t_i} \left(\int_{t_{i-1}}^{t} \dot{f}(\theta_0, X_u) dW_u \right) dt$$

$$+ T^{-1/2}u \sum_{i=1}^{n} f'(\theta_0, X_{t_{i-1}}) \int_{t_{i-1}}^{t_i} \left(\int_{t_{i-1}}^{t} [f(\theta_0, X_u) \dot{f}(\theta_0, X_u) \right.$$
$$\left. + \tfrac{1}{2}\ddot{f}(\theta_0, X_u)] du \right) dt$$
(by Itô formula)
$$=: S_{21}(u) + S_{22}(u).$$

Define $\bar{S}_{2j}(u) := \frac{1}{1+u^2} S_{2j}(u), \ \ j = 1, 2.$ It is easy to show that $\sup_{u \in U_1} |\bar{S}_{22}(u)| \overset{P}{\to} 0$ as $\frac{T}{n^{2/3}} \to 0$. As in the estimation of $\bar{S}_1$, we can show that for $p \geq 1$,

$$E|\bar{S}_{21}(u)|^{2p} \leq C \left(\frac{T}{n} \right)^p,$$

$$E|\bar{S}_{21}(u_2) - \bar{S}_{21}(u_1)|^{2p} \leq C \left(\frac{T}{n} \right)^{2p} |u_2 - u_1|^{2p}$$

Hence using Lemma 2.1, $\sup_{u \in U_1} |\bar{S}_{21}(u)| \overset{P}{\to} 0$ as $\dfrac{T}{n} \to 0$.

Thus $\sup_{u \in U_1} \dfrac{|\tilde{\rho}_{n,T}(u)|}{1 + u^2} \overset{P}{\to} 0$ as $T \to \infty$ and $\dfrac{T}{n^{2/3}} \to 0$.

On the other hand, for any $\epsilon > 0$, if $\delta > 0$ is small enough, by (A2) we have

$$\lim_{T \to \infty, \frac{T}{n} \to 0} P \left\{ \sup_{u \in U_1} \frac{1}{1 + u^2} |u^2 \Gamma - \Gamma_{n,T}| < \epsilon \right\} = 1.$$

Note that

$$\tilde{\gamma}_{n,T} := \tilde{\rho}_{n,T} - \frac{1}{2}(\Gamma_{n,T} - u^2 \Gamma).$$

Hence, for any $\epsilon > 0$, for small $\delta > 0$, $\lim_{T \to \infty, \frac{T}{n} \to 0} P\{V_{n,T}^c\} = 0$

where $V_{n,T} := \{ \sup_{u \in U_1} \dfrac{1}{1 + u^2} |\tilde{\gamma}_{n,T}(u)| < \epsilon \}.$

Let $\epsilon < \eta$. On the event $V_{n,T}$, if $|u| \leq \delta T^{1/2}$, then

$$\log \tilde{Z}_{n,T}(u) \le |u| |\tilde{\Delta}_{n,T}| - \frac{1}{2} u^2 \Gamma + |\tilde{\gamma}_{n,T}(u)|$$

$$\le |u| |\tilde{\Delta}_{n,T}| - \frac{1}{2} u^2 \Gamma + \epsilon(1 + u^2)$$

$$\le |u| |\tilde{\Delta}_{n,T}| - \frac{1}{2} u^2 \Gamma + \eta u^2 + \epsilon$$

$$\le |u| |\tilde{\Delta}_{n,T}| - \eta u^2 + \epsilon.$$

Let $U_2 := \{u : q \le |u| \le \delta T^{1/2}\}$ where q is a positive number. Then

$$P\left\{ \sup_{u \in U_2} \tilde{Z}_{n,T}(u) \ge \exp(-\frac{\eta q^2}{2}) \right\}$$

$$\le P(V_{n,T}^c) + P\left\{ \sup_{u \in U_2} \left(|u| |\tilde{\Delta}_{n,T}| - \eta u^2 \right) + \epsilon \ge -\frac{\eta q^2}{2} \right\}$$

$$\le o(1) + P\left\{ q|\tilde{\Delta}_{n,T}| - \eta q^2 + \epsilon \ge -\frac{\eta q^2}{2} \right\} + P\left\{ |\tilde{\Delta}_{n,T}| > 2\eta q \right\}$$

$$\le 2P\left\{ |\tilde{\Delta}_{n,T}| > \frac{\eta q}{2} - \frac{\epsilon}{q} \right\} + o(1).$$

Let χ and τ be arbitrary positive numbers. For large q, $\exp(\frac{-\eta q^2}{2}) < \tau$ and

$$\limsup_{T \to \infty, \frac{T}{n^{2/3}} \to 0} P\left\{ |\tilde{\Delta}_{n,T}| > \frac{\eta q}{2} - \frac{\epsilon}{q} \right\} < \frac{\chi}{3}.$$

Then $\displaystyle \limsup_{T \to \infty, \frac{T}{n^{2/3}} \to 0} P\left\{ \sup_{u \in U_2} \tilde{Z}_{n,T}(u) \ge \tau \right\} \le \chi.$

Define $U_3 := \{u : |u| \ge \delta T^{1/2}\}$ and $H_1 := \{y : |y| \ge \delta\}$. Then for $v > 0$, we have

$$\limsup_{T \to \infty, \frac{T}{n^{2/3}} \to 0} P\left\{ \sup_{u \in U_3} \tilde{Z}_{n,T}(u) \ge \tau \right\}$$

$$= \limsup_{T \to \infty, \frac{T}{n^{2/3}} \to 0} P\left\{ \sup_{y \in H_1} \left[\tilde{l}_{n,T}(\theta_0 + y) - \tilde{l}_{n,T}(\theta_0) \right] \ge T^{-1} \log \tau \right\}$$

$$\le \limsup_{T \to \infty, \frac{T}{n^{2/3}} \to 0} P\left\{ \sup_{y \in H_1} [\tilde{l}_{n,T}(\theta_0 + y) - l(\theta_0 + y) + \tilde{l}_{n,T}(\theta_0) - l(\theta_0)] \ge v \right\}$$

$$+ \limsup_{T \to \infty, \frac{T}{n^{2/3}} \to 0} P\left\{ \sup_{y \in H_1} [l(\theta_0 + y) - l(\theta_0)] \ge T^{-1} \log \tau - v \right\}$$

If v is small the second term in the r.h.s. is zero. The first term tends to zero from Lemma 3.5. Therefore, for $\tau > 0$ and $\chi > 0$, if q is large

$$\limsup_{T \to \infty, \frac{T}{n^{2/3}} \to 0} P\left\{ \sup_{|u| \ge q} \tilde{Z}_{n,T}(u) > \tau \right\} \le \chi. \qquad \square$$

9.4 Asymptotics of Approximate Estimators and Bernstein-von Mises Type Theorems

Let $C_0(\mathbb{R})$ be the Banach space of real valued continuous functions on $\mathbb{R}$ vanishing at the infinity with sup-norm. Let

$$U_{n,T} = \left\{ u : \theta_0 + T^{-1/2}u \in \Theta \right\}.$$

For $u \in U_{n,T}, Z_{n,T}(u)$ and $\tilde{Z}_{n,T}(u)$ have been defined and extend it to an element of $C_0(\mathbb{R})$ whose maximal points are contained in $U_{n,T}$. Using the weak convergence of the random field $Z_{n,T}(u)$ (Lemma 3.2, Lemma 3.3 and Lemma 3.4), we obtain the following results. Below $\Delta \sim \mathcal{N}(0, \Gamma)$ as defined in Lemma 3.2.

Theorem 4.1 *Under the conditions (A1) - (A7), we have*

$$Z_{n,T}(\cdot) \overset{D}{\to} Z(\cdot) \quad \text{in } C_0(\mathbb{R}) \text{ as } T \to \infty \text{ and } \frac{T}{n^{2/3}} \to 0$$

$$\text{where } Z(\cdot) = \exp(u\Delta - \frac{1}{2}u^2\Gamma),$$

i.e., for any continuous functional g on $C_0(\mathbb{R})$

$$E\left[g(Z_{n,T}(\cdot))\right] \to E[g(Z(\cdot))] \text{ as } T \to \infty \text{ and } \frac{T}{n^{2/3}} \to 0.$$

In particular for the AMLE1, AMPE1, ABE1 and APD1, as $T \to \infty$ and $\frac{T}{n^{2/3}} \to 0$,

(a) $T^{1/2}(\hat{\theta}_{n,T} - \theta_0) \overset{D}{\to} \Gamma^{-1}\Delta \sim \mathcal{N}(0, \Gamma^{-1})$

(b) $T^{1/2}(\bar{\theta}_{n,T} - \theta_0) \overset{D}{\to} \Gamma^{-1}\Delta,$

(c) $T^{1/2}(\tilde{\theta}_{n,T} - \theta_0) \overset{D}{\to} \Gamma^{-1}\Delta,$

(d) $\int_{-\infty}^{\infty} \left| p_{n,T}^*(t|X_0^{n,h}) - (\frac{\Gamma}{2\pi})^{1/2} \exp(-\frac{1}{2}\Gamma t^2) \right| dt \overset{P}{\to} 0.$

Proof: We use Lemma 3.2, Lemma 3.3 and Lemma 3.4 to prove the theorem. With the aid of these lemmas, following the arguments in the proof of Theorem III.1.1 of Ibragimov and Hasminskii (1981) or Theorem 3.4.1 of Kutoyants (1984a), (a) follows. Following the arguments in the proof of Theorem III.2.1 of Ibragimov and Hasminskii (1981) or Theorem 3.4.2 of Kutoyants (1984a) (c) follows. Folowing the arguments in of Wolfowitz (1975), (b) follows. Following the arguments in Theorem 1 in Ghosal, Ghosh and Samanta (1995), (d) follows. □

Corollary 4.2 *Under the assumptions of Theorem 4.1, we have*
$T^{1/2}(\widetilde{\theta}_{n,T} - \hat{\theta}_{n,T}) \overset{P}{\to} 0$ *as* $T \to \infty$ *and* $\frac{T}{n^{2/3}} \to 0$,
i.e., AMLE1 and ABE1 are asymptotically equivalent.

Proof Using Lemmas 3.2, Lemma 3.3 and Lemma 3.4, all the conditions of Corollary 1 in Ghosal, Ghosh and Samanta (1995) are satisfied. Hence the result follows from the said Corollary. □

Theorem 4.3 *Under the conditions (A1) - (A8), we have*

$$\widetilde{Z}_{n,T}(\cdot) \overset{D}{\to} Z(\cdot) \ \text{in} \ C_0(\mathbb{R}) \ \text{as} \ T \to \infty \ \text{and} \ \frac{T}{n^{2/3}} \to 0$$

$$\text{where} \ Z(u) = \exp(u\Delta - \frac{1}{2}u^2\Gamma),$$

i.e, for any continuous functional g on $C_0(\mathbb{R})$,

$$E\left[g(\widetilde{Z}_{n,T}(\cdot))\right] \to E\left[g(Z(\cdot))\right] \ \text{as} \ T \to \infty \ \text{and} \ \frac{T}{n^{2/3}} \to 0.$$

In particular for the AMLE2, AMPE2, ABE2 and APD2, as $T \to \infty$ and $\frac{T}{n^{2/3}} \to 0$,

(a) $T^{1/2}(\hat{\hat{\theta}}_{n,T} - \theta_0) \overset{D}{\to} \Gamma^{-1}\Delta \sim \mathcal{N}(0, \Gamma^{-1})$

(b) $T^{1/2}(\overline{\overline{\theta}}_{n,T} - \theta_0) \overset{D}{\to} \Gamma^{-1}\Delta,$

(c) $T^{1/2}(\widetilde{\widetilde{\theta}}_{n,T} - \theta_0) \overset{D}{\to} \Gamma^{-1}\Delta,$

(d) $\displaystyle\int_{-\infty}^{\infty} \left| \widetilde{p}^*_{n,T}(s|X_0^{n,h}) - (\frac{\Gamma}{2\pi})^{1/2}\exp(-\frac{1}{2}\Gamma s^2) \right| ds \overset{P}{\to} 0$

Proof: Using the weak convergence of the random field $\widetilde{Z}_{n,T}(u)$ (Lemma 3.6, Lemma 3.7 and Lemma 3.8) arguments are same as those for the proof of Theorem 6.4.1. □

Corollary 4.4 *Under the assumptions of Theorem 4.2, we have*
$T^{1/2}(\hat{\hat{\theta}}_{n,T} - \widetilde{\widetilde{\theta}}_{n,T}) \overset{P}{\to} 0$ *as* $T \to \infty$ *and* $\frac{T}{n^{2/3}} \to 0$,
i.e., AMLE2 and ABE2 are asymptotically equivalent.

Proof Using Lemma 3.6, Lemma 3.7 and Lemma 3.8 all the conditions Corollary 1 in Ghosal, Ghosh and Samanta (1995). Hence the result follows from the said Corollary. □

9.5 Example: Logistic Diffusion

Consider the stochastic analogue of the logistic growth model:

$$dX_t = \alpha X_t(1 - X_t/\beta)dt + \sigma X_t dW_t, \quad X_0 = x_0 > 0, \quad t \in [0, T]$$

for positive parameters α, β and σ. This diffusion is useful for modeling the growth of populations. The instantaneous population of some species X_t grows, in the absence of any restraints, exponentially fast in t with growth rate per individual equal to α. The actual evolution of the population is cut back by the saturation inducing the term $(1 - X_t/\beta)$. The constant β is called the carrying capacity of the environment and usually represents the maximum population that can be supported by the resources of the environment. The parameter σ represents the effect of the noise on the dynamics of X.

After setting $Y_t = -\log(X_t)/\sigma$, we obtain the SDE

$$dY_t = \left\{ \frac{\sigma}{2} - \frac{\alpha}{\sigma} + \frac{\alpha}{\sigma\beta} \exp(-\sigma X_t) \right\} dt + dW_t, \quad Y_0 = -\frac{\log(x_0)}{\sigma}, \quad t \in [0, T].$$

The original process X does not hit 0 in finite time for any values of the parameters with probability 1.

Remarks

(1) Theorem 4.1 (d) and Theorem 4.3 (d) give two versions of the *Bernstein-von Mises type theorem* for discretely observed diffusions.

(2) For the conditional least squares estimator, Prakasa Rao (1983) using the Cramer's approach assumed stronger conditions on the drift function for obtaining asymptotic normality, e.g., existence of third derivative of drift function with respect to the parameter. Here we assume existence of only second derivative through the Hajek-Le Cam approach. We also relax the RIED condition by using the MIED condition.

(3) The problem of obtaining the rates of convergence to normality of the AMLEs, the AMPEs, the ABEs and the CLSE now remain open.

The rates of convergence of approximate posterior densities to normal density also remain open. Note that in the linear case we have shown in Chapters 4 and 5 that AMLE2 has a faster rate of convergence than AMLE1. We conjecture that the estimators and the posterior based on the second type of approximate likelihood would have faster rates of convergence than those based on the first type of approximate likelihood.

(4) Generalization of the results of this Chapter to multiparameter case is worth investigating.

(5) It remains open if one can relax the MIED condition to obtain the limit distributions of different estimators.

Estimating Function for Discretely Observed Homogeneous Diffusions

10.1 Introduction

This chapter shows that discretization after the application of Itô formula in the Girsanov likelihood produces estimators of the drift which have faster rates of convergence than the Euler estimator for stationary ergodic diffusions and is free of approximating the stochastic integral. The discretization schemes are related to the Hausdorff moment problem. *Interalia* we use a new stochastic integral which will be of independent interest. We show strong consistency, asymptotic normality and a Berry-Esseen bound for the corresponding approximate maximum likelihood estimators of the drift parameter from high frequency data observed over a long time period.

Parameter estimation in diffusion processes based on discrete observations is one of the hottest topic in finance in recent days, see Ait-Sahalia (2002), Ait-Sahalia and Mykland (2003, 2004), Elerian et al. (2001), Duffie and Glynn (2004), Bibby and Sorensen (1995), Kessler (1997), Clement (1997), Brandt and Santa-Clara (2002), Genon-Catalot (1990). Here our aim is to study estimators with higher order accuracy.

This chapter is adapted from Bishwal (2007a).

Suppose the diffusion process satisfies the Itô stochastic differential equation

$$
\begin{aligned}
dX_t &= f(\theta, X_t)dt + dW_t, \ t \geq 0 \\
X_0 &= \xi
\end{aligned}
\tag{1.1}
$$

where $\{W_t, t \geq 0\}$ is a one dimensional standard Wiener process, $\theta \in \Theta$, Θ is a compact subset of $\mathbb{R}$, f is a known real valued function defined on $\Theta \times \mathbb{R}$, the unknown parameter θ is to be estimated on the basis of discrete observations of the process $\{X_t, t \geq 0\}$. Let θ_0 be the true value of the parameter which is in the interior of Θ. We assume that the process $\{X_t, t \geq 0\}$ is observed at $0 = t_0 < t_1 < \ldots < t_n = T$ with $\Delta t_i := t_i - t_{i-1} = \frac{T}{n} = h$ for some fixed real number $d > 0$. We estimate θ from the high frequency ($h \to 0$) data $\{X_{t_0}, X_{t_1}, \ldots, X_{t_n}\}$.

We start with least squares method. The conditional least squares estimator (CLSE) of θ is defined as

$$\theta_{n,T} := \arg\min_{\theta \in \Theta} Q_{n,T}(\theta)$$

where $Q_{n,T}(\theta) := \sum_{i=1}^{n} \frac{\left[X_{t_i} - X_{t_{i-1}} - f(\theta, X_{t_{i-1}})h\right]^2}{\Delta t_i}$.

This estimator was first studied by Dorogovcev (1976) who obtained its weak consistency under some regularity conditions as $T \to \infty$ and $\frac{T}{n} \to 0$. Kasonga (1988) obtained the strong consistency of the CLSE under some regularity conditions as $n \to \infty$ assuming that $T = dn^{1/2}$ for some fixed real number $d > 0$. Bishwal (2002b) obtained strong consistency and conditional asymptotic normality of the CLSE for nonlinear nonhomogeneous diffusions.

Florens-Zmirou (1989) studied minimum contrast estimator, based on an Euler-Maruyama type first order approximate discrete time scheme of the SDE (1.1) which is given by

$$Z_{t_i} - Z_{t_{i-1}} = f(\theta, Z_{t_{i-1}})(t_i - t_{i-1}) + W_{t_i} - W_{t_{i-1}}, \ i \geq 1, \ Z_0 = X_0.$$

The log-likelihood function of $\{Z_{t_i}, 0 \leq i \leq n\}$ is given by

$$L_{n,T}(\theta) := C \sum_{i=1}^{n} \frac{\left[Z_{t_i} - Z_{t_{i-1}} - f(\theta, Z_{t_{i-1}})h\right]^2}{\Delta t_i}$$

where C is a constant independent of θ. A contrast for the estimation of θ is derived from the above log-likelihood by substituting $\{Z_{t_i}, 0 \leq i \leq n\}$ with $\{X_{t_i}, 0 \leq i \leq n\}$. The resulting contrast is

$$H_{n,T}(\theta) := C \sum_{i=1}^{n} \frac{\left[X_{t_i} - X_{t_{i-1}} - f(\theta, X_{t_{i-1}})h\right]^2}{\Delta t_i}$$

and the resulting minimum contrast estimator, called the Euler-Maruyama estimator, is given by

$$\bar{\theta}_{n,T} := \arg\min_{\theta \in \Theta} H_{n,T}(\theta)$$

Florens-Zmirou (1989) showed consistency of this estimator as $T \to \infty$ and $\frac{T}{n} \to 0$ and asymptotic normality as $T \to \infty$ and $\frac{T}{n^{2/3}} \to 0$. Yoshida (1992) extended Florens-Zmirou (1989) to multidimensional case.

If continuous observation of $\{X_t\}$ on the interval $[0, T]$ were available, then the likelihood function of θ would be

$$L_T(\theta) := \exp\left\{\int_0^T f(\theta, X_t)dX_t - \frac{1}{2}\int_0^T f^2(\theta, X_t)dt\right\} \tag{1.2}$$

Since we have discrete data, we have to approximate the likelihood to obtain the MLE. Taking Itô type approximation of the stochastic integral and rectangle rule approximation of the ordinary integral in (1.2) we obtain the approximate likelihood function

$$\hat{L}_{n,T}(\theta) := \exp\left\{ \sum_{i=1}^{n} f(\theta, X_{t_{i-1}})(X_{t_i} - X_{t_{i-1}}) - \frac{h}{2}\sum_{i=1}^{n} f^2(\theta, X_{t_{i-1}}) \right\}. \quad (1.3)$$

Itô-Euler approximate maximum likelihood estimator (IEAMLE) based on $\hat{L}_{n,T}$ is defined as

$$\hat{\theta}_{n,T} := \arg\max_{\theta \in \Theta} \hat{L}_{n,T}(\theta).$$

Weak consistency and asymptotic normality of this estimator were obtained by Yoshida (1992) as $T \to \infty$ and $\frac{T}{n^{2/3}} \to 0$.

Note that the CLSE, the Euler-Maruyama estimator and the IEAMLE are the same estimator (see Shoji (1997)).

Prime denotes derivative with respect to θ and dot denotes derivative with respect to x. Let $\Phi(x)$ denote standard normal distribution function.

In order to obtain better estimators, in terms of having faster rates of convergence, first we discuss the algorithm proposed in Mishra and Bishwal (1995). Note that the Itô and the Fisk-Stratonovich (FS, hence forth) integrals are connected by

$$\int_0^T f(\theta, X_t)dX_t = \int_0^T f(\theta, X_t) \, o \, dX_t - \frac{1}{2}\int_0^T \dot{f}(\theta, X_t)dt. \quad (1.4)$$

where o is the Itô's circle put for FS integral (see Ikeda and Watanabe (1989)). We transform the Itô integral in (1.2) to FS integral and apply FS type approximation of the stochastic integral and rectangular rule type approximation of the Lebesgue integrals and obtain the approximate likelihood

$$\tilde{L}_{n,T}(\theta) := \exp\left\{ \frac{1}{2}\sum_{i=1}^{n}[f(\theta, X_{t_{i-1}}) + f(\theta, X_{t_i})](X_{t_i} - X_{t_{i-1}}) \right.$$

$$\left. -\frac{h}{2}\sum_{i=1}^{n} \dot{f}(\theta, X_{t_{i-1}}) - \frac{h}{2}\sum_{i=1}^{n} f^2(\theta, X_{t_{i-1}}) \right\} \quad (1.5)$$

The Fisk-Stratonovich approximate maximum likelihood estimator (FSAMLE) based on $\tilde{L}_{n,T}$ is defined as

$$\tilde{\theta}_{n,T} := \arg\max_{\theta \in \Theta} \tilde{L}_{n,T}(\theta).$$

FSAMLE was proposed in Mishra and Bishwal (1995) and its wick consistency and rate of convergence were studied. Bishwal (2006b) studied local asymptotic normality (LAN) property of the model.

We propose a new algorithm to obtain several approximate likelihoods. Using Itô formula, we can write

$$L_T(\theta) := \exp\left\{A(\theta, X_T) - A(\theta, X_0) - \frac{1}{2}\int_0^T [f^2(\theta, X_t) + \dot{f}(\theta, X_t)]dt\right\} \quad (1.6)$$

where

$$A(\theta, x) := \int^x f(\theta, y)dy.$$

Observe that this form of likelihood is free of any stochastic integral. Taking a rectangular approximation of the Lebesgue integral in (1.6), we obtain another approximate likelihood

$$\bar{L}_{n,T}(\theta) := \exp\left\{A(\theta, X_T) - A(\theta, X_0) - \frac{h}{2}\sum_{i=1}^n [f^2(\theta, X_{t_{i-1}}) + \dot{f}(\theta, X_{t_{i-1}})]\right\}.$$
$$(1.7)$$

The corresponding Itô-Lebesgue approximate maximum likelihood estimator (ILAMLE) is given by

$$\bar{\theta}_{n,T} := \arg\max_{\theta \in \Theta} \bar{L}_{n,T}(\theta).$$

The advantage of this estimator is its faster rate of convergence over the Euler estimator. Note that the rate of convergence is pulled down by approximating the stochastic integral. But this estimator is free of stochastic integral approximation. Denote

$$g(\theta, X_t) := f^2(\theta, X_t) + \dot{f}(\theta, X_t). \quad (1.8)$$

Thus

$$L_T(\theta) = \exp\left\{A(\theta, X_T) - A(\theta, X_0) - \frac{1}{2}\int_0^T g(\theta, X_t)dt\right\} \quad (1.9)$$

We obtain several discrete approximations of the Lebesgue integral in the definition (1.9).

Define a weighted sum of squares

$$M_{n,T} := \frac{T}{n}\left\{\sum_{i=1}^n w_{t_i}g(\theta, X_{t_{i-1}}) + \sum_{i=2}^{n+1} w_{t_i}g(\theta, X_{t_{i-1}})\right\}. \quad (1.10)$$

where $w_{t_i} \geq 0$ is a weight function.

Denote

$$I_{n,T} := \frac{T}{n}\sum_{i=1}^n g(\theta, X_{t_{i-1}}), \quad (1.11)$$

$$J_{n,T} := \frac{T}{n}\sum_{i=2}^{n+1} g(\theta, X_{t_{i-1}}) = \frac{T}{n}\sum_{i=1}^n g(\theta, X_{t_i}). \quad (1.12)$$

General weighted approximate log-likelihood (WALL) is defined as

$$\tilde{l}_{n,T,G} := A(\theta, X_T) - A(\theta, X_0) - \frac{1}{2}M_{n,T}. \tag{1.13}$$

With $w_{t_i} = 1$, we obtain the forward WALL as

$$\tilde{l}_{n,T,F} := A(\theta, X_T) - A(\theta, X_0) - \frac{1}{2}I_{n,T}. \tag{1.14}$$

With $w_{t_i} = 0$, we obtain the backward WALL as

$$\tilde{l}_{n,T,B} := A(\theta, X_T) - A(\theta, X_0) - \frac{1}{2}J_{n,T}. \tag{1.15}$$

With $w_{t_i} = 0.5$, the simple symmetric WALL is defined as

$$\tilde{l}_{n,T,z} := A(\theta, X_T) - A(\theta, X_0) - \frac{1}{4}[I_{n,T} + J_{n,T}] \tag{1.16}$$

which in turn is equal to

$$A(\theta, X_T) - A(\theta, X_0) - \frac{T}{4n}\sum_{i=2}^{n} g(\theta, X_{t_{i-1}}) + \frac{T}{2n}(g(\theta, X_{t_0}) + g(\theta, X_{t_n})).$$

With the weight function

$$w_{t_i} = \begin{cases} 0 & : \quad i = 1 \\ \frac{i-1}{n} & : \quad i = 2, 3, \cdots, n \\ 1 & : \quad i = n+1 \end{cases}$$

the weighted symmetric WALL defined as

$$\tilde{l}_{n,T,w} := A(\theta, X_T) - A(\theta, X_0) - \frac{T}{4n}\sum_{i=2}^{n} g(\theta, X_{t_{i-1}}) + \frac{T}{4n}\sum_{i=1}^{n} g(\theta, X_{t_{i-1}}). \tag{1.17}$$

Note that estimator (1.13) is analogous to the trapezoidal rule in numerical analysis. One can instead use the midpoint rule to define another WALL

$$\tilde{l}_{n,T,M} := A(\theta, X_T) - A(\theta, X_0) - \frac{T}{2n}\sum_{i=1}^{n} g\left(\theta, \frac{X_{t_{i-1}} + X_{t_i}}{2}\right). \tag{1.18}$$

One can use the Simpson's rule to define another WALL where the approximant is a convex combination of the approximants in (1.16) and (1.18)

$$\tilde{l}_{n,T,S} := A(\theta, X_T) - A(\theta, X_0)$$
$$- \frac{T}{12n}\sum_{i=1}^{n}\left\{g(\theta, X_{t_{i-1}}) + 4g\left(\frac{X_{t_{i-1}} + X_{t_i}}{2}\right) + g(\theta, X_{t_i})\right\}. \tag{1.19}$$

In general, one can generalize Simpson's rule as

$$\tilde{l}_{n,T,GS} := A(\theta, X_T) - A(\theta, X_0)$$
$$- \frac{T}{6n} \sum_{i=1}^{n} \left\{ \alpha \frac{g(\theta, X_{t_{i-1}}) + g(\theta, X_{t_i})}{2} + (1-\alpha)g\left(\frac{X_{t_{i-1}} + X_{t_i}}{2}\right) \right\} \quad (1.20)$$

for any $0 \leq \alpha \leq 1$.

The case $\alpha = 0$ produces the WALL (1.18). The case $\alpha = 1$ produces the WALL (1.17). The case $\alpha = \frac{1}{3}$ produces the WALL (1.19).

We propose a very general form of the quadrature based WALL as

$$\tilde{l}_{n,T,w} := A(\theta, X_T) - A(\theta, X_0) - \frac{T}{2n} \sum_{i=1}^{n} \sum_{j=1}^{m} \left[(1 - s_j)g(\theta, X_{t_{i-1}}) + s_j g(\theta, X_{t_i}) \right] p_j$$

$$(1.21)$$

where p_j, $j \in \{1, 2, \cdots, m\}$ is a probability mass function of a discrete random variable S on $0 \leq s_1 < s_2 < \cdots < s_m \leq 1$ with $P(S = s_j) := p_j$, $j \in \{1, 2, \cdots, m\}$.

Denote the k-th moment of the random variable S as $\mu_k := \sum_{j=1}^{m} s_j^k p_j$, $k = 1, 2, \cdots$.

If one chooses the probability distribution as uniform distribution for which the moments are a harmonic sequence $(\mu_1, \mu_2, \mu_3, \mu_4, \mu_5, \mu_6, \cdots) = (\frac{1}{2}, \frac{1}{3}, \frac{1}{4}, \frac{1}{5}, \frac{1}{6}, \frac{1}{7} \cdots)$ then there is no change in rate of convergence than the second order. If one can construct a probability distribution for which the harmonic sequence is truncated at a point, then there is an improvement of rate of convergence at the point of truncation.

Conjecture: *Given a positive integer m, construct a probability mass function p_j, $j \in \{1, 2, \cdots, m\}$ on $0 \leq s_1 < s_2 < \cdots < s_m \leq 1$ such that*

$$\sum_{j=1}^{m} s_j^r p_j = \frac{1}{r+1}, \quad r \in \{0, \cdots, m-2\} \quad (1.22)$$

$$\sum_{j=1}^{m} s_j^{m-1} p_j \neq \frac{1}{m}. \quad (1.23)$$

Neither the probabilities p_j nor the atoms, s_j, of the distribution are specified in advance.

This problem is related to the truncated Hausdorff moment problem. I obtain examples of such probability distributions and use them to get higher order accurate (up to sixth order) WALLs.

The order of approximation error (rate of convergence) of an estimator is $n^{-\nu}$ where

$$\nu := \inf \left\{ k : \mu_k \neq \frac{1}{1+k}, \ \mu_j = \frac{1}{1+j}, j = 1, 2, \cdots, k-1 \right\}. \quad (1.24)$$

We construct probability distributions satisfying these moment conditions and obtain estimators of the rate of convergence up to order 6.

Probability $p_1 = 1$ at the point $s_1 = 0$ gives the WALL (1.11) for which $\mu_1 = 0$. Note that $\mu_1 \neq \frac{1}{2}$. Thus $\nu = 1$.

Probability $p_1 = 1$ at the point $s_1 = 1$ gives the WALL (1.12) for which $\mu_1 = 1$. Note that $\mu_1 \neq \frac{1}{2}$. Thus $\nu = 1$.

Probabilities $(p_1, p_2) = (\frac{1}{2}, \frac{1}{2})$ at the respective points $(s_1, s_2) = (0, 1)$ produces the WALL $\tilde{l}_{n,T,Z}$ for which $(\mu_1, \mu_2) = (\frac{1}{2}, \frac{1}{4})$. Thus $\nu = 2$.

Probability $p_j = 1$ at the point $s_j = \frac{1}{2}$ produce the WALL $\tilde{l}_{n,T,M}$ for which $(\mu_1, \mu_2) = (\frac{1}{2}, \frac{1}{2})$. Thus $\nu = 2$.

Probabilities $(p_1, p_2) = (\frac{1}{4}, \frac{3}{4})$ at the respective points $(s_1, s_2) = (0, \frac{2}{3})$ produce the asymmetric WALL

$$\tilde{l}_{n,T,3} := A(\theta, X_T) - A(\theta, X_0)$$
$$-\frac{T}{4n} \sum_{i=1}^{n} \left[g(\theta, X_{t_{i-1}}) + 3g(\theta, \frac{X_{t_{i-1}} + 2X_{t_i}}{3}) \right] \tag{1.25}$$

for which $(\mu_1, \mu_2, \mu_3) = (\frac{1}{2}, \frac{1}{3}, \frac{2}{9})$. Thus $\nu = 3$.

Probabilities $(p_1, p_2) = (\frac{3}{4}, \frac{1}{4})$ at the respective points $(s_1, s_2) = (\frac{1}{3}, 1)$ produce asymmetric WALL

$$\tilde{l}_{n,T,4} := A(\theta, X_T) - A(\theta, X_0) - \frac{T}{4n} \sum_{i=1}^{n} \left[3g(\theta, \frac{2X_{t_{i-1}} + X_{t_i}}{3}) + g(\theta, X_{t_i}) \right] \tag{1.26}$$

for which $(\mu_1, \mu_2, \mu_3) = (\frac{1}{2}, \frac{1}{3}, \frac{10}{36})$. Thus $\nu = 3$.

Probabilities $(p_1, p_2, p_3) = (\frac{1}{6}, \frac{2}{3}, \frac{1}{6})$ at the respective points $(s_1, s_2, s_3) = (0, \frac{1}{2}, 1)$ produce the estimator $\tilde{l}_{n,T,5}$ for which $(\mu_1, \mu_2, \mu_3, \mu_4) = (\frac{1}{2}, \frac{1}{3}, \frac{1}{4}, \frac{5}{25})$. Thus $\nu = 4$.

Probabilities $(p_1, p_2, p_3, p_4) = (\frac{1}{8}, \frac{3}{8}, \frac{3}{8}, \frac{1}{8})$ at the respective points $(s_1, s_2, s_3, s_4) = (0, \frac{1}{3}, \frac{2}{3}, 1)$ produce the symmetric WALL

$$\tilde{l}_{n,T,5} := A(\theta, X_T) - A(\theta, X_0) - \frac{T}{8n} \sum_{i=1}^{n} \Big[g(\theta, X_{t_{i-1}}))$$
$$+3g(\theta, \frac{2X_{t_{i-1}} + X_{t_i}}{3}) + 3g(\theta, \frac{X_{t_{i-1}} + 2X_{t_i}}{3}) + g(\theta, X_{t_i}) \Big] s$$
$$\tag{1.27}$$

for which $(\mu_1, \mu_2, \mu_3, \mu_4) = (\frac{1}{2}, \frac{1}{3}, \frac{1}{4}, \frac{11}{54})$. Thus $\nu = 4$.

Probabilities $(p_1, p_2, p_3, p_4, p_5) = (\frac{1471}{24192}, \frac{6925}{24192}, \frac{1475}{12096}, \frac{2725}{12096}, \frac{5675}{24192}, \frac{1721}{24192})$ at the respective points $(s_1, s_2, s_3, s_4, s_5) = (0, \frac{1}{5}, \frac{2}{5}, \frac{3}{5}, \frac{4}{5}, 1)$ produce the asymmetric WALL

$$\tilde{l}_{n,T,7} := A(\theta, X_T) - A(\theta, X_0)$$

$$-\frac{T}{24192n}\sum_{i=1}^{n}\left[1471g(\theta, X_{t_{i-1}}) + 6925g(\theta, \tfrac{X_{t_{i-1}}+X_{t_i}}{5}) + 2950g(\theta, \tfrac{2X_{t_{i-1}}+2X_{t_i}}{5})\right.$$

$$\left.+5450g(\theta, \tfrac{3X_{t_{i-1}}+3X_{t_i})}{5}) + 5675g(\theta, \tfrac{4X_{t_{i-1}}+4X_{t_i}}{5}) + 1721g(\theta, X_{t_i})\right] \qquad (1.28)$$

for which $(\mu_1, \mu_2, \mu_3, \mu_4, \mu_5, \mu_6) = (\tfrac{1}{2}, \tfrac{1}{3}, \tfrac{1}{4}, \tfrac{1}{5}, \tfrac{841}{5040})$. Thus $\nu = 5$.

Probabilities $(p_1, p_2, p_3, p_4, p_5) = (\tfrac{7}{90}, \tfrac{16}{45}, \tfrac{2}{15}, \tfrac{16}{45}, \tfrac{7}{90})$ at the respective points $(s_1, s_2, s_3, s_4, s_5) = (0, \tfrac{1}{4}, \tfrac{1}{2}, \tfrac{3}{4}, 1)$ produce the symmetric WALL $\tilde{l}_{n,T,8}$ given by

$$\tilde{l}_{n,T,8} := A(\theta, X_T) - A(\theta, X_0)$$

$$-\frac{T}{90n}\sum_{i=1}^{n}\left[7g(\theta, X_{t_{i-1}} + 32g(\theta, \tfrac{3X_{t_{i-1}}+X_{t_i}}{4}) + 12g(\theta, \tfrac{X_{t_{i-1}}+X_{t_i}}{2})\right.$$

$$\left.+32g(\theta, \tfrac{X_{t_{i-1}}+3X_{t_i}}{4}) + 7g(\theta, X_{t_i})\right] \qquad (1.29)$$

for which $(\mu_1, \mu_2, \mu_3, \mu_4, \mu_5, \mu_6) = (\tfrac{1}{2}, \tfrac{1}{3}, \tfrac{1}{4}, \tfrac{1}{5}, \tfrac{1}{6}, \tfrac{110}{768})$. Thus $\nu = 6$.

Probabilities $(p_1, p_2, p_3, p_4, p_5) = (\tfrac{19}{288}, \tfrac{75}{288}, \tfrac{50}{288}, \tfrac{50}{288}, \tfrac{75}{288}, \tfrac{19}{288})$ at the respective points
$(s_1, s_2, s_3, s_4, s_5) = (0, \tfrac{1}{5}, \tfrac{2}{5}, \tfrac{3}{5}, \tfrac{4}{5}, 1)$ produce symmetric WALL

$$\tilde{l}_{n,T,9} := A(\theta, X_T) - A(\theta, X_0)$$

$$-\frac{T}{288n}\sum_{i=1}^{n}\left[19g(\theta, X_{t_{i-1}} + 75g(\theta, \tfrac{4X_{t_{i-1}}+X_{t_i}}{5}) + 50(\theta, \tfrac{3X_{t_{i-1}}+2X_{t_i}}{5})\right.$$

$$\left.+50g(\theta, \tfrac{2X_{t_{i-1}}+3X_{t_i}}{5}) + 75g(\theta, \tfrac{X_{t_{i-1}}+4X_{t_i}}{5}) + 19g(\theta, X_{t_i})\right]. \qquad (1.30)$$

for which $(\mu_1, \mu_2, \mu_3, \mu_4, \mu_5, \mu_6) = (\tfrac{1}{2}, \tfrac{1}{3}, \tfrac{1}{4}, \tfrac{1}{5}, \tfrac{1}{6}, \tfrac{3219}{22500})$. Thus $\nu = 6$.

Define the estimators

$$\tilde{\theta}_{n,T,r} := \arg\max_{\theta \in \Theta}\tilde{l}_{n,T,r}(\theta), \quad r = 1, 2, \cdots, 9.$$

The WALL $\tilde{l}_{n,T,Z}$ is based on the arithmetic mean of $I_{n,T}$ and $J_{n,T}$. One can use geometric mean and harmonic mean instead. The geometric mean based symmetric WALL (which is based on the ideas of partial autocorrelation) is defined as

$$\tilde{l}_{n,T,G} := A(\theta, X_T) - A(\theta, X_0) - \frac{T}{2n}\sqrt{I_{n,T}\,J_{n,T}} \qquad (1.31)$$

The harmonic mean based symmetric WALL is defined as

$$\tilde{l}_{n,T,H} := A(\theta, X_T) - A(\theta, X_0) - \frac{T/n}{\frac{1}{I_{n,T}} + \frac{1}{J_{n,T}}} \qquad (1.32)$$

Note that

$$\tilde{l}_{n,T,H} \leq \tilde{l}_{n,T,G} \leq \tilde{l}_{n,T,Z}. \tag{1.33}$$

We shall use the following notations : $\Delta X_i := X_{t_i} - X_{t_{i-1}}, \Delta W_i := W_{t_i} - W_{t_{i-1}}$, C is a generic constant independent of h, n and other variables (it may depend on θ). We assume the following conditions:

(A1) $|f(\theta, x)| \leq K(\theta)(1 + |x|)$,
$\qquad |f(\theta, x) - f(\theta, y)| \leq K(\theta)|x - y|$.
(A2) $|f(\theta, x) - f(\phi, y)| \leq J(x)|\theta - \phi|$ for all $\theta, \phi \in \Theta, x, y \in \mathbb{R}$ where

$$\sup_{\theta \in \Theta} |K(\theta)| = K < \infty, \quad E|J(X_0)|^m < \infty \text{ for some } m > 16.$$

(A3) The diffusion process X is stationary and ergodic with invariant measure ν, i.e., for any ψ with $E[\psi(\cdot)] < \infty$

$$\frac{1}{n} \sum_{i=1}^{n} \psi(X_{t_i}) \to E_\nu[\psi(X_0)] \text{ a.s. as } n \to \infty, T \to \infty \text{ and } h \to 0.$$

(A4) $E|\xi|^m < \infty$ for some $m > 16$.
(A5) $E|f(\theta, \xi) - f(\theta_0, \xi)|^2 = 0$ iff $\theta = \theta_0$.
(A6) f is twice continuously differentiable function in x with

$$E \sup_t |\dot{f}(X_t)|^2 < \infty, \quad E \sup_t |\ddot{f}(X_t)|^2 < \infty.$$

(A7) $|\dot{f}'(\theta, x)| \leq C(1 + |x|^c)$ for any $x \in \mathbb{R}$ and $\theta \in \Theta$.
The Fisher information

$$0 < I(\theta) := \int_{-\infty}^{\infty} (f'(\theta, x))^2 d\nu(x) < \infty$$

and for any $\delta > 0$, or any compact $\Xi \subset \Theta$,

$$\inf_{\theta_0 \in \Xi} \sup_{|\theta - \theta_0| > \delta} E_{\theta_0} |f'(\theta, X_0) - f'(\theta_0, X_0)|^2 > 0.$$

(A8) The Malliavin covariance of the process is nondegenerate.
In this chapter, we study strong consistency, asymptotic normality, and a Berry-Esseen type bound for the AMLEs.

10.2 Rate of Consistency

To obtain the strong consistency of the AMLE, we need the following two lemmas. The proof of the first lemma is standard.

Lemma 2.1. Suppose the random function D_n satisfy the following conditions:
(C1) With probability one, $D_n(\theta) \to D(\theta)$ uniformly in $\theta \in \Theta$ as $n \to \infty$.

(C2) The limiting nonrandom function D is such that

$$D(\theta_0) \geq D(\theta) \text{ for all } \theta \in \Theta.$$

(C3) $D(\theta) = D(\theta_0)$ iff $\theta = \theta_0$.
Then $\theta_n \to \theta_0$ a.s. as $n \to \infty$, where

$$\theta_n = \sup_{\theta \in \Theta} D_n(\theta).$$

Lemma 2.2 Under for any $m \geq 1$,

$$E \left| \sum_{i=1}^{n} f(\theta_0, X_{t_{i-1}}) \Delta t_i - \int_0^T f(\theta_0, X_t) dt \right|^{2m} \leq C \left(\frac{T^2}{n} \right)^{2m}.$$

Proof. By Itô formula, we have

$$f(\theta_0, X_t) - f(\theta_0, X_{t_{i-1}})$$

$$= \int_{t_{i-1}}^t \dot{f}(\theta_0, X_u) dX_u + \frac{1}{2} \int_{t_{i-1}}^t \ddot{f}(\theta_0, X_u) du$$

$$= \int_{t_{i-1}}^t \dot{f}(\theta_0, X_u) dW_u + \int_{t_{i-1}}^t [\dot{f}(\theta_0, X_u) f(\theta_0, X_u) + \frac{1}{2} \ddot{f}(\theta_0, X_u)] du$$

$$=: \int_{t_{i-1}}^t \dot{f}(\theta_0, X_u) dW_u + \int_{t_{i-1}}^t B(\theta_0, X_u) du.$$

Thus

$$E \left| \sum_{i=1}^{n} f(\theta_0, X_{t_{i-1}}) \Delta t_i - \int_0^T f(\theta_0, X_t) dt \right|^{2m}$$

$$\leq E | \sum_{i=1}^{n} \int_{t_{i-1}}^{t_i} [f(\theta_0, X_t) - f(\theta_0, X_{t_{i-1}})] \, dt |^{2m}$$

$$= E | \sum_{i=1}^{n} \int_{t_{i-1}}^{t_i} \left[\dot{f}(\theta_0, X_u) dW_u \right.$$

$$\left. + \int_{t_{i-1}}^t f'(\theta_0, X_{t_{i-1}}) B(\theta_0, X_u) du \right] dt |^{2m}$$

$$\leq 2^m \left\{ E | \sum_{i=1}^{n} \int_{t_{i-1}}^{t_i} \int_{t_{i-1}}^t \dot{f}(\theta_0, X_u) dW_u dt |^{2m} \right.$$

$$\left. + E | \sum_{i=1}^{n} \int_{t_{i-1}}^{t_i} \int_{t_{i-1}}^t f'(\theta_0, X_{t_{i-1}}) B(\theta_0, X_u) du dt |^{2m} \right\}$$

$$=: 2^m (M_1 + M_2)^m$$

where

$$M_1 := E | \sum_{i=1}^{n} \int_{t_{i-1}}^{t_i} \int_{t_{i-1}}^{t} \dot{f}(\theta_0, X_u) dW_u dt |^2$$

and

$$M_2 := | \sum_{i=1}^{n} \int_{t_{i-1}}^{t_i} \int_{t_{i-1}}^{t} f'(\theta_0, X_{t_{i-1}}) A(\theta_0, X_u) du dt |^2.$$

Observe that with $S_{i,t} := \int_{t_{i-1}}^{t} f'(\theta_0, X_{t_{i-1}}) \dot{f}(\theta_0, X_u) dW_u,\ 1 \leq i \leq n$,

$$M_1 = \sum_{i=1}^{n} E \left(\int_{t_{i-1}}^{t_i} S_{i,t} dt \right)^2 + \sum_{j \neq i=1}^{n} E \left(\int_{t_{i-1}}^{t_i} S_{i,t} dt \right) \left(\int_{t_{j-1}}^{t_j} S_{j,t} dt \right)$$

$$\leq \sum_{i=1}^{n} (t_i - t_{i-1}) \int_{t_{i-1}}^{t_i} E(S_{i,t}^2) dt$$

(the last term being zero due to orthogonality of the integrals)

$$\leq \sum_{i=1}^{n} (t_i - t_{i-1}) \int_{t_{i-1}}^{t_i} \left\{ \int_{t_{i-1}}^{t} E \left[f'(\theta_0, X_{t_{i-1}}) \dot{f}(\theta_0, X_u) \right]^2 du \right\} dt$$

$$\leq C \frac{T}{n} \sum_{i=1}^{n} \int_{t_{i-1}}^{t_i} (t - t_{i-1}) dt \quad \text{(by (A4) and (A3))}$$

$$\leq C \frac{T}{n} \sum_{i=1}^{n} (t_i - t_{i-1})^2$$

$$= C \frac{T^3}{n^2}.$$

On the other hand, with $U_{i,t} := \int_{t_{i-1}}^{t} f'(\theta_0, X_{t_{i-1}}) A(\theta_0, X_u) du,\ 1 \leq i \leq n$,

$$M_2 = E | \sum_{i=1}^{n} \int_{t_{i-1}}^{t_i} \int_{t_{i-1}}^{t} f'(\theta_0, X_{t_{i-1}}) A(\theta_0, X_u) du dt |^2$$

$$= E | \sum_{i=1}^{n} \int_{t_{i-1}}^{t_i} U_{i,t} dt |^2$$

$$= \sum_{i=1}^{n} \left(\int_{t_{i-1}}^{t_i} U_{i,t} dt \right)^2 + \sum_{j \neq i=1}^{n} E \left(\int_{t_{i-1}}^{t_i} U_{i,t} dt \right) \left(\int_{t_{j-1}}^{t_j} U_{i,t} dt \right)$$

$$\leq \sum_{i=1}^{n} (t_i - t_{i-1}) E \left(\int_{t_{i-1}}^{t_i} U_{i,t} dt \right)^2 + \sum_{j \neq i=1}^{n} \left\{ E \left(\int_{t_{i-1}}^{t_i} U_{i,t} d \right)^2 \right.$$

$$\left. \times E \left(\int_{t_{j-1}}^{t_j} U_{j,t} dt \right)^2 \right\}^{1/2}$$

$$\leq \sum_{i=1}^{n}(t_i - t_{i-1}) \int_{t_{i-1}}^{t_i} E(U_{i,t}^2)dt$$

$$+ \sum_{j\neq i=1}^{n} \left\{ (t_i - t_{i-1}) \int_{t_{i-1}}^{t_i} E(U_{i,t}^2)dt \; (t_j - t_{j-1}) \int_{t_{j-1}}^{t_j} E(U_{j,t}^2)dt \right\}^{1/2}$$

But $E(U_{i,t}^2) \leq C(t - t_{i-1})^2$ using (A4) and (A3). On substitution, the last term is dominated by

$$C\sum_{i=1}^{n}(t_i - t_{i-1})^4 + C \sum_{j\neq i=1}^{n} (t_i - t_{i-1})^2(t_j - t_{j-1})^2$$

$$= C\frac{T^4}{n^3} + C\frac{n(n-1)T^4}{2n^4} \leq C\frac{T^4}{n^2}.$$

Thus

$$E\left| \sum_{i=1}^{n} f(\theta, X_{t_{i-1}})\Delta t_i - \int_{0}^{T} f(\theta, X_t)dt \right|^2$$

$$\leq 2^m(M_1 + M_2)^m$$

$$\leq C_m \left(\frac{T^2}{n} \right)^m.$$

$\square$

Lemma 2.3 As $T \to 0$ and $T/n \to 0$

$$\sup_{\theta \in \Theta} \left[\frac{1}{n}\sum_{i=1}^{n} f(\theta, X_{t_{i-1}}) - \frac{1}{T}\int_{0}^{T} f(\theta, X_t)dt \right] \longrightarrow 0 \; a.s.$$

Proof Use of Lemma 2.2, ergodicity, Arzella-Ascoli theorem, and Borel-Cantelli theorem proves the lemma.

With the aid of Lemma 2.3 and Lemma 2.1 we obtain the strong consistency result.

Proposition 2.3
$$\bar{\theta}_{n,T} \to \theta_0 \; a.s.$$

as $T \to 0$ and $T/n \to 0$.

The following is a general theorem on approximate maximum likelihood estimation.

Theorem 2.4 (Huzak (2001): Let Θ be an open subset of the Euclidean space $\mathbb{R}^d$, let $(\omega, \mathcal{F}, \mathbb{P})$ be a probability space, and let $(\mathcal{F}_n, n \in \mathbb{N})$ be a family of sub-σ-algebras of $\mathcal{F}$. Moreover, let $(\gamma_n; n \in \mathbb{N})$ be a sequence of positive numbers such that $\lim_n \gamma_n = 0$, and let $L, L_n : \Omega \times \Theta \to \mathbb{R}$, $n \in \mathbb{N}$, be functions satisfying the following assumptions:

(A1): For all $\theta \in \Theta$, $L(\omega, \theta)$ is $\mathcal{F}$-measurable and $L_n(\omega, \theta)$ is $\mathcal{F}_n$-measurable, $n \in \mathbb{N}$ is. For all $\omega \in \Omega$, $L(\omega, \theta)$ and $L_n(\omega, \theta)$ are twice continuously differentiable in θ.

(A2): For all $\omega \in \Omega$, $L(\omega, \theta)$ has a unique point of global maximum $\hat{\theta} \equiv \hat{\theta}(\omega)$ in θ and $D^2 L(\hat{\theta}) < 0$.

(A3): For any relatively compact set $\mathcal{K} \subset \Theta$,

$$\sup_{\theta \in \mathcal{K}} |D^l L_n(\theta) - D^l L(\theta)| = O_{\mathbb{P}}(\gamma_n), \ n \in \mathbb{N}, \ l = 1, 2.$$

Then there exists a sequence $(\hat{\theta}_n, n \in \mathbb{N})$ of Θ-valued random variables such that

(i) $\lim_n \mathbb{P}\{DL_n(\hat{\theta}_n) = 0\} = 1$.

(ii) $\hat{\theta}_n \to^{\mathbb{P}} \hat{\theta}$ as $n \to \infty$.

(iii) If $\tilde{\theta}_n; n \in \mathbb{N}$ is any other sequence of random variables which satisfy (i) and (ii), then $\lim_n \mathbb{P}\{\tilde{\theta}_n = \hat{\theta}_n\} = 1$;

(iv) The sequence $\gamma_n^{-1}(\hat{\theta}_n - \hat{\theta})$, $n \in \mathbb{N}$ is bounded in probability.

An application of the Theorem 2.4 to our WALLs provide the following stochastic bound on the difference between the discrete AMLEs and the continuous MLE. Details are omitted.

Euler Estimator has the following rate:

Proposition 2.5 For fixed T,

$$\left|\hat{\theta}_{n,T} - \theta_T\right| = O_P\left(\frac{1}{\sqrt{n}}\right).$$

The new estimators have the following rates:

Proposition 2.6 For fixed T,

$$(a) \qquad \left|\bar{\theta}_{n,T} - \theta_T\right| = O_P\left(\frac{1}{n}\right).$$

$$(b) \quad \left|\bar{\theta}_{n,T,M} - \theta_T\right| = O_P\left(\frac{1}{n^2}\right).$$

$$(c) \quad \left|\tilde{\theta}_{n,T,Z} - \theta_T\right| = O_P\left(\frac{1}{n^2}\right).$$

$$(d) \quad \left|\tilde{\theta}_{n,T,3} - \theta_T\right| = O_P\left(\frac{1}{n^3}\right).$$

$$(e) \quad \left|\tilde{\theta}_{n,T,4} - \theta_T\right| = O_P\left(\frac{1}{n^3}\right).$$

$$(f) \quad \left|\tilde{\theta}_{n,T,5} - \theta_T\right| = O_P\left(\frac{1}{n^4}\right).$$

$$(g) \quad \left|\tilde{\theta}_{n,T,7} - \theta_T\right| = O_P\left(\frac{1}{n^5}\right).$$

$$(h) \quad \left|\tilde{\theta}_{n,T,8} - \theta_T\right| = O_P\left(\frac{1}{n^6}\right).$$

$$(i) \quad \left|\tilde{\theta}_{n,T,9} - \theta_T\right| = O_P\left(\frac{1}{n^6}\right).$$

10.3 Berry-Esseen Bound

We will use Lemma 2.1.1 and the following lemma in obtaining the Berry-Esseen bound.

The following lemma which is an immediate consequence of Theorem 7 in Yoshida (1997).

Lemma 3.1 We have

$$\sup_{x \in \mathbb{R}} \left| P_{\theta_0} \left\{ \frac{1}{\sqrt{TI(\theta_0)}} \int_0^T f(\theta_0, X_t) dW_t \leq x \right\} - \Phi(x) \right| \leq CT^{-1/2}.$$

Proposition 3.2

$$\sqrt{T}(\bar{\theta}_{n,T} - \theta_0) \to^{\mathcal{D}} \mathcal{N}(0, I^{-1}(\theta_0)).$$

Theorem 3.3

$$\sup_{x \in \mathbb{R}} \left| P_\theta \left\{ \sqrt{TI(\theta_0)}(\bar{\theta}_{n,T} - \theta_0) \leq x \right\} - \Phi(x) \right| = O\left(T^{-1/2} \bigvee \frac{T^4}{n^2}\right).$$

Proof Let $l_{n,T}(\theta) := \log \bar{L}_{n,T}(\theta)$.
By Taylor expansion, we have

$$l'_{n,T}(\bar{\theta}_{n,T}) = l'_{n,T}(\theta_0) + (\bar{\theta}_{n,T} - \theta_0)l''_{n,T}(\theta^*_{n,T})$$

where $\left|\theta^*_{n,T} - \theta\right| \leq \left|\bar{\theta}_{n,T} - \theta_0\right|$.
Since $l'_{n,T}(\bar{\theta}_{n,T}) = 0$, hence we have

$$\sqrt{TI(\theta_0)}(\bar{\theta}_{n,T} - \theta_0)$$

$$= -\frac{\frac{1}{\sqrt{TI(\theta_0)}} l'_{n,T}(\theta_0)}{\frac{1}{TI(\theta_0)} l''_{n,T}(\theta^*_{n,T})}$$

$$= \left[\frac{1}{\sqrt{TI(\theta_0)}} \left\{ (A'(\theta, X_T) - A'(\theta, X_0)) - \frac{h}{2} \sum_{i=1}^{n} [2f(\theta, X_{t_{i-1}})f'(\theta, X_{t_{i-1}}) \right. \right.$$

$$\left. \left. + \dot{f}'(\theta, X_{t_{i-1}})] \right\} \right] \left[\frac{1}{TI(\theta_0)} \sum_{i=1}^{n} f''(\theta^*_{n,T}, X_{t_{i-1}}) \right]^{-1}$$

$$=: [Z_{n,T}] [V_{n,T}]^{-1} .$$

Note that

$$V_{n,T} = \frac{1}{TI(\theta_0)} \sum_{i=1}^{n} f''(\theta^*_{n,T}, X_{t_{i-1}}) \Delta t_i = \frac{1}{TI(\theta_0)} \sum_{i=1}^{n} f'(\theta^*_{n,T}, X_{t_{i-1}})^2 \Delta t_i.$$

But $E(I_T - 1)^2 \leq CT^{-1}$ from Theorem 2.3.1
Hence $E(V_{n,T} - 1)^2 = E[(V_{n,T} - I_T) + (I_T - 1)]^2 \leq C(T^{-1} \bigvee \frac{T^2}{n^2})$
since

$$\sup_{\theta \in \Theta} E[V_{n,T} - I_T]^2 \leq C \frac{T^2}{n^2}.$$

Thus

$$\sup_{x \in \mathbb{R}} \left| P_\theta \left\{ \sqrt{TI(\theta)}(\bar{\theta}_{n,T} - \theta) \leq x \right\} - \Phi(x) \right|$$

$$= \sup_{x \in \mathbb{R}} \left| P_\theta \left\{ \frac{Z_{n,T}}{V_{n,T}} \leq x \right\} - \Phi(x) \right|$$

$$= \sup_{x \in \mathbb{R}} |P_\theta \{Z_{n,T} \leq x\} - \Phi(x)| + P_\theta \{|V_{n,T} - 1| \geq \epsilon\} + \epsilon$$

$$\leq C(T^{-1/2} \bigvee \frac{T^2}{n}) + \epsilon^{-2} C(T^{-1} \bigvee \frac{T^2}{n^2}) + \epsilon.$$

since

$$\sup_{x \in \mathbb{R}} |P_\theta \{Z_{n,T} \leq x\} - \Phi(x)|$$

$$\leq \sup_{x \in \mathbb{R}} |P_\theta \{Z_T \leq x\} - \Phi(x)| + P_\theta \{|Z_{n,T} - Z_T| \geq \epsilon\} + \epsilon$$

$$\leq C(T^{-1/2} \bigvee \frac{T^2}{n}) + \epsilon^{-2} E |Z_{n,T} - Z_T|^2 + \epsilon$$

$$\leq (T^{-1/2} \bigvee \frac{T^2}{n}) + \epsilon^{-2} C(T^{-1} \bigvee \frac{T^3}{n^2}) + \epsilon.$$

Choosing $\epsilon = T^{-1/2}$, we have the result. $\qquad\qquad\square$

For the Euler estimator we obtain the following Berry-Esseen bound.

Proposition 3.4

$$\sup_{x\in\mathbb{R}}\left|P_\theta\left\{\sqrt{TI(\theta)}(\hat\theta_{n,T}-\theta)\le x\right\}-\Phi(x)\right|=O\left(\max\left(T^{-1/2},\frac{T^2}{n},\frac{T^3}{n^2}\right)\right).$$

10.4 Examples

(a) Ornstein-Uhlenbeck Process

The Ornstein-Uhlenbeck process satisfies the Itô stochastic differential equation

$$dX_t = \theta\,X_t\,dt + dW_t,\ t\ge 0, \tag{1.1}$$

where $\{W_t\}$ is a standard Wiener process with the filtration $\{\mathcal{F}_t\}_{t\ge0}$, X_0 has stationary distribution $\mathcal{N}(0,\,-1/2\theta)$ and $\theta<0$ is the unknown parameter.
Here

$$A(\theta,X_T)-A(\theta,X_0)=\theta\int^{X_T}y\,dy-\theta\int^{X_0}y\,dy=\frac{\theta}{2}[X_T^2-X_0^2].$$

$$\frac{h}{2}\sum_{i=1}^n[f^2(\theta,X_{t_{i-1}})+\dot f(\theta,X_{t_{i-1}})]=\frac{h}{2}\sum_{i=1}^n\theta^2 X_{t_{i-1}}^2+\frac{T}{2}\theta.$$

Hence

$$l'_{n,T}=\frac{1}{2}[X_T^2-X_0^2]-h\theta\sum_{i=1}^n X_{t_{i-1}}^2-\frac{T}{2}.$$

Thus

$$\bar\theta_{n,T}=\frac{\frac{1}{2}[X_T^2-X_0^2]-\frac{T}{2}}{h\sum_{i=1}^n X_{t_{i-1}}^2}.$$

This is a better estimator than Euler estimator as shown in Chapter 8.

(b) Gompertz Diffusions

Gomtertz diffusion model describes the *in vivo* tumor growth. The drift parameter describes the intrinsic growth rate (mitosis rate) of the tumor. We introduce some new approximate maximum likelihood estimators of the drift parameter in the Gompertz diffusion model based on discretely sampled data and study their properties.

The Gompertz process has been recently used in tumor growth modeling. In view of this, it becomes necessary to estimate the unknown parameters in the model from discrete data. The knowledge of distribution of the estimator may be applied to evaluate the distribution of other important growing parameters used to assess the tumor treatment modalities. Some of these parameters are the platue of deterministic Gomperzian model, $X_\infty = \exp(\alpha/\beta)$, tumor growth delay, and the first time the growth curve of the Gompertz diffusion reaches X_∞.

Let $(\Omega, \mathcal{F}, \{\mathcal{F}_t\}_{t\geq 0}, P)$ be a stochastic basis on which is defined the Gompertz diffusion model $\{X_t\}$ satisfying the Itô stochastic differential equation

$$dX_t = (\alpha X_t - \beta X_t \log X_t)\, dt + \sigma X_t dW_t, \ t \geq 0, \ X_0 = x_0 \qquad (4.1)$$

where $\{W_t\}$ is a standard Wiener process with the filtration $\{\mathcal{F}_t\}_{t\geq 0}$ and $\theta < 0$ is the unknown parameter to be estimated on the basis of discrete observations of the process $\{X_t\}$ at times $0 = t_0 < t_1 < \cdots t_n = T$ with $t_i - t_{i-1} = \frac{T}{n}, i = 1, 2 \cdots, n$. We assume two types of high frequency data with long observation time:
1) $T \to \infty, n \to \infty, \frac{T}{\sqrt{n}} \to 0$,   2) $T \to \infty, n \to \infty, \frac{T}{n^{2/3}} \to 0$.

Here X_t is the tumor volume which is measured at discrete time, α is the intrinsic growth rate of the tumor, β is the tumor growth acceleration factor, and σ is the diffusion coefficient.

The knowledge of the distribution of the estimator may be applied to evaluate the distribution of other important growing parameters used to access tumor treatment madalities. Some of these parameters are the plateau of the model $X_\infty = \exp(\frac{\alpha}{\beta})$, tumor growth decay, and the first time the growth curve of the model reaches X_∞. We assume that the growth deceleration factor β does not change, while the variability of environmental conditions induces fluctuations in the intrinsic growth rate (mitosis rate) α.

Let the continuous realization $\{X_t, 0 \leq t \leq T\}$ be denoted by X_0^T. Let P_θ^T be the measure generated on the space (C_T, B_T) of continuous functions on $[0, T]$ with the associated Borel σ-algebra B_T generated under the supremum norm by the process X_0^T and let P_0^T be the standard Wiener measure. It is well known that when θ is the true value of the parameter P_θ^T is absolutely continuous with respect to P_0^T and the Radon-Nikodym derivative (likelihood) of P_θ^T with respect to P_0^T based on X_0^T is given by

$$L_T(\theta) := \frac{dP_{\alpha,\beta}^T}{dP_0^T}(X_0^T) = \exp\left\{ \frac{1}{\sigma^2} \int_0^T (\alpha - \beta \log X_t)X_t^{-1}dX_t \right.$$
$$\left. - \frac{1}{2\sigma^2} \int_0^T (\alpha - \beta \log X_t)^2 dt \right\}. \qquad (4.2)$$

Consider the log-likelihood function, which is given by

$$\gamma_T(\alpha, \beta, \sigma) := \frac{1}{\sigma^2}\left\{ \int_0^T (\alpha - \beta \ln X_t)X_t^{-1}dX_t - \frac{1}{2}\int_0^T (\alpha - \beta \ln X_t)^2 dt \right\}.$$
$$(4.3)$$

A solution of the estimating equation $\gamma_T'(\alpha, \beta) = 0$ provides the conditional maximum likelihood estimate (MLE)

$$\hat{\alpha}_T := \frac{(\int_0^T X_t^{-1} dX_t)(\int_0^T \ln^2 X_t dt) - (\int_0^T X_t^{-1} \ln X_t dX_t)(\int_0^T \ln X_t dt)}{T \int_0^T \ln^2 X_t dt - (\int_0^T \ln X_t dt)^2}. \tag{4.4}$$

$$\hat{\beta}_T := \frac{(\int_0^T X_t^{-1} dX_t)(\int_0^T \ln X_t dt) - T(\int_0^T X_t^{-1} \ln X_t dX_t)}{T \int_0^T \ln^2 X_t dt - (\int_0^T \ln X_t dt)^2}. \tag{4.5}$$

$$\hat{\sigma}_T^2 := \frac{\sum_{i=1}^n (X_{t_i} - X_{t_{i-1}})^2}{\sum_{i=1}^n X_{t_{i-1}}^2 (t_i - t_{i-1})}. \tag{4.6}$$

As an alternative to maximum likelihood method and to obtain estimators with higher order accuracy our aim is to use contrast functions. Using Itô formula, the score function $\gamma_T(\theta)$ can be written as

$$\gamma_T(\alpha) = \frac{1}{2} \ln X_T + \int_0^T (\beta \ln X_t + \frac{\sigma^2}{2}) dt. \tag{4.7}$$

Consider the estimating function

$$M_T(\alpha) = \int_0^T (\alpha \ln X_t + \frac{1}{2}) dt \tag{4.8}$$

and the minimum contrast estimate (MCE)

$$\tilde{\alpha}_T := \frac{\beta}{T} \left\{ \int_0^T \ln X_t dt \right\}. \tag{4.9}$$

Define a weighted sum of squares

$$M_{n,T} := \frac{T}{n} \left\{ \sum_{i=1}^n w_{t_i} \ln X_{t_{i-1}} + \sum_{i=2}^{n+1} w_{t_i} \ln X_{t_{i-1}} \right\}. \tag{4.10}$$

where $w_{t_i} \geq 0$ is a weight function.

Denote the discrete increasing functions

$$I_{n,T} := \frac{T}{n} \sum_{i=1}^n \ln X_{t_{i-1}}, \tag{4.11}$$

$$J_{n,T} := \frac{T}{n} \sum_{i=2}^{n+1} \ln X_{t_{i-1}} = \frac{T}{n} \sum_{i=1}^n \ln X_{t_i}. \tag{4.12}$$

General weighted AMCE is defined as

$$\tilde{\alpha}_{n,T} := \left\{ \frac{2}{n} M_{n,T} \right\}. \tag{4.13}$$

With $w_{t_i} = 1$, we obtain the forward AMCE as

$$\tilde{\alpha}_{n,T,F} := \left\{ \frac{2}{n} I_{n,T} \right\}^{-1}. \tag{4.14}$$

With $w_{t_i} = 0$, we obtain the backward AMCE as

$$\tilde{\alpha}_{n,T,B} := \left\{ \frac{2}{n} J_{n,T} \right\}^{-1}. \tag{4.15}$$

Analogous to the estimators for the discrete AR (1) model, we define the simple symmetric and weighted symmetric estimators (see Fuller (1996)):

With $w_{t_i} = 0.5$, the simple symmetric AMCE is defined as

$$\tilde{\alpha}_{n,T,z} := \left\{ \frac{1}{n} [I_{n,T} + J_{n,T}] \right\}^{-1} = \left\{ \frac{2}{n} \sum_{i=2}^{n} \ln X_{t_{i-1}} + 0.5(\ln X_{t_0} + \ln X_{t_n}) \right\}. \tag{4.16}$$

With the weight function

$$w_{t_i} = \begin{cases} 0 & : \ i = 1 \\ \frac{i-1}{n} & : \ i = 2, 3, \cdots, n \\ 1 & : \ i = n+1 \end{cases}$$

the weighted symmetric AMCE is defined as

$$\tilde{\alpha}_{n,T,w} := -\left\{ \frac{2}{n} \sum_{i=2}^{n} \ln X_{t_{i-1}} + \frac{1}{n} \sum_{i=1}^{n} \ln X_{t_{i-1}} \right\}. \tag{4.17}$$

Note that estimator (4.13) is analogous to the trapezoidal rule in numerical analysis. One can instead use the midpoint rule to define another estimator

$$\tilde{\alpha}_{n,T,A} := \left\{ \frac{2}{n} \sum_{i=1}^{n} \ln \left(\frac{X_{t_{i-1}} + X_{t_i}}{2} \right) \right\}. \tag{4.18}$$

One can use the Simpson's rule to define another estimator where the denominator is a convex combination of the denominators in (4.11) and (4.12)

$$\tilde{\alpha}_{n,T,S} := \left\{ \frac{1}{3n} \sum_{i=1}^{n} \left\{ \ln X_{t_{i-1}} + 4\ln \left(\frac{X_{t_{i-1}} + X_{t_i}}{2} \right) + \ln X_{t_i} \right\} \right\}. \tag{4.19}$$

In general, one can generalize Simpson's rule as

$$\tilde{\alpha}_{n,T,GS} := -\left\{ \frac{2}{n} \sum_{i=1}^{n} \left\{ \alpha \frac{\ln X_{t_{i-1}} + \ln X_{t_i}}{2} + (1-\alpha)\ln \left(\frac{X_{t_{i-1}} + X_{t_i}}{2} \right) \right\} \right\} \tag{4.20}$$

for any $0 \le \alpha \le 1$.

The case $\alpha = 0$ produces the estimator (4.18). The case $\alpha = 1$ produces the estimator (4.17). The case $\alpha = \frac{1}{3}$ produces the estimator (4.19).

We propose a very general form of the quadrature based estimator as

$$\tilde{\alpha}_{n,T,w} := \left\{ \frac{2}{n} \sum_{i=1}^{n} \sum_{j=1}^{m} \ln \left[(1 - s_j) X_{t_{i-1}} + s_j X_{t_i} \right] p_j \right\} \qquad (4.21)$$

where p_j, $j \in \{1, 2, \cdots, m\}$ is a probability mass function of a discrete random variable S on $0 \leq s_1 < s_2 < \cdots < s_m \leq 1$ with $P(S = s_j) := p_j$, $j \in \{1, 2, \cdots, m\}$.

Denote the k-th moment of the random variable S as $\mu_k := \sum_{j=1}^{m} s_j^k p_j$, $k = 1, 2, \cdots$.

If one chooses the probability distribution as uniform distribution for which the moments are a harmonic sequence $(\mu_1, \mu_2, \mu_3, \mu_4, \mu_5, \mu_6, \cdots) = (\frac{1}{2}, \frac{1}{3}, \frac{1}{4}, \frac{1}{5}, \frac{1}{6}, \frac{1}{7}, \cdots)$ then there is no change in rate of convergence than second order. If one can construct a probability distribution for which the harmonic sequence is truncated at a point, then there is a rate of convergence improvement at the point of truncation.

References

1. Adler, R., Muller, P. and Rozovskii, B.L. (1996): *Stochastic Modeling in Physical Oceonography*, Birkhauser, Boston - Berlin.
2. Aihara, S.I. (1992): Regularized maximum likelihood estimate for an infinite dimensional parameter in stochastic parabolic systems, *SIAM J. Cont. Optim.* **30**, 745-764.
3. Aihara, S.I. (1994): Maximum likelihood estimate for discontinuous parameter in stochastic hyperbolic systems, *Acta Applicandae Mathematicae* **35**, 131-151.
4. Aihara, S.I. (1995): Identification of a discontinuous parameter in stochastic parabolic systems, To appear in *Applied Math. Optimization.*
5. Aihara, S.I. and Bagchi, A. (1988): Parameter identification for stochastic diffusion equations with unknown boundary conditions, *App. Math. Optimization* **17**, 15-36.
6. Aihara, S.I. and Bagchi, A. (1989): Infinite dimensional parameter identification for stochastic parabolic systems, *Statist. Probab. Letters* **8**, 279-287.
7. Aihara, S.I. and Bagchi, A. (1991): Parameter identification for hyperbolic stochastic systems, *J. Math. Anal. Appl.* **160**, 485-499.
8. Ait-Sahalia, Y. (1996): Nonparametric pricing of interest rate derivative securities, *Econometrica* **64**, 527-560.
9. Ait-Sahalia, Y. (1999): Transition densities for interest rates and other nonlinear diffusions, *J. Finance* **54**, 1361-1395.
10. Ait-Sahalia, Y. (2002): Maximum likelihood estimation of discretely-sampled diffusions: a closed form approximation approach, *Econometrica* **70**, 223-262.
11. Ait-Sahalia, Y. and Hansen, L. (2006): *Handbook of Financial Econometrics*, North-Holland, Amsterdam, forthcoming.
12. Ait-Sahalia, Y. and Mykland, P. (2003): The effects of random and discrete sampling when estimatiing continuous time diffusion, *Econometrica*, **71**, 483-549.
13. Ait-Sahalia, Y. and Mykland, P. (2004): Estimating diffusions from discretely and possibly randomly spaced data: a general theory, *Annals of Statistics*, **32**, 2186-2222.
14. Alos, E., Mazet, O and Nualart, D. (2000): Stochastic calculus with respect to fractional Brownian motion with Hurst parameter less than $\frac{1}{2}$, *Stoch. Proc. Appl.*, **86** (1), 121-139.
15. Alos, E., Mazet, O. and Nualart, D. (2001): Stochastic calculus with respect to Gaussian processes, *Ann. Probab.* **29**, 766-801.

16. Anscombe, F.J. (1952): Large sample theory of sequential estimation, *Proc. Camb. Phil. Soc.*, **48**, 600-607.

17. Arato, M. (1978): On the statistical examination of continuous state Markov processes, I, II, III, IV, *Selected Transl. in Math. Statist. and Probability*, Vol. **14**, 203-287, American Mathematical Society, Providence.

18. Arato, M. (1982): *Linear Stochastic Systems with Constant Coefficients : A Statistical Approach*, Lecture Notes in Control and Information Sciences **45**, Springer-Verlag, New York.

19. Arato, M., Kolmogorov, A.N. and Sinai, Ya. G. (1962): On estimation of parameters of complex stationary Gaussian processes, *Doklady of Academy of Sciences* **146**(4), 747-750.

20. Arnold, L. (1974): *Stochastic Differential Equations*, Wiley, New York.

21. Bagchi, A. and Borkar, V. (1984): Parameter identification in infinite dimensional linear systems, *Stochastics* **12**, 201-213.

22. Basawa, I.V. and Prakasa Rao, B.L.S. (1980): *Statistical Inference for Stochastic Processes*, Academic Press, New York-London.

23. Basawa, I.V. and Scott, D.J. (1983): *Asymptotic Optimal Inference for Non-ergodic Models*, Lecture Notes in Statistics **17**, Springer-Verlag, Berlin.

24. Barndorff-Neilson, J.E. and Sørensen, M. (1994): A review of some aspects of asymptotic likelihood theory for stochastic processes, *Int. Statist. Review*, **62**, 133-165.

25. Bauer, K. (1980): On asymptotic properties of estimates of the drift coefficient parameter of a diffusion process, *Theor. Probab. Appl.* **25**, 429-431.

26. Bellach, B. (1980): Consistency, asymptotic normality and asymptotic efficiency of the maximum likelihood estimator in linear stochastic differential equations, *Math. Operationsforch. Statist., Ser. Statistics*, **11**, 227-266.

27. Bellach, B. (1983): Parameter estimation in linear stochastic differential equations and their asymptotic properties, *Math. Operationsforch. Statist. Ser. Statistics*, **14**, 141-191.

28. Beran, R. (1994): *Statistics for Long-Memory Processes*, Chapman & Hall, New York.

29. Bergstrom, A.R. (ed.) (1976): *Statitical Inference in Continuous Time Economic Models*, North-Holland, Amsterdam.

30. Bergstrom, A.R. (1988): The history of continuous time econometric models, *Econometric Theory* **4**, 365-383.

31. Bibby, B.M. and Sørensen, M. (1995a): Martingale estimation functions for discretely observed diffusion processes, *Bernoulli* **1**, 17-39.

32. Bibby, B.M. and Sørensen, M. (1995b): On estimation for discretely observed diffusions : A review, *Theory of Stochastic Processes* **2** (18), 49-56.

33. Bibby, B.M. and Sørensen, M. (1997): A hyperbolic diffusion model for stock prices, *Finance Stoch.* **1**, 25-41.

34. Bishwal, J.P.N. (1999a): Large deviations inequalities for the maximum likelihood estimator and the Bayes estimators in nonlinear stochastic differential equations, *Statistics and Probability Letters* **43**(2), 207-215.

35. Bishwal, J.P.N. (1999b): Bayes and sequential estimation in Hilbert space valued stochastic differential equations, *Journal of the Korean Statistical Society* **28** (1), 93-106.

36. Bishwal, J.P.N. (1999c): Asymptotic theory of approximate Bayes estimator for diffusion processes from discrete observations, In: *Statistical Inference for*

Diffusion Type Processes, (B.L.S. Prakasa Rao) Arnold Publishers, London, 221-224.

37. Bishwal, J.P.N. (2000a): Sharp Berry-Esseen bound of the maximum likelihood estimator in the Ornstein-Uhlenbeck process, *Sankhyā, Ser. A* **62** (1), 1-10.

38. Bishwal, J.P.N. (2000b): Rates of convergence of the posterior distributions and the Bayes estimators in the Ornstein-Uhlenbeck process, *Random Operators and Stochastic Equations* **8** (2), 47-65.

39. Bishwal, J.P.N. (2001): Accuracy of normal approximation for the maximum likelihood and the Bayes estimators in the Ornstein-Uhlenbeck process using random normings, *Statistics and Probability Letters* **52** (4), 427-439.

40. Bishwal, J.P.N. (2002a): The Bernstein-von Mises theorem and spectral asymptotics of Bayes estimators in parabolic SPDEs, *Journal of the Australian Mathematical Society* **72** (2), 289-300.

41. Bishwal, J.P.N. (2002b): Conditional least squares estimation for nonhomogeneous diffusions, *Statistica Neerlandica* (to appear).

42. Bishwal, J.P.N. (2003a): Maximum likelihood estimation in partially observed stochastic differential system driven by a fractional Brownian motion, *Stochastic Analysis and Applications* **21** (5), 995-1007.

43. Bishwal, J.P.N. (2003b): Large deviations and Berry-Esseen inequalities estimators in the nonlinear nonhomogeneous diffusions, *REVSTAT* (to appear).

44. Bishwal, J.P.N. (2004a): Uniform rate of weak convergence of the minimum contrast estimator in the Ornstein-Uhlenbeck process, *Methodology and Computing in Applied Probability* (to appear).

45. Bishwal, J.P.N. (2004b): Asymptotic theory of estimation in nonergodic diffusion type processes, To appear in *Mathematical Methods of Statistics*.

46. Bishwal, J.P.N. (2005a): Rates of convergence of posterior distributions and Bayes estimators in nonlinear ergodic diffusions, *Stochastic Processes and their Applications* (to appear).

47. Bishwal, J.P.N. (2005b): Non-martingale estimating function for discretely sampled diffusions, *Applied Stochastic Models in Business and Industry* (to appear).

48. Bishwal, J.P.N. (2006a): Rates of weak convergence of the approximate minimum contrast estimator for the discretely observed Ornstein-Uhlenbeck process, *Statistics and Probability Letters* **76** (13), 1397-1409.

49. Bishwal, J.P.N. (2006b): Sequential maximum likelihood estimation in semimartingales, *Journal of Statistics and Applications* **1** (2-4), 15-25.

50. Bishwal, J.P.N. (2007a): A new estimating function for discretely sampled diffusions, *Random Operators and Stochastic Equations* **15** (1), 65-88.

51. Bishwal, J.P.N. (2007b): Sequential maximum likelihood estimation in nonlinear diffusion type processes, *Brazilian Journal of Probability and Statistics* (to appear).

52. Bishwal, J.P.N. and Bose, A. (1995): Speed of convergence of the maximum likelihood estimator in the Ornstein-Uhlenbeck process, *Calcutta Statistical Association Bulletin* **45**, 245-251.

53. Bishwal, J.P.N. and Bose, A. (2001): Rates of convergence of approximate maximum likelihood estimators in the Ornstein-Uhlenbeck process, *Computers & Mathematics with Applications* **42** (1-2), 23-38.

54. Billingsley, P. (1961): *Statistical Inference for Markov Processes*, Chicago University Press, Chicago.

248   References

55. Black, F. and Scholes, M. (1973): The pricing of options and corporate liabilities, *Journal of Political Economy* **81**, 637-659.
56. Borkar, V. and Bagchi, A. (1982): Parameter estimation in continuous time stochastic processes, *Stochastics* **8**, 193-212.
57. Borokov, A.A. (1973): on the rate of convergence for the invariance principle, *Theory Probab. Appl.* **18**, 217-234.
58. Borwanker, J.D., Kallianpur, G. and Prakasa Rao, B.L.S. (1971): The Bernstein-von Mises theorem for Markov processes, *Ann. Math. Statist.* **42** (1971), 1241-1253.
59. Bose (Basu), A. (1983a): Asymptotic theory of estimation in non-linear stochastic differential equations for the multiparameter case, *Sankhyā Ser. A* **45**, 56-65.
60. Bose (Basu), A. (1983b): The Bernstein-von Mises theorem for a certain class of diffusion processes, *Sankhyā Ser. A* **45**, 150-160.
61. Bose, A. (1985): Rate of convergence of the maximum likelihood estimator in the Ornstein-Uhlenbeck process, Technical Report No. 4/85, Stat-Math Unit, Indian Statistical Institute, Calcutta.
62. Bose, A. (1986a): Berry-Esseen bound for the maximum likelihood estimator in the Ornstein-Uhlenbeck process, *Sankhyā Ser. A* **48**, 181-187.
63. Bose, A. (1986b): *Asymptotic study of estimators in some discrete and continuous time models*, Ph. D. Thesis, Indian Statistical Institute, Calcutta.
64. Brandt, M.W. and Santa-Clara, P. (2002): Simulated likelihood estimation of diffusions with an application to exchange rate dynamics, *Journal of Financial Economics*, **63**, 161-210.
65. Brown, B.M. and Hewitt, J.T. (1975a): Asymptotic likelihood theory for diffusion processes, *J. Appl. Prob.* **12**, 228-238.
66. Brown, B.M. and Hewitt, J.T. (1975b): Inference for the diffusion branching process, *J. Appl. Prob.* **12**, 588-594.
67. Broze, L., Scaillet, O. and Zakoian, J.M. (1998): Quasi indirect inference for diffusion processes, *Econometric Theory* **14**, 161-186.
68. Campillo, F. and Le Gland, F. (1989): MLE for partially observed diffusions: direct maximization vs. the EM algorithm, *Stochastic Process. Appl.* **33**, 245-274.
69. Carmona, P. and Coutin, L. (1999): Stochastic integration with respect to fractional Brownian motion, Preprint, Universite paul Sabatier.
70. Comte, F. (1996): Simulation and estimation of long memory continuous time models, *J. Time Series Analysis* **17**, 19-36.
71. Comte, F. and Renault, E. (1996): Long memory continuous time models, *J. Econometrics* **73**, 101-149.
72. Comte, F. and Renault, E. (1998): Long memory in continuous-time stochastic volatility models, *Mathematical Finance* **8**, 291-324.
73. Coutin, L. and Decreusefond, L. (1999a): Stochastic differential equations driven by fractional Brownian motion, *Preprint*.
74. Coutin, L. and Decreusefond, L. (1999b): Abstract nonlinear filtering in the presence of fractional Brownian motion, *Ann. Applied Probab.* **9** (4), 1058-1090.
75. Gripenberg, G. and Norros, I. (1997): on the prediction of fractional Brownian motion, *J. Appl. Probab.*, **33**, 400-410.
76. Clement, E. (1993): Inference statistique des processes de diffusion, Research Report No. 9404, CREST-INSEE, Universite Paris VI.
77. Clement, E. (1995): Estimation de processus de diffusion par methodes simulees, Research Report No.9507, CREST-INSEE, Universite Paris VI.

78. Clement, E. (1997a): Estimation of diffusion processes by simulated moments methods, *Scand. J. Statist.* **24**(3), 353-369.

79. Clement, E. (1997b): Bias correction for the estimation of discretized diffusion processes from an approximated likelihood function, *Theor. Probab. Appl.* **42**, 283-288.

80. Curtain, R. and Prichard, A. (1978): *Infinite Dimensional Linear Systems Theory*, Springer-Verlag, Berlin.

81. Dacunha-Castelle, D. and Florens-Zmirou, D. (1986): Estimation of the coefficients of a diffusion from discrete observations, *Stochastics* **19**, 263-284.

82. Dai, W. and Heyde, C.C. (1996): Itô formula with respect to fractional Brownian motion and its applications, *J. Appl. Maths. Stoch. Anal.* **9**, 439-448.

83. Decreusefond, L. and Ustunel, A.S. (1998): Fractional Brownian motion: theory and applications, In: *Fractional Differential Systems: Models, Methods and Applications, ESAIM Proceedings* **5**, 75-86.

84. Decreusefond, L. and Ustunel, A.S. (1999): Stochastic analysis of the fractional Brownian motion, *Potential Analysis* **10**, 177-214.

85. Djehiche, B. and Eddahbi, M. (1999): Hedging options with market models modulated by fractional Brownian motion, Preprint, Royal Institute of Technology, Sweden.

86. Duncan, T.E., Hu, Y. and Pasik-Duncan, B. (2000): Stochastic calculus for fractional Brownian motion I. Theory, *SIAM J. Control. Optim.*

87. Dieterich, W., Fulde, P. and Peschel, I. (1980): Theoretical models for superionic conductors, *Adv. Physics* **29**, 527-605.

88. Dietz, H.M. (1989): Asymptotic properties of maximum likelihood estimators in diffusion type models, Preprint No. 228, Humboldt-Universitat Zu Berlin, Sektion Mathematik.

89. Dietz, H.M. (1992): A non-Markovian relative to the Ornstein-Uhlenbeck process and some of its local statistical properties, *Scand. J. Statist.* **19**, 363-379.

90. Dietz, II.M. and Kutoyants, Yu. A. (1997): A class of minimum-distance estimators for diffusion processes with with ergodic properties, *Statist. Decisions* **15**, 211-227.

91. Dohnal, G. (1987): On estimating the diffusion coefficient, *J. Appl. Prob.* **24**, 105-114.

92. Doob, J.L. (1948): Application of the theory of martingales, *Coll. Int. du C.N.R.S. Paris*, 22-28.

93. Doob, J.L. (1953): *Stochastic Processes*, John Wiley, New York.

94. Dorogovcev, A. Ja. (1976): The consistency of an estimate of a parameter of a stochastic differential equation, *Theory Prob. Math. Statist.* **10**, 73-82.

95. Duffie, J.D. (1996): *Dynamic Asset Pricing Theory*, Princeton University Press, Princeton, NJ.

96. Duffie, J.D. and Glynn, P. (2004): Estimation of Continuous time Markov processes samples at random time intervals, *Econometrica*, **72**, 1773-1808.

97. Duncan, T.E., Hu, Y. and Pasik-Duncan, B. (2000): Stochastic calculus for fractional Brownian motion I. Theory, *SIAM J. Control. Optim.*

98. Dupuis, P. and Kushner, H.J. (1989): Stochastic approximations and large deviations : upper bounds and w.p.1 convergence, *SIAM J. Control Optim.* **27**, 1108-1135.

99. Elerian, O. (1998): A note on the existence of a closed form conditional transition density for the Milstein scheme, Economics Discussion Paper, Nuffield College, Oxford.

100. Elerian, O., Chib, S. and Shephard, N. (2001): Likelihood inference for discretely observed nonlinear diffusions, *Econometrica*, **69**, 959-993.
101. Elliot, R.J. (1982): *Stochastic Calculus and Applications*, Springer-Verlag, New York.
102. Fan, J, Jiang, J., Zhang, C. and Zhou, Z. (2003): Time-dependent diffucion models for term structure dynamics, *Statistica Sinica*, **13**, 965-992.
103. Feigin, P.D. (1976): Maximum likelihood estimation for continuous time stochastic processes, *Adv. Appl. Prob.* **8**, 712-736.
104. Feigin, P.D. (1979): Some comments concerning curious singularity, *J. Appl. Prob.* **16**, 440-444.
105. Feller, W. (1957): *An Introduction to Probability Theory and its Applications*, Vol. I, Wiley, New York.
106. Fisk, D.L. (1963): Quasimartingales and stochastic integrals, Technical Report No. 1, Dept. of Mathematics, Michigan State University, East Lansing.
107. Florens-Landais, D. and Pham, H. (1999): Large deviations in estimation of an Ornstein-Uhlenbeck model, *J. Appl. Probab.* **36**, 60-77.
108. Friedman, A. (1975): *Stochastic Differential Equations*, Vol. I, Academic Press, New York.
109. Florens-Zmirou, D. (1989): Approximate discrete-time schemes for statistics of diffusion processes, *Statistics*, **20**, 547-557.
110. Florens-Zmirou, D. (1991): Statistics on crossings of discretized diffusions and local time, *Stoch. Proc. Appl.* **39**, 139-151.
111. Florens-Zmirou, D. (1993): On estimating the diffusion coefficient from discrete observations, *J. Appl. Probab.* **30**, 790-804.
112. Freedman, D. (1999): On the Bernstein-von Mises theorem with infinite dimensional parameters, *Ann. Statist.* **27**, 1119-1140.
113. Fuller, W.A. (1996): *Introduction to Statistical Time Series*, Second Edition, Wiley, New York.
114. Gallant, A.R. and Long, J.R. (1997): Estimating stochastic differential equations efficiently by minimum chi-squared, *Biometrika* **84**, 125-141.
115. Gard, T.C. (1988): *An Introduction to Stochastic Differential Equations*, Marcel Dekker.
116. Gelb, A. (ed.) (1974): *Applied Optimal Estimation*, MIT Press, Cambridge, MA.
117. Genon-Catalot, V. (1987): *Observations Partielles de Diffusions: Traitement Statistique dans L' Asymptotique de la variance*, Thesis, Universite Paris Sud, Orsay, Cedex.
118. Genon-Catelot, V. (1990): Maximum contrast estimation for diffusion processes from discrete observations, *Statistics* **21**, 99-116.
119. Genon-Catalot, V. and Jacod, J. (1993): On the estimation of the diffusion coefficient for multidimensional diffusion processes, *Ann. Inst. Henri Poincaré, Probabilités et Statistiques* **29**, 119-151.
120. Genon-Catalot, V. and Jacod, J. (1994): On the estimation of diffusion coefficient for diffusion processes, *Scand. J. Statist.* **21**, 193-221.
121. Genon-Catalot, V., Jeantheau, T. and Laredo, C. (1998a): Parameter estimation for discretely observed stochastic volatility models, *Bernoulli* **4**(1), 1-18.
122. Genon-Catalot, V., Jeantheau, T. and Laredo, C. (1998b): Limit theorems for for discretely observed stochastic volatility models, *Bernoulli* **4**(3), 283-303.
123. Ghosal, S. (1994): Asymptotic Properties of the posterior distributions and study of some nonregular cases, Ph.D. Thesis, Indian Statistical Institute, Calcutta.

124. Ghosal, S. , Ghosh, J.K. and Samanta, T. (1995): On convergence of posterior distributions, *Ann. Statist.* **23**, 2145-2152.

125. Ghosh, J.K., Ghosal, S. and Samanta, T. (1994): Stability and convergence of the posterior in some nonregular problems, In: *Statistical Decision Theory and Related Topics V*, (Eds.) Gupta, S.S. and Berger, J.O., Proceedings of the Fifth Purdue International Symposium, Springer-Verlag, New York, 183-200.

126. Ghosal, S., Ghosh, J.K. and Van der Vaart, A. (2000): Convergence rates of posterior distributions, *Ann. Statist.*, **28** (2), 500-531.

127. Gikhman, I.I. and Skorohod, A.V. (1972): *Stochastic Differential Equations*, Springer-Verlag, Berlin.

128. Gouriéroux, C., Monfort, A. and Renault, E. (1993): Indirect inference, *Journal of Applied Econometrics*, **8**, 585-618.

129. Gouriéroux, C. and Monfort, A. (1994): Indirect inference for stochastic differential equations, Discussion paper 9443, CREST-INSEE, University of Paris VI.

130. Grambsch, P. (1983): Sequential sampling based on the observed Fisher information to guarantee the accuracy of the maximum likelihood estimator, *Ann. Statist.* **11**, 68-77.

131. Greenwood, P.E. and Shiryayev, A.N. (1992): Asymptotic minimaxity of a sequential estimator of a first order autoregressive model, *Stoch. Stoch. Reports* **38**, 49-65.

132. Greenwood, P.E. and Welfelmeyer, W. (1993): Asymptotic minimax results for stochastic process families with critical points, *Stoch. Proc. Appl.* **44**, 107-116.

133. Gripenberg, G. and Norros, I. (1997): on the prediction of fractional Brownian motion, *J. Appl. Probab.*, **33**, 400-410.

134. Gushchin, A.A. (1995): On asymptotic optimality of estimators of parameters under the LAQ condition, *Theory. Probab. Appl.* **40**, 261-272.

135. Gushchin, A.A. and Küchler, U. (1999): Asymptotic inference in linear stochastic differential equation with time delay, *Bernoulli* **5**, 1059-1098.

136. Guttorp, P. and Kulperger, R. (1984): Statistical inference for some Volterra population processes in a random environment, *Canadian J. Statist.* **12**, 289-302.

137. Hall, P. and Heyde, C.C. (1980): *Martingale Limit Theory and its Applications*, Academic Press, New York.

138. Hansen, L. and Scheinkman, J.A. (1995): Back to the future: generating moment implications for continuous-time Markov processes, *Econometrica* **63**, 1995 (with Lars Hansen).

139. Harison, V. (1992): The Bernstein-von Mises theorem for a certain class of Gaussian diffusion processes, *Publications du Service de Mathematiques*, Vol. 6, Universite d' Antananarivo, 1-7.

140. Harison, V. (1996): Drift estimation of a certain class of diffusion processes from discrete observations, *Comput. Math. Applic.* **31**, 121-133.

141. Heyde, C.C. (1998): *Quasi-likelihood and its Applications*, Springer-Verlag, Berlin.

142. Hipp, C. and Michel, R. (1976): On the Bernstein-von Mises approximation of posterior distributions, *Ann. Statist.* **4**, 972-980.

143. Hopfner, R. and Kutoyants, Y. (2002): On a problem of statistical inference for null recurrent diffusions, *Statist. Inf. Stochastic Process* **6** (1), 25-42.

144. Holden, H., Øksendal, B. Ubøe, J. and Zhang, T. (1996): *Stochastic Partial Differential Equations*, Birkhäuser, Boston.

252   References

145. Huang, T.M. (2000): Convergence rates of posterior distributions and adaptive estimation, Technical Report No. 711, Department of Statistics, Carnegie Mellon University.

146. Huebner, M. (1993): *Parameter Estimation of Stochastic Differential Equations,* Ph.D. Thesis, University of Southern California, Los Angeles.

147. Huebner, M. (1997): A characterization of asymptotic behaviour of maximum likelihood estimators for stochastic PDE's, *Math. Methods. Statist.* **6**, 395-415.

148. Huebner, M. (1999): Asymptotic properties of the maximum likelihood estimator for stochastic PDEs disturbed by small noise, *Statistical Inference for Stochastic Processes,* **2**, 57-68.

149. Huebner, M., Khasminskii, R.Z. and Rozovskii, B.L. (1992): Two examples of parameter estimation for stochastic partial differential equations, In : S. Cambanis, J.K. Ghosh, R.L. Karandikar, P.K. Sen (Eds.) *Stochastic Processes, Freschrift in honour of G. Kallianpur,* Springer-Verlag, Berlin, 149-160.

150. Huebner, M. and Rozovskii, B.L. (1995): On the asymptotic properties of maximum likelihood estimators for parabolic stochastic PDEs, *Prob. Theor. Rel. Fields* **103**, 143-163.

151. Hutton, J.E. and Nelson, P.E. (1984): Interchanging the order of differentiation and stochastic integration, *Stochastic Procsss. Appl.* **18**, 371-377.

152. Huzak, M. (2001): A general theorem on approximate maximum likelihood estimation, *Glasnik Matematicki,* **36**, 139-153.

153. Ibragimov, I.A. and Khasminskii, R.Z. (1981): *Statistical Estimation: Asymptotic Theory,* Springer-Verlag, Berlin.

154. Ikeda, N. and Watanabe, S. (1989): *Stochastic Differential Equations and Diffusion Processes,* Second Edition, North-Holland, Amsterdam (Kodansha Ltd., Tokyo).

155. Itô, K. and McKean, H.P. Jr. (1974): *Diffusion Processes and their Sample Paths,* Springer-Verlag, Berlin.

156. Itô, K. (1984): *Foundations of Stochastic Differential Equations in Infinite Dimensional Spaces,* CBMS-NSF Reginal Conference Series in Applied Mathematics, Vol. 49, SIAM, Philadeplphia, Pennsylvania.

157. Ivanov, A.V. (1976): The Berry-Esseen inequality for the distribution of the least squares estimate, *Mathematical Notes* **20**, 721-727.

158. Jacobsen, M. (2002): Discretely observed diffusions: classes of estimating functions and small Δ-optimality, *Bernoulli* **8**, 643-668.

159. Jacod, J. and Shiryayev, A.N. (1987): *Limit Theorems for Stochastic Processes,* Springer-Verlag, Berlin.

160. Jacod, J. (1993): Random sampling in estimation problems for continuous Gaussian processes with independent increments, *Stoch. Proc. Appl.* **44**, 181-204.

161. Jacod, J. (2001): Inference for stochastic processes, *Prepubication Universite Paris VI.*

162. James, M. and Le Gland, F. (1995): Consistent parameter estimation for partially observed diffusions with small noise, *Applied Math. Optim.* **32**, 47-72.

163. Jankunas, A. and Khasminskii, R.Z. (1997): Estimation of the parameters of linear homogeneous stochastic differential equations, *Stoch. Proc. Appl.* **72**, 205-219.

164. Jeganathan, P. (1982): On the asymptotic theory of statistical inference when the limit of the log-likelihood ratios is mixed normal, *Sankhyā Ser. A* **44**, 173-212.

165. Jeganathan, P. (1995): Some aspects of asymptotic theory with applications to time series models, *Econometric Theory* **11**, 818-887.
166. Jennrich, R.I. and Bright, P.B. (1976): Fitting systems of linear differential equations using computer generated exact derivatives. *Technometrics* **18**, 385-392.
167. Jensen, B. and Poulsen, R. (1999): A comparison of approximate techniques for transition densities of diffusion processes, Working Paper Series No. 38, Center for Analytical Finance, University of Aaarhus - Aarhus School of Business.
168. Jones, R.H. (1984): Fitting multivariate models to unequally spaced data, In: Time Series Analysis of Irregularity Observed Data (Ed. E. Parzen), *Lecture Notes in Statistics* **25** Springer-Verlag, New York, 155-188.
169. Kallianpur, G. (1980): *Stochastic Filtering Theory*, Springer-Verlag, New York.
170. Kallianpur, G. and Selukar, R.S. (1991): Parameter estimation in linear filtering, *J. Multivariate Anal.* **39**, 284-304.
171. Kallianpur, G. and Selukar, R.S. (1993): Estimation of Hilbert space valued parameter by the method of sieves, In :*Statistics and Probability, A R.R. Bahadur Festschrift*(Eds.) J.K.Ghosh, S.K. Mitra, K.R. Parthasarathy and B.L.S. Prakasa Rao, Wiley Eastern, 325-347.
172. Kallianpur, G. and Xiong, X. (1995): *Infinite Dimensional Stochastic Differential Equations*, Lecture Notes-Monograph Series, **26**, Institute of Mathematical Statistics, California.
173. Karandikar, R.L. (1983): Interchanging the order of stochastic integration and ordinary differentiation, *Sankhyā Ser. A* **45**, 120-124.
174. Karatzas, I. and Shreve, S.E. (1987): *Brownian Motion and Stochastic Calculus*, Springer-Verlag, New York.
175. Kasonga, R.A. (1988): The consistency of a non linear least squares estimator from diffusion processes, *Stoch. Proc. Appl.* **30**, 263-275.
176. Kasonga, R.A. (1990): Parameter estimation by deterministic approximation of a solution of a stochastic differential equation, *Commun. Statist. - Stoch. Models* **6**, 59-67.
177. Kessler, M. (1997): Estimation of an ergodic diffusion from discrete observations, *Scand. J. Statist.* **24**(2), 211-229.
178. Kessler, M. (2000): Simple and explicit estimating functions for a discretely observed diffusion process, *Scand. J. Statist.* **27**, 65-82.
179. Kessler, M. and Sørensen, M. (1999): Estimating equations based on eigen function for a discretely observed diffusion process, *Bernoulli* **5**, 299-314.
180. Khasminskii, R.Z., Krylov, N. and Moshchuk, N. (1999): On estimation of parameters of linear stochastic differential equations, *Prob. Theor. Rel. Fields* **113** (3), 443-472.
181. Kim, Y.T. (1996): Parameter estimation for an infinite dimensional stochastic differential equations, *J. Korean Statistical Society* **25**, No. 2, 161-173.
182. Kleptsyna, M.L., Kloeden, P.E. and Anh, V.V. (1998a): Linear filtering with fractional Brownian motion, *Stoch. Anal. Appl.*, **16**, 907-914.
183. Kleptsyna, M.L., Kloeden, P.E. and Anh, V.V. (1998b): Nonlinear filtering with fractional Brownian motion, *Problems of Information Transmission* **34** (2), 1-12.
184. Kleptsyna, M.L., Kloeden, P.E. and Anh, V.V. (1998c): Existence and uniqueness theorem for fBm stochastic differential equations, *Problems of Information Transmission*, **34** (4), 51-61.
185. Kleptsyna, M.L. and Le Breton, A. (2002): Statistical analysis of the fractional Ornstein Uhlenbeck type process, *Statist. Inf. Stochastic Process* **5** (3), 229-248.

254   References

186. Kleptsyna, M.L., Le Breton, A. and Roubaud, M.C. (1999): An elementary approach to filtering in systems with fractional Brownian motion observation noise, In: (Eds.) B. Grigelionis *et al.*, *Prob. Theory and Math. Stat.*, VSP, The Netherlands.

187. Kleptsyna, M.L., Le Breton, A. and Roubaud, M.C. (1999): General approach to filtering with fractional Brownian noises- application to linear systems, *Preprint*.

188. Kolmogorov, A.N. (1940): Wiener skewline and other interesting curves in Hilbert space, *Doklady Akad. Nauk* **26**, 115-118.

189. Kloeden, P.E. and Platen, E. (1995): *Numerical Solution of Stochastic Differential Equations*, Second Print, Springer-Verlag, Berlin.

190. Kloeden, P.E., Platen, E. and Schurz, H. (1994): *Numerical Solution of SDE Through Computer Experiments*, Springer-Verlag, Berlin.

191. Kloeden, P.E., Schurz, H., Platen, H. and Sørensen, M. (1996): On the effects of discretization of estimators of drift parameters for diffusion processes, *J. Appl. Prob.* **33**, 1061-1076.

192. Konecky, F. (1990): Maximum likelihood estimator from a partially observed diffusions in the case of small measurement noise, *Statist. Decisions* **8**, 115-130.

193. Koski, T. and Loges, W. (1985): Asymptotic statistical inference for a stochastic heat flow problem, *Statist. Probab. Letters* **3**, 185-189.

194. Koski, T. and Loges, W. (1986): On minimum contrast estimation for Hilbert space valued stochastic differential equations, *Stochastics* **16**, 217-225.

195. Krylov, N.V. (1995): *An Introduction to the Theory of Diffusion Processes*, American Mathematical Society, Providence.

196. Küchler, U. and Sørensen, M. (1994a): Exponential families of stochastic processes and Levy processes, *J. Statist. Planning Inf.* **39**, 211-237.

197. Küchler, U. and Sørensen, M. (1994b): Exponential families of stochastic processes with time continuous likelihood functions, *Scand. J. Statist.* **21**, 421-431.

198. Küchler, U. and Sørensen, M. (1997): *Exponential Family of Stochastic Process*, Springer-Verlag, New York.

199. Kulinich, G.L. (1975): On estimation of the drift parameter of a stochastic diffusion equation, *Theory Prob. Appl.* **20**, 384-387.

200. Kunita, H. (1990): *Stochastic Flows and Stochastic Differential Equations*, Cambridge Univ. Press, New York.

201. Kutoyants, Yu. A. (1977): Estimation of the trend parameter of a diffusion process in the smooth case, *Theory Probab. Appl.* **22**, 399-406.

202. Kutoyants, Yu. A. (1978): Estimation of a parameter of a diffusion process, *Theory Probab. Appl.* **23**, 641-649.

203. Kutoyants, Yu. A. (1984a): *Parameter Estimation for Stochastic Processes* (Translated and edited by B.L.S. Prakasa Rao), Research and Expositions in Mathematics, Vol. 6, Heldermann-Verlag, Berlin.

204. Kutoyants, Yu. A. (1984b): Parameter estimation for diffusion type processes of observations, *Math. Operationsforch. U. Statist., Ser. Statist.* **15**, 541-551.

205. Kutoyants, Yu. A. (1994a): *Identification of Dynamical Systems with Small Noise*, Kluwer, Dordrecht.

206. Kutoyants, Yu. A. (1994b): Nonconsistent estimation by diffusion type observations, *Statist. Probab. Letters.* **20**, 1-7.

207. Kutoyants, Yu. A. (2003): *Statistical Inference for Ergodic Diffusion Processes*, Springer-Verlag, New York-Berlin.

208. Kutoyants, Yu. A. and Pohlman, H. (1994): Parameter estimation for Kalman-Bucy filter wih small noise, *Statistics* **25**, 307-323.

209. Lanksa, V. (1979): Minimum contrast estimation in diffusion processes, *J. Appl. Prob.* **16**, 65-75.

210. Laredo, C. (1990): Sufficient condition for asymptotic sufficiency of incomplete observations of a diffusion process, *Ann. statist.* **18**, 1158-1171.

211. Larsen, K.S. and Sørensen, M. (2003): Diffusion models for exchange rates in a target zone, *Mathematical Finance* (to appear).

212. Le Breton, A. (1976): On continuous and discrete sampling for parameter estimation in diffusion type processes, *Mathematical Programming Study* **5**,124-144.

213. Le Breton, A. (1977): Parameter estimation in a linear stochastic differential equation, *Transcactions of the Seventh Prague Conference on Information on Theory, Statistical Decision Functions and Random Processes and of the 1974 EMS, Vol. A*, Academica, Prague, 353-366.

214. Le Breton, A. (1998a): Filtering and parameter estimation in a simple linear system driven by fractional Brownian motion, *Statist. Probab. Letters* **38**, 263-274.

215. Le Breton, A. (1998b): A Girsanov type approach to filtering in a simple linear system driven by fractional Brownian motion, *C. R. Acad. Sci. Paris, Series I* **326**, 997-1002.

216. Le Breton, A. and Musiela, M. (1984): A study of a one dimensional bilinear differential model for stochastic processes, *Prob. Math. Statist.* **4**, 91-107.

217. Le Cam, L. (1953): On some asymptotic properties of maximum likelihood estimates and related Bayes estimates, *Univ. Calif. Publ. Statist.* **1**, 277-330.

218. Le Cam, L. (1973): Convergence of estimates under dimensionality restrictions, *Ann. Statist.* **1**, 38-53.

219. Le Cam, L. (1986): *Asymptotic Methods in Statistical Decision Theory*, Springer-Verlag, Berlin-New York.

220. Le Cam, L. and Yang, G.L. (1990): *Asymptotics in Statistics*, Springer-Verlag, New York.

221. Lee, T.S. and Kozin, F. (1977): Almost sure asymptotic likelihood theory for diffusion processes, *J. Appl. Prob.* **14**, 527-537.

222. Leon, J.R. and Ludena, C. (2000): Estimating the diffusion coefficient for diffusions driven by fBm, *Statistical Inference for Stochastic Processes* **8**, 1-10.

223. Levy, P. (1948): Processus stochastiques et movement Brownien, Paris.

224. Lin, S.J. (1995): Stochastic analysis of fractional Brownian motions, *Stochastics and Stochastics Reports* **55**, 121-140.

225. Liptser, R.S. and Shiryayev, A.N. (1989): *Theory of Martingales*, Kluwer, Dordrecht.

226. Norros, I., Valkeila, E. and Virtamo, J. (1999): An elementary approach to a Girsanov formula and other analytical results on fractional Brownian motion, *Bernoulli* **5**, 571-587.

227. Peltier, R.F. and Levy Vehel, J. (1994): A new method for estimating the parameter of fractional Brownian motion, *INRIA Research Report No. 2396*.

228. Nualart, D. and Rovira, C. (2000): Large deviations for stochastic Volterra equations, *Bernoulli* **6** (2), 339-355.

229. Pipiras, V. and Taqqu, M.S. (2000): Integration questions related to fractional Brownian motion, Preprint, Boston University.

230. Levanony, D., Shwartz, A. and Zeitouni, O. (1993): Uniform decay and equicontinuity for normalized parameter dependent Itô integrals, *Stoch. Stoch. Reports* **43**, 9-28.

256   References

231. Levanony, D., Shwartz, A. and Zeitouni, O. (1994): Recursive identification in continuous-time stochastic processes, *Stoch. Proc. Appl.* **49**, 245-275.

232. Linkov, Yu. N. (1990): Local asymptotic mixed normality for distributions of processes of diffusion type, *Theory Prob. Math. Statist.* **40**, 67-73.

233. Linkov, Yu. N. (1993): *Asymptotic Methods of Statistics for Stochastic Processes*, Naukova, Dumka (In Russian); English Translation: (2001) American Mathematical Society, Providence, R.I.

234. Liptser, R.S. and Shiryayev, A.N. (1977): *Statistics of Random Processes I: General Theory* Springer-Verlag, Berlin.

235. Liptser, R.S. and Shiryayev, A.N. (1978): *Statistics of Random Processes II: Applications* Springer-Verlag, Berlin.

236. Liptser, R.S. and Shiryayev, A.N. (1989): *Theory of Martingales*, Kluwer, Dordrecht.

237. Lo, A.W. (1988): Maximum likelihood estimation for generalized Itô processes with discretely sampled data, *Econometric Theory* **4**, 231-247.

238. Loges, W. (1984): Girsanov theorem in Hilbert space and an application to the statistics of Hilbert space-valued stochastic differential equations, *Stoch. Proc. Appl.* **17**, 243-263.

239. Lototsky, S.V. and Rozovskii, B. L. (1999): Spectral asymptotics of some functionals arising in statistical inference for SPDEs, *Stoch. Process Appl.* **79**, 69-94.

240. Luschgy, H. (1992): Local asymptotic mixed normality for semimartingale experiments, *Probab. Theor. Rel. Fields*, **92**, 151-176.

241. Lyons, T. (1994): Differential equations driven by rough signals (I): An extension of an inequality of L.C Young, *Mathematical Research Letters*, **1**, 451-464.

242. Mandelbrot, B. (1997): *Fractals and Scaling in Finance: Discontinuity, Concentration, Risk*, Springer, New York.

243. Mandelbrot, B. and Van Ness, J.W. (1968): Fractional Brownian motions, fractional noises and applications, *SIAM Review* **10** (1968), 422-437.

244. Mao, X. (1995): *Stochastic Differential Equations and Applications* Horwood Publishing, Chichester.

245. McKean, Jr. H.P. (1969): *Stochastic Integrals*, Academic Press, New York.

246. McLeish, D.L. and Kolkiewicz, A.W. (1997): Fitting diffusion models in finance, In: *Selected Proceedings of the Simposium on Estimating Functions* (Eds.) Basawa, I.V., Godambe, V.G. and Taylor, R.L., IMS Lecture Notes - Monograph Series **32**, 327-350.

247. McShane, E.J. (1974): *Stochastic Calculus and Stochastic Models*, Academic Press, New York.

248. Merton, R.C. (1973): Theory of rational option pricing, *Bell J. Econ. Management Sci.*, **4**, 141-183.

249. Metivier, M. and Pellaumail, J. (1980): *Stochastic Integration*, Academic Press, New York.

250. Michel, R. and Pfanzagl, J. (1971): The accuracy of the normal approximation for minimum contrast estimate, *Zeit. Wahr. Verw. Gebiete* **18**, 73-84.

251. Milshtein, G.N. (1995): *Numerical Integration of Stochastic Differential Equations*, Kluwer, Norwell, MA.

252. Mishra, M.N. (1989): The Bernstein- von Mises theorem for a class of non-homogeneous diffusion processes, *Statist. Decisions*, **7**, 153-165.

253. Mishra, M.N. and Bishwal, J.P.N. (1995): Approximate maximum likelihood estimation for diffusion processes from discrete observations, *Stochastics and Stochastics Reports* **52**, 1-13.

254. Mishra, M.N. and Prakasa Rao, B.L.S. (1985a): On the Berry-Esseen theorem for maximum likelihood estimator for linear homogeneous diffusion processes, *Sankhyā, Ser A* **47**, 392-398.

255. Mishra, M.N. and Prakasa Rao, B.L.S. (1985b): Asymptotic study of maximum likelihood estmation for nonhomogeneous diffusion processes, *Statist. Decisions* **3**, 193-203.

256. Mishra, M.N. and Prakasa Rao, B.L.S. (1987): Rate of convergence in the Bernstein-von Mises theorem for a class of diffusion processes, *Stochastics* **22**, 59-75.

257. Mishra, M.N. and Prakasa Rao, B.L.S. (1991): Bounds on the equivalence of Bayes and maximum likelihood estimators for a class of diffusion processes, *Statistics* **22**, 613-625.

258. Mohapl, J. (1994): Maximum likelihood estimation in linear infinite dimensional models, *Commun. Statist. - Stoch. Models* **10**, 781-794.

259. Mohapl, J. (1996): On estimation in planar Ornstein-Uhlenbeck process, Preprint, Dept. of Statistics and Actuarial Sciences, University of Waterloo, Canada.

260. Monroe (1972): Processes that can be embedded in Bownian motion, *Ann. Math. Statist.*

261. Musiela, M. (1976): Risks of maximum likelihood estimators of parameters of an Itô type processes, *Bulletin de L' Academic Polonaise des Sciences, Serie de Sci. Math.* **24**, 1035-1039.

262. Nualart, D. (1995): *Malliavin Calculus and Related Topics*, Springer-Verlag, New York.

263. Nualart, D. and Rovira, C. (2000): Large deviations for stochastic Volterra equations, *Bernoulli* **6** (2), 339-355.

264. Musiela, M. (1984): *Processses de Diffusions: Aspects Probabilistics et Statistiques*, These l' Universite' Scientifique et Medicale de Grenoble, France.

265. Novikov, A. (1972): Sequential estimation of the parameters of diffusion type processes, *Mathematical Notes* **12**, 812-818.

266. Øksendal, O. (1995): *Stochastic Differential Equations*, 4th edition, Springer-Verlag, New York-Berlin.

267. Overbeck, L. and Ryden, T. (1997): Estimation in the Cox-Ingersoll-Ross model, *Econometric Theory* **13**, 430-461.

268. Pedersen, A.R. (1994): Quasi likelihood inference for discretely observed diffusion processes, Research Rep. No. 295, Dept. of Theoretical Statistics, Univ. of Åarhus.

269. Pedersen, A.R. (1995a): A new approach to maximum likelihood estimation for stochastic differential equations based on discrete observations, *Scand. J. Statist.* **22**, 55-71.

270. Pedersen, A.R. (1995b): Consistency and asymptotic normality of an approximate maximum likelihood estimator for discretely observed diffusion processes, *Bernoulli* **1**, 257-279.

271. Peltier, R.F. and Levy Vehel, J. (1994): A new method for estimating the parameter of fractional Brownian motion, *INRIA Research Report No. 2396*.

272. Penev, S.I. (1985): Parametric statistical inference for multivariate diffusion processes using discrete observations, *Mathematics and Education in Mathematics*, Proc. Fourteenth Spring Conf. Bulg. Math., Sunny Beach, Sophia, 501-510.

273. Pfanzagl, J. (1971): The Berry-Esseen bound for minimum contrast estimators, *Metrika* **17**, 82-91.

274. Piterbarg, L. and Rozovskii, B.L. (1995): Maximum likelihood estimators in the equations of physical oceanography, In: *Stochastic Modelling in Physical Oceanography*, (Eds.) R.Adler, P.Muller, and B.Rozovskii, Birkhauser, Boston.

275. Piterbarg, L. and Rozovskii, B.L. (1996): On asymptotic problems of parameter estimation in stochastic PDEs: discrete time sampling, *Math. Method. Statist.* **6** (2), 200-223.

276. Poulsen, R. (1999): Approximate maximum likelihood estimation of discretely observed diffusion processes, Working Paper No. 29, Centre for Ananytical Finance, University of Aarhus - Aaarhus School of Businesss.

277. Prakasa Rao, B.L.S. (1978): On the rate of convergence of the Bernstein-von Mises approximation for Markov processes, *Serdica Bulg. Math. Publ.* **4**, 36-42.

278. Prakasa Rao, B.L.S. (1980): The Bernstein - von Mises theorem for a class of diffusion processes, *Theory of Random Processes* **9**, 95-101 (in Russian).

279. Prakasa Rao, B.L.S. (1981): On Bayes estimation for diffusion fields, *In: Statistics: Applications and New Directions, Proc. ISI Golden Jubilee Conferences*, ed. J.K. Ghosh and J. Roy, 504-511.

280. Prakasa Rao, B.L.S. (1982): Maximum probability estimation for diffusion processes, *In : Statistics and Probability, Essays in honour of C.R. Rao* (Eds.) G. Kallianpur, P.R. Krishnaiah, J.K. Ghosh, North-Holland, Amsterdam, 557-590.

281. Prakasa Rao, B.L.S. (1983): Asymptotic theory for non-linear least squares estimator for diffusion processes, *Math. Operationsforch Statist. Ser. Statistics* **14**, 195-209.

282. Prakasa Rao, B.L.S. (1984): On the exponential rate of convergence of the least squares estimator in the nonlinear regression model with Gaussian errors, *Statist. Probab. Letters* **2**, 139-142.

283. Prakasa Rao, B.L.S. (1985): Estimation of the drift for diffusion processes, *Statistics* **16**, 263-275.

284. Prakasa Rao, B.L.S. (1988a): Law of the iterated logarithm for fluctuation of posterior distributions for a class of diffusion processes and a sequential test of power one. *Theor. Prob. Appl.* **33**, 269-275.

285. Prakasa Rao, B.L.S. (1988b): Statistical inference from sampled data for stochastic processes, In: *Statistical Inference from Stochastic Processes*, (Ed.) N.U.Prabhu *Contemporary Mathematics* **80**, 249-284, American Mathematical Society, Providence, Rhode-Island.

286. Prakasa Rao, B.L.S. (1999a): *Statistical Inference for Diffusion Type Processes*, Arnold, London.

287. Prakasa Rao, B.L.S. (1999b): *Semimartingales and Their Statistical Inference*, CRC Press, Boca Raton, Florida.

288. Prakasa Rao, B.L.S. and Rubin, H. (1981): Asymptotic theory of estimation in nonlinear stochastic differential equations, *Sankhyā Ser. A* **43**, 170-189.

289. Prakasa Rao, B.L.S. and Rubin, H. (1998): Uniform approximation for families of stochastic integrals, Sankhyā, Ser. A **60**, 57-73.

290. Protter, P. (1990): *Stochastic Integration and Differential Equations: A New Approach*, Springer-Verlag, Berlin.

291. Renyi, A. (1963): On stable sequences of events, *Sankhyā Ser. A* **25**, 293-302.

292. Revuz, D. and Yor, M. (1991): *Continuous Martingales and Brownian Motion*, Springer-Verlag.

293. Rogers, L.C.G. (1997): Arbitrage with fractional Brownian motion, *Math. Finance* **7**, 95-105.

294. Rogers, L.C.G. and Williams, D. (1987): *Diffusions, Markov Processes and Martingales Vol. 2 : Itô Calculus*, John Wiley, New York.

295. Rootzen, H. (1980): Limit distributions for the error in approximation of stochastic integrals, *Ann. Proab.* **8**, 241-251.

296. Roussas, G.G. (1972): *Contiguity of Probability Measures*, Cambridge University Press.

297. Robinson, P.M. (1977): Estimation of a time series model from unequally spaced data, *Stoch. Proc. Appl.* **6**, 9-24.

298. Rogers, L.C.G. (1997): Arbitrage with fractional Brownian motion, *Math. Finance* **7**, 95-105.

299. Ruzmaikina, A.A. (2000): Stochastic calculus for fractional Brownian motion, *J. Statistical Physics* **100**, 1049-1069.

300. Samarodnitsky, G. and Taqqu, M.S. (1994): *Stable Non-Gaussian Random Processes: Stochastic Models with Infinite Variance*, Chapman and Hall, London.

301. Shiryaev, A.N. (1998): On arbitrage and replication for fractal models, *Preprint, MaPhySto, Aaarhus.*

302. Shiryayev, A.N. (1999): *Essentials of Stochastic Finance: Facts, Models and Theory*, World Scientific, Singapore-New Jersey-London.

303. Schmetterer, L. (1974): *Introduction to Mathematical Statistics*, Springer-Verlag, Berlin.

304. Schwartz, L. (1965): On Bayes procedures, *Z. Wahrscheinlichkeitstheorie* **4**, 10-26.

305. Shiryayev, A.N. and Spokoiny, V.G. (2000): *Statistical Experiments and Decisions: Asymptotic Theory*, World Scientific, Singapore.

306. Shen, X. and Wasserman, L. (1998): Rates of convergence of posterior distributions, Technical Report No. 678, Department of Statistics, Carnegie Mellon University.

307. Shoji, I. (1997): A note on asymptotic properties of estimator derived from the Euler method for diffusion processes at discrete times, *Statist. Probab. Letters* **36**, 153-159.

308. Singer, H. (1993): Continuous time dynamical systems with sampled data, errors of measurement and unobserved components, *J. Time Series Analysis* **14**, 527-544.

309. Sørensen, B.E. (1992): Continuous record asymptotics in systems of stochastic differential equations, *Econometric Theory* **8**, 28-51.

310. Sørensen, H. (2001): Discretely observed diffusions: approximation of the continuous time score function, *Scand. J. Statist.* **28**, 113-121.

311. Sørensen, H. (2004): Parametric inference for diffusion processes observed at discrete points in time: a survey, *Int. Statist. Review* **72**, 337-354.

312. Sørensen, M. (1983): On maximum likelihood estimation in randomly stopped diffusion type processes, *Int. Statist. Review* **51**, 93-110.

313. Sørensen, M. (1997): Estimating functions for discretely observed diffusions: A review, In: *Selected Proceedings of the Simposium on Estimating Functions* (Eds.) Basawa, I.V., Godambe, V.G. and Taylor, R.L., IMS Lecture Notes - Monograph Series **32**, 305-325.

314. Sørensen, M. (1999): On asymptotics of estimating functions, *Braz. J. Probab. Statist.* **13**, 111-136.

315. Spokoiny, V.G. (1992): Lower bounds of minimax risk in estimation problems without LAN condition, *Stoch. Stoch. Reports* **38**, 67-93.
316. Stoyanov, J. (1984): Problems of estimation in continuous-discrete stochastic models, *Proc. Seventh Conf. on Prob. Theory, Statistical Decision Functions and Random Processes*, Brasov, Rumania, 363-373.
317. Stoyanov, J. and Vladeva, D. (1982): Estimation of unknown parameter of continuous time stochastic processes by observations at random moments, *C.R. Acad. Bulgar. Sci.* **35**, 153-156.
318. Strasser, H. (1975): The asymptotic equivalence of Bayes and maximum likelihood estimation, *J. Multivariate Anal.* **5**, 206-226.
319. Strasser, H. (1976): Asymptotic properties of posterior distributions, *Zeit. Wahr. Verw. Gebiete* **35**, 269-282.
320. Strasser, H. (1977): Improved bounds for Bayes and maximum likelihood estimation, *Theory Prob. Appl.* **22**, 349-361.
321. Strasser, H. (1985): *Mathematical Theory of Statistics*, Walter de Gruyter, Berlin-New York.
322. Stratonovich, R.L. (1964): Novaya forma zapisi stokhasticheskikh integralov i uravenni, *Vestnik Moskov Univ. Ser I Mat. Mekh.*, **1**, 3-12. Translated as: A new representation for stochastic integrals and equations, *SIAM J. Control* (1966), 362-371.
323. Stroock, D.W. and Varadhan, S.R.S. (1979): *Multidimensional Diffusion Process*, Springer-Verlag, Berlin.
324. Sweeting, T.J. and Adekola, A.O. (1987): Asymptotic posterior normality for stochastic processes revisited, *J. Royal Statist. Soc. B 49*, 215-222.
325. Taraskin, A.F. (1974): On the asymptotic normality of vector valued stochastic integrals and estimates of a multidimensional diffusion process, *Theory Prob. Math. Statist.* **2**, 209-224.
326. Taraskin, A.F. (1980): The asymptotic statistical problems for fields of diffusion type, *In : Stochastic Differential Systems* (Ed. B. Grigelionis) *Lecture Notes in Control and Information Sciences* **25**, Springer-Verlag, Berlin.
327. Tikhov, M.S. (1978): On optimal sequential estimation plans for nonquadratic loss functions, *Theory Prob. Appl.*, **23**, 132-138.
328. Tikhov, M.S. (1980): On the sequential estimation of the drift coefficient of a diffusion process with quadratic and nonquadratic losses, *Theory Prob. Appl.* **25**, 607-614.
329. Touti, A. (1993): Two theorems on convergence in distribution for stochastic integrals and statistical applications, *Theory Prob. Applic.* **38**, 95-117.
330. Trofimov, E.I. (1982): A stochastic equation for the maximum likelihood estimate of the parameter of a process of diffusion type, *Russian Math. Surveys* **37**(4), 221-222.
331. Tsitovich, I.I. (1977): On estimating the drift of a diffusion process from observations in a domain, *Theory Prob. Appl.* **22**, 851-858.
332. Tudor, C. and Tudor, M. (1995): Approximation schemes for Itô-Volterra stochastic equations, *Boletin de la Sociedad Matematica Mexicana, Tercera Serie* (Ser. 3) **1**, 73-95.
333. Tyurin, K. and Phillips, P.C.B. (1999): The occupation density of a fractional Brownian motion and some of its applications, Working paper, Yale University.
334. Van Kampen, N.G. (1981): *Stochastic Processes in Physics and Chemistry*, North Holland, Amsterdam.

335. Walsh, J.B. (1986): An introduction to stochastic partial differential equations, *Lecture Notes in Mathematics* **1180**, 265-439, Springer, Berlin.

336. Wolfowitz, J. (1975): Maximum probability estimators in the classical and in the 'almost smooth' case, *Theory Probab. Appl.* **20**, 363-371.

337. Yor, M. (1976): Sur quelques approximations d'integrals stochastiques, Seminaire de Probabilites XI, Lecture Notes in Mathematics 581, Springer-Verlag.

338. Yoshida, N. (1988): Robust *M*-estimators in diffusion processes, *Ann. Inst. Statist. Math.* **40**, 799-820.

339. Yoshida, N. (1990): Asymptotic behaviour of *M*-estimator and related random field for diffusion process, *Ann. Inst. Statist. Math.* **42**, 221-251.

340. Yoshida, N. (1992): Estimation for diffusion processes from discrete observations, *J. Multivariate Anal.* **41**, 220-242.

341. Yoshida, N. (1997): Malliavin calculus and asymptotic expansion for martingales, *Probab. Theory Relat. Fields*, **109**, 301-342.

342. Zahle, M. (1998): Integration with respect to fractal functions and stochastic calculus, *Prob. Theory. Rel. Fields* **111**, 333-374.

Index

Lecture Notes in Mathematics

For information about earlier volumes
please contact your bookseller or Springer
LNM Online archive: springerlink.com

Vol. 1833: D.-Q. Jiang, M. Qian, M.-P. Qian, Mathematical Theory of Nonequilibrium Steady States. On the Frontier of Probability and Dynamical Systems. IX, 280 p, 2004.

Vol. 1834: Yo. Yomdin, G. Comte, Tame Geometry with Application in Smooth Analysis. VIII, 186 p, 2004.

Vol. 1835: O.T. Izhboldin, B. Kahn, N.A. Karpenko, A. Vishik, Geometric Methods in the Algebraic Theory of Quadratic Forms. Summer School, Lens, 2000. Editor: J.-P. Tignol (2004)

Vol. 1836: C. Năstăsescu, F. Van Oystaeyen, Methods of Graded Rings. XIII, 304 p, 2004.

Vol. 1837: S. Tavaré, O. Zeitouni, Lectures on Probability Theory and Statistics. Ecole d'Eté de Probabilités de Saint-Flour XXXI-2001. Editor: J. Picard (2004)

Vol. 1838: A.J. Ganesh, N.W. O'Connell, D.J. Wischik, Big Queues. XII, 254 p, 2004.

Vol. 1839: R. Gohm, Noncommutative Stationary Processes. VIII, 170 p, 2004.

Vol. 1840: B. Tsirelson, W. Werner, Lectures on Probability Theory and Statistics. Ecole d'Eté de Probabilités de Saint-Flour XXXII-2002. Editor: J. Picard (2004)

Vol. 1841: W. Reichel, Uniqueness Theorems for Variational Problems by the Method of Transformation Groups (2004)

Vol. 1842: T. Johnsen, A. L. Knutsen, K_3 Projective Models in Scrolls (2004)

Vol. 1843: B. Jefferies, Spectral Properties of Noncommuting Operators (2004)

Vol. 1844: K.F. Siburg, The Principle of Least Action in Geometry and Dynamics (2004)

Vol. 1845: Min Ho Lee, Mixed Automorphic Forms, Torus Bundles, and Jacobi Forms (2004)

Vol. 1846: H. Ammari, H. Kang, Reconstruction of Small Inhomogeneities from Boundary Measurements (2004)

Vol. 1847: T.R. Bielecki, T. Björk, M. Jeanblanc, M. Rutkowski, J.A. Scheinkman, W. Xiong, Paris-Princeton Lectures on Mathematical Finance 2003 (2004)

Vol. 1848: M. Abate, J. E. Fornaess, X. Huang, J. P. Rosay, A. Tumanov, Real Methods in Complex and CR Geometry, Martina Franca, Italy 2002. Editors: D. Zaitsev, G. Zampieri (2004)

Vol. 1849: Martin L. Brown, Heegner Modules and Elliptic Curves (2004)

Vol. 1850: V. D. Milman, G. Schechtman (Eds.), Geometric Aspects of Functional Analysis. Israel Seminar 2002-2003 (2004)

Vol. 1851: O. Catoni, Statistical Learning Theory and Stochastic Optimization (2004)

Vol. 1852: A.S. Kechris, B.D. Miller, Topics in Orbit Equivalence (2004)

Vol. 1853: Ch. Favre, M. Jonsson, The Valuative Tree (2004)

Vol. 1854: O. Saeki, Topology of Singular Fibers of Differential Maps (2004)

Vol. 1855: G. Da Prato, P.C. Kunstmann, I. Lasiecka, A. Lunardi, R. Schnaubelt, L. Weis, Functional Analytic Methods for Evolution Equations. Editors: M. Iannelli, R. Nagel, S. Piazzera (2004)

Vol. 1856: K. Back, T.R. Bielecki, C. Hipp, S. Peng, W. Schachermayer, Stochastic Methods in Finance, Bressanone/Brixen, Italy, 2003. Editors: M. Fritelli, W. Runggaldier (2004)

Vol. 1857: M. Émery, M. Ledoux, M. Yor (Eds.), Séminaire de Probabilités XXXVIII (2005)

Vol. 1858: A.S. Cherny, H.-J. Engelbert, Singular Stochastic Differential Equations (2005)

Vol. 1859: E. Letellier, Fourier Transforms of Invariant Functions on Finite Reductive Lie Algebras (2005)

Vol. 1860: A. Borisyuk, G.B. Ermentrout, A. Friedman, D. Terman, Tutorials in Mathematical Biosciences I. Mathematical Neurosciences (2005)

Vol. 1861: G. Benettin, J. Henrard, S. Kuksin, Hamiltonian Dynamics – Theory and Applications, Cetraro, Italy, 1999. Editor: A. Giorgilli (2005)

Vol. 1862: B. Helffer, F. Nier, Hypoelliptic Estimates and Spectral Theory for Fokker-Planck Operators and Witten Laplacians (2005)

Vol. 1863: H. Führ, Abstract Harmonic Analysis of Continuous Wavelet Transforms (2005)

Vol. 1864: K. Efstathiou, Metamorphoses of Hamiltonian Systems with Symmetries (2005)

Vol. 1865: D. Applebaum, B.V. R. Bhat, J. Kustermans, J. M. Lindsay, Quantum Independent Increment Processes I. From Classical Probability to Quantum Stochastic Calculus. Editors: M. Schürmann, U. Franz (2005)

Vol. 1866: O.E. Barndorff-Nielsen, U. Franz, R. Gohm, B. Kümmerer, S. Thorbjønsen, Quantum Independent Increment Processes II. Structure of Quantum Lévy Processes, Classical Probability, and Physics. Editors: M. Schürmann, U. Franz, (2005)

Vol. 1867: J. Sneyd (Ed.), Tutorials in Mathematical Biosciences II. Mathematical Modeling of Calcium Dynamics and Signal Transduction. (2005)

Vol. 1868: J. Jorgenson, S. Lang, $Pos_n(R)$ and Eisenstein Series. (2005)

Vol. 1869: A. Dembo, T. Funaki, Lectures on Probability Theory and Statistics. Ecole d'Eté de Probabilités de Saint-Flour XXXIII-2003. Editor: J. Picard (2005)

Vol. 1870: V.I. Gurariy, W. Lusky, Geometry of Müntz Spaces and Related Questions. (2005)

Vol. 1871: P. Constantin, G. Gallavotti, A.V. Kazhikhov, Y. Meyer, S. Ukai, Mathematical Foundation of Turbulent Viscous Flows, Martina Franca, Italy, 2003. Editors: M. Cannone, T. Miyakawa (2006)

Vol. 1872: A. Friedman (Ed.), Tutorials in Mathematical Biosciences III. Cell Cycle, Proliferation, and Cancer (2006)

Vol. 1873: R. Mansuy, M. Yor, Random Times and Enlargements of Filtrations in a Brownian Setting (2006)

Vol. 1874: M. Yor, M. Émery (Eds.), In Memoriam Paul-André Meyer - Séminaire de Probabilités XXXIX (2006)

Vol. 1875: J. Pitman, Combinatorial Stochastic Processes. Ecole d'Eté de Probabilités de Saint-Flour XXXII-2002. Editor: J. Picard (2006)

Vol. 1876: H. Herrlich, Axiom of Choice (2006)

Vol. 1877: J. Steuding, Value Distributions of L-Functions (2007)

Vol. 1878: R. Cerf, The Wulff Crystal in Ising and Percolation Models, Ecole d'Eté de Probabilités de Saint-Flour XXXIV-2004. Editor: Jean Picard (2006)

Vol. 1879: G. Slade, The Lace Expansion and its Applications, Ecole d'Eté de Probabilités de Saint-Flour XXXIV-2004. Editor: Jean Picard (2006)

Vol. 1880: S. Attal, A. Joye, C.-A. Pillet, Open Quantum Systems I, The Hamiltonian Approach (2006)

Vol. 1881: S. Attal, A. Joye, C.-A. Pillet, Open Quantum Systems II, The Markovian Approach (2006)

Vol. 1882: S. Attal, A. Joye, C.-A. Pillet, Open Quantum Systems III, Recent Developments (2006)

Vol. 1883: W. Van Assche, F. Marcellàn (Eds.), Orthogonal Polynomials and Special Functions, Computation and Application (2006)

Recent Reprints and New Editions